CW01486659

THE LESSONS OF YUGOSLAVIA

RESEARCH ON RUSSIA AND EASTERN EUROPE

Series Editor: Metta Spencer

RESEARCH ON RUSSIA AND EASTERN EUROPE
VOLUME 3

THE LESSONS OF YUGOSLAVIA

EDITED BY

METTA SPENCER

Department of Sociology, University of Toronto, Canada

2000

JAI
An Imprint of Elsevier Science

Amsterdam – London – New York – Oxford – Paris – Shannon – Tokyo

ELSEVIER SCIENCE Inc.
655 Avenue of the Americas
New York, NY 10010, USA

First edition 2000

Library of Congress Cataloging in Publication Data
A catalog record from the Library of Congress has been applied for.

ISBN: 0-7623-0280-1

∞ The paper used in this publication meets the requirements of ANSI/NISO Z39.48-1992 (Permanence of Paper).
Printed in The Netherlands.

CONTENTS

PART II WARS IN SLOVENIA, CROATIA, AND BOSNIA

PART III AFTER THE BREAKUP AND DAYTON

PART IV KOSOVO: THE POST-WAR WAR

CONCLUSION

LIST OF CONTRIBUTORS

Milica Z. Bookman

5907 Riviera Drive, Coral Gables, FL 33146, Dept Econ, St Joseph's University, 5600 City Ave, Philadelphia 19131–1395, USA
E-mail: mbookman@sju.edu

Michael Chossudovksy

21 First Avenue, Paris-Vaudreuil, QC J7V3T5 Canada
E-mail: chosso@travel-net.com

Timothy Donais

Institut Linguistique Adenet 33, Grand Rue Jean Moulin, 34000, Montpellier, France
E-mail: t_donais@hotmail.com

Randy Hodson

Ohio State University, USA

C. G. Jacobsen

KATTEN, 4818, Faervik, Sildevik, Norway
E-mail: cgj@magi.com

David Last

Department of Politics-Economics, Royal Military College, Kingston, Ontario, Canada
E-mail: last-d@rmc.ca

Corey Levine

33 First Avenue, Ottawa, Ontario K15 2G1, Canada
E-mail: cmjlevine@yahoo.com

Sonja Licht

Omladinskih Brigada 216/13, Beograd, Yugoslavia, YU 11070
slicht@sfj.opennet.org

Garth Massey   University of Wyoming

Srecko Mihailovic   Inst. Soc. Sci. Centre for Pol., Narodnog
Fronta 45, PO B 927, 11000 Belgrade
E-mail: srecko@EUnet.yu

Jan Øberg   TFF Vegagatan 25, S–224 57 Lund,
Sweden
E-mail: tff@transnational.org

Margarita Papandreou   1 Romilias Str., Kastri, Greece, GR 14671,
E-mail: MPAP.ATH.FORTHNET.GR
@Forthnet.GR

Robert K Schaeffer   2391 Grandview Terrace, Manhattan,
Kansas, 66502, USA
Robschaeff@aol.com

Dusko Sekulic   The Flinders University, 60 Bella View
Rd, Flagstaff Hill, 5159 SA, Australia
E-mail: sodusko@psy1.ssn.flinders.edu.au.

Darko Silovic   10 East End Avenue, Apt. 9C, New York,
NY 10021, USA
E-mail: dsilovic@compuserve.com

Ken Simons   710 Bathurst St., Toronto, Ontario,
M5S 2R4, Canada
E-mail: ksimons@interlog.com

Metta Spencer   155 Marlee Avenue, Apt. 201, Toronto,
Ontario, M6B 4B5 Canada
E-mail: mspencer@web.net

Dorie Wilsnack   c/o International BPT Office, Ringstr 9a,
D-32427, Minden, Germany.
E-mail: doriew@igc.apc.org

Mitja Žagar   Institute for Ethnic Studies, Erjavceva 26,
S1–1000 Ljubljana, Slovenia
E-mail: mitja.zagar@guest.arnes.si

"To the memory of Alison and George Ignatieff, and to all the victims of the tragedy that befell Yugoslavia."

PREFACE

Science for Peace is a Canadian organization consisting of several hundred academics who address key issues of peace, justice, environment, and sustainable development. In March 1997 that group organized a conference in Toronto to reflect on the conflicts surrounding the break-up of the former Yugoslavia. This meeting was made possible by a generous grant from the John Holmes Fund, which is administered by the Canadian Centre for Foreign Policy Development in Ottawa. Panel discussions took place, with short presentations by the participants and follow-up conversations over a three-day period among peace activists, diplomats, governmental experts, scholars, leaders of non-governmental organizations, and professional military peacekeepers.

As one of the organizers of this meeting, it was my intention at the time to publish extended versions of the papers within a few months after the conference. Largely because the Yugoslav conflicts were not really over, many of the participants were too busy to expand their remarks in a timely way, and some had to give up the attempt altogether. Nevertheless, it was possible to find other contributors who had not attended the conference. In a sense, this proved to be fortuitous when later the Kosovo conflict culminated in another round of warfare; our regrettable delay had enabled us to address the Kosovo problem along with the other conflicts that had preceded it.

Nevertheless, I must point out that the resulting collection of papers were not all prepared at the same time. Some of them had been updated a year or so before we went to press, while others - especially those focusing on Kosovo - were completed much later. In any case, so many changes take place during the publication of any edited book that events always will have overtaken some parts of particular papers. Nevertheless, we believe many scholars and peacemakers will find this compilation particularly valuable, including as it does the reflections of some who sought, without success, to prevent the tragedies in Kosovo.

The 1997 conference was dedicated to the memory of Alison and George Ignatieff, two Canadians who had departed from our midst only a few years before and who were dear friends to many members of Science for Peace. George Ignatieff had been Canadian ambassador to Yugoslavia after World War II, and his sons Michael and Andrew recall with warmth the childhood years they spent in Belgrade. The Ignatieffs also served the Canadian people in other important diplomatic posts, and after retiring George became Chancellor of the University of Toronto and then president of Science for Peace. His active work for peace inspired us all. We were fortunate that Andrew was able to participate in organizing the conference that honored his parents.

I also want to acknowledge some of the people who worked on that conference and who advised and assisted me in preparing and editing this book. Dozens of people contributed, but I can name only a few. I am especially grateful to Steven Lee of the Canadian Centre for Foreign Policy Development, to the staff of Innis College, and to Patricia Albanese, Nurit Amor, John Bacher, Walter Dorn, Slobodan Drakulic, Andre Gunder Frank, L. Terrell Gardner, Kristina Hemon, Andrew Ignatieff, Sue McClelland, Susan McClelland, James Milner, Sylvanna Papaleo, Lara Paul, Miro Prpic, Kenneth Simons, Jean Smith, Suzanne Soto, Daniela Stor, and John Valleau.

Metta Spencer,
Science for Peace
Toronto, Canada
June 2000

PART I

HISTORIC AND STRUCTURAL DEVELOPMENTS

1. WHAT HAPPENED IN YUGOSLAVIA?

Metta Spencer

What killed Yugoslavia? Since this entire book will offer numerous responses to that question, the present introduction will not seriously attempt to anticipate those insights or to propose any analysis. Instead, it is simply meant as a bare-bones chronological sketch of the most significant historical events in that down-hill progression. The other contributors will elaborate and analyze specific aspects of the saga in their own chapters. Some readers will not need to refresh their memory with this chronology but will go directly into the main body of the book, referring only when necessary to an even more concise display of the sequence of significant events. Those of you who do skip this chapter will be able to refer to the 'time line' synopsis presented in the appendix of the book.

ORIGINS OF A MULTICULTURAL SOCIETY

The Roman province of Illyricum occupied a territory corresponding, more or less, to the short-lived twentieth century country, Yugoslavia. But if that land was politically unified during the Roman period and again recently, it was divided throughout the intervening time. The boundary between Western Christendom and Byzantium ran through the region and became the boundary between the Ottoman and Habsburg (later the Austro-Hungarian) Empires. These divisions long ago became significant in cultural terms for, although the people were almost all Slavs who spoke dialects of the same language, about half were Catholic who wrote in the Latin script and were ruled by Austrian or Hungarian monarchs, while the rest (mainly Serbs and Montenegrins) were Orthodox

Research on Russia and Eastern Europe, Volume 3, pages 3–45.
2000 by Elsevier Science Inc.
ISBN: 0-7623-0280-1

Christians who used Cyrillic and were ruled for centuries by the Ottomans until the disintegration of that empire was far advanced.

Before the Ottomans came, however, the Serbian Empire had covered in the 14th century most of today's Greece, Albania, and Macedonia, as well as Serbia, the Sandzak, and Montenegro, with Kosovo at its heart. It was in 1389 that Turkish Ottomans defeated the Serbs' army at the battle of Kosovo Polje – a historic turning point of lasting symbolic significance to the vanquished people. By the mid-fifteenth century, the Turks had conquered Serbia proper and a large part of the region, which they then ruled for 500 years. There was no forcible conversion to Islam, but most people in Bosnia and the Sandzak area did adopt that faith. By the time Yugoslavia dissolved in the 1990s, Muslims comprised around one-fifth of the country's total population of 23 million. The regions that we know as Slovenia and Croatia remained predominantly Catholic; Serbia remained Orthodox Christian; and Bosnia and Hercegovina was a patchwork of all three major faiths.

Most Yugoslav inhabitants identified themselves as belonging to one of the five 'constituent nations' that supposedly voluntarily created the 'South Slavic' state: Slovenes, Croats, Serbs, Montenegrins, and Macedonians. There were also more than twenty 'nationalities' that were regarded as minority groups with certain guaranteed rights, but (unlike the constituent nations) no right to secede. These include the Albanians, Hungarians, Roma (Gypsies), and Turks. Two of these nationalities – the Albanians of Kosovo ('Kosovars'), and the Muslims of Bosnia (whom I shall call Bosniaks) – demanded recognition as 'constituent nations' and the latter group would achieve that status in 1963 by constitutional change.

The Hungarians of Vojvodina and the Albanians of Kosovo, on the other hand, were not given such recognition, since they are not Slavic peoples. The latter group are Muslims who belong to a community that predominates in the adjacent state, Albania, and are numerous in Macedonia. The Albanian language, though Indo-European, is entirely distinct from the tongues spoken elsewhere in the region, and is the only surviving representative of the so-called Thraco-Illyrian group of languages, which was spoken by the inhabitants of the Balkan Peninsula before the Roman period.[1] (So say the Albanians. However, Serb historians commonly dispute this account.) Serbs had constituted about half of the two million inhabitants of Kosovo at the end of World War II, but the government encouraged them to move to prosperous cities in Serbia and, after 1974, the Albanian majority also successfully pressured Serbs to leave. This migration, combined with the high birth rate of the Kosovars, meant that by the 1990s, the latter group constituted about 90% of Kosovo's population.[2]

Yugoslavia's other ethnic communities, such as Jews, were considered minority groups. Finally, some 5% of the population called themselves only

'Yugoslavs,' either as an expression of their political commitment to unity or because they belonged to families of mixed ancestry.[3] In Bosnia-Herzegovina fully 16% of the children were from mixed marriages, though in the other republics intermarriage was less common.[4]

The idea of Yugoslavia was first suggested in Croatia in the 1830s by a movement called the 'Illyrianists,' who wanted all south Slavs to unify culturally and politically to defend themselves against aggressive Ottoman, German, Italian, and Hungarian nationalisms. It was then assumed that the culturally advanced Croatian city Zagreb would be the centre of this proposed state. The idea remained only a dream for almost a century, for there was no opportunity to form a new Yugoslav state until the end of World War I, when the Ottoman and Austro-Hungarian Empires were both defeated and dismembered by the allied victors.

By then, the Ottoman Empire had long been in decline. Serbia had been an independent country since 1878, and in 1912 it had joined a coalition with Bulgaria, Greece, and Montenegro to attack Ottoman Turkey. This, the First Balkan War, ended by Turkey's ceding almost all its European possessions. A new state, Albania, was created, but it was too weak to control all the regions where Albanians lived. Serbia annexed the Albanian regions of Macedonia and Kosovo, which, five centuries after the Ottomans had captured it, remained symbolically the 'cradle of the Serb nation.' The next year, in the Second Balkan War, Serbia and Greece seized most of the spoils of war from Bulgaria. Serbia doubled its size and territory, incorporating more non-Serbs than it could either assimilate or eliminate. The Muslim populations particularly suffered from the brutality of these wars.[5]

It was the Western allies who created Yugoslavia from the fragments of the two empires they had vanquished. At its founding in 1918 Yugoslavia was known as the 'Kingdom of Serbs, Croats and Slovenes'. The Serbian Prince Alexander, who had sided with the war's victors, agreed to unite his territory with the former Habsburg lands with Belgrade as its capital.[6] It had a territory of about 248,000 square km and a population of about 12 million.[7]

The Croats were immediately disappointed, for the new state gained control of their funds and replaced their old Habsburg institutions with Serbian ones.[8] No efforts to integrate the old states through compromise were successful. No parliamentary coalitions could be created that transcended the concerns of nationality groups. The former Habsburg subjects resisted the dominant Serbs, sometimes violently, and there were already conflicts between Serbs and Albanians in Kosovo.[9] In 1928 King Alexander became dictator of the country, giving it the new name 'Yugoslavia'.

Some Croatians wanted full independence and were prepared to use violent means of breaking up the country. In particular, the fascist Ustasha-Croat

Revolutionary Organization, led by Ante Pavelic, received support from Mussolini. However, their assassination of King Alexander in 1934 proved to be counterproductive, for it unified public opinion against their cause. Prince Paul, who succeeded Alexander as regent, tried to placate the Croat nationalists by offering considerable autonomy to Croatia. This would have done nothing to satisfy Yugoslavia's Hungarians, Albanians, Macedonians, and Bosniaks, who had grievances of their own.[10] In the 1920s the Albanians of Kosovo had carried out an unsuccessful guerrilla war against the federal Yugoslav state.

As World War II approached, Prince Paul found it necessary to enter into agreements with Hitler, but his negotiations succeeded in keeping the German troops out of Yugoslavia – at least until certain of his military officers seized power with the intention of extracting more concessions from Germany. Their actions failed, for Hitler simply invaded Yugoslavia and kept the route open for his army to wage their campaign in Greece. He gave portions of Yugoslavia to Germany, Italy, Hungary, Bulgaria, and Albania. Then he established a fascist rule by Ante Pavelic and his Ustashas over the rest of the country, which included some of Serbia, Croatia, and Bosnia-Herzegovina. The Ustashas planned to kill a third of the Serbs, expel another third, and convert the others to Catholicism.[11] They came close, killing 85,000 people in one death camp, Jasenovac.[12]

Around 6.4% of Yugoslavia's population died during or just after World War II, which was fought mainly in Bosnia-Herzegovina and Montenegro, with heaviest loss of life in those areas.[13] About 600,000 people were also killed in Croatia in that war.[14]

Of approximately one million Yugoslavs killed during World War II, probably more than half died at the hands of other Yugoslavs.[15] For Yugoslavia, World War II actually consisted of several civil wars that had little to do with the larger war being fought elsewhere.[16] These civil wars were three-way fights between the Ustashas and the competing resistance movements of two military leaders, Josip Broz Tito and Draza Mihailovic. Tito and Mihailovic alike opposed the German rulers of Serbia from 1941 onward, though during most years there were not many Germans in Yugoslavia.

Tito, the communist who led the Partisans, recruited fighters from all Yugoslav peoples, whereas Mihailovic claimed to lead a group of strictly Serbian guerrilla fighters called Chetniks. In fact, he had little control over the Chetniks, who perpetrated atrocities of their own. However, because the Ustashas were in charge of a state organization, the extent of their murders and ethnic cleansing outstripped the Chetniks.[17] As a royalist, Mihailovic had expected aid from Britain. However, Winston Churchill opted in Tito's favor instead. Faced with a choice between fighting against Tito's Partisans or the Germans, Mihailovic eventually decided to fight the former, and in fact there was no significant battle between Chetniks and Germans or Italians. Tito won; after the war he put Mihailovic on trial and executed him.

TITO'S YUGOSLAVIA AFTER WORLD WAR II

Tito's long and impressive career as dictator of Yugoslavia was just beginning. King Peter II was deposed and a republic was proclaimed. Tito suppressed internal opposition, nationalized Yugoslav industry, and launched a planned economy. He was a deeply committed Communist who was especially talented at reconciling and balancing nationalist conflicts. His top priority was to keep ethnic rivalry from tearing apart the new state. For example, he reduced Serbian influence in the federal institutions of Belgrade, while promoting Serbs to positions of power in Croatia, where they were a minority.[18] Unlike Germany, where the guilty past was discussed in public almost constantly, in Yugoslavia the discussion of World War II conflicts and injustices was forbidden.[19] Tito insisted that the past was past and that it was time to get on with the present and future. 'Brotherhood and unity' was a slogan that guided his every action and he tried to build that principle into a new constitution, which the communist victors adopted in 1946.

According to that constitution Yugoslavia was a highly decentralized country comprising six republics, plus two relatively autonomous provinces within Serbia – Vojvodina and Kosovo. Borders between republics could be changed only on the basis of consensus. Each republic (but not Vojvodina or Kosovo) supposedly had the right of self-determination, though this was a matter of dispute. Secession certainly was not expected and no procedures were provided by which it could occur.[20] On the other hand, Yugoslavia was – at least in theory – the most decentralized country in Europe, a condition that Tito fostered for the sake of limiting inter-ethnic fights.[21] He gambled on his own skill at power-brokering among the communist elites of the republics and in general he succeeded;[22] one could almost get the impression that Yugoslavia was a society without conflict.[23] However, his less talented successors would be seriously handicapped by the decentralized structures he left behind.

Yugoslavia was originally a leading member of the Cominform, but Tito showed considerable independence from the beginning. In 1948 he displeased Stalin, mainly on the basis of regional issues. Yugoslavia supported the partisans in the Greek civil war and Tito wanted to create a Balkan confederation, which would have included Albania. Stalin forced Yugoslavia's expulsion from the Cominform on the charge of deviating from the correct Communist line.

There was every basis for expecting a Soviet invasion, and Tito prepared for the worst. Yet no such thing happened, and Tito found himself in a position to play East against West, to the advantage of his own country. He accepted billions of dollars of loans and non-repayable aid from the West without changing his own program to suit the donors. The United States did not, however, provide Marshall Plan funding for Yugoslavia, nor did the Soviet Union admit that country to its own economic bloc, the Council for Mutual Economic Assistance ('COMECON'). However, during the 1970s the economy suffered from foreign debt, inefficiency, and inflation.

Over time, Tito made Yugoslavia into the most open Communist country, though he kept considerable control over intellectual life, mainly so as to prevent any re-emergence of nationalism.

Tito was a major political figure on the world stage because he worked closely with President Nasser of Egypt and Prime Minister Jawaharlal Nehru of India in trying to create a neutral bloc. He helped found the Nonaligned Movement and the Group of 77, which stood apart from either superpower bloc.[24] His personal leadership gave Yugoslavia a much more prominent place in the international arena than it would otherwise have occupied. Much of this influence was leveraged, however, and did not persist long after the cold war ended, much to the dismay and shock of some Yugoslav politicians, who believed that their influence on world affairs was secure.

Tito was able to restrain ethnic conflict for decades, not by challenging the power of nations, but rather by artfully balancing each one against the others. His approach rested on a conception of ethnic equality that is quite unlike the liberal Western notion, which protects the rights of individuals, not of groups.

Each major Yugoslav ethnic community was a 'constituent nation' entitled to its own republic. This was the basis for a decentralized federal system of governance that limited the power of the federal government. Although each nationality was allowed to express its culture (indeed, all nationalities were supposed to be represented at every major public occasion),[25] *nationalism* was prohibited and even punished.

The Yugoslav constitution was re-written time after time – the earliest one (1946) was replaced in 1953, and again in 1964 and 1974,[26] which was the last version prepared under Tito's leadership. It greatly decentralized the already decentralized system and would later become a factor in the collapse of the country. Republics were suddenly declared to be nation-states, each with complete autonomy, except that its constitution must not contradict the federal one.[27] Many of the functions carried out by parliaments in other countries were to be carried out in Yugoslavia by self-managing organizations that were supposed to create 'compacts' instead of laws.

According to the 1974 constitution, citizens were not to elect parliamentary deputies directly. Instead, they were to elect 'delegations' that would then elect the parliamentarians. And after electing the deputies, the delegations continued to control them, and could replace them if they did not follow their orders.[28] Thus federal parliamentarians did not have the autonomy required for effective legislative work, but had to spend much of their time waiting in the corridor for their instructions to arrive from delegations back at home.[29]

The 1974 constitution allowed any decision to be blocked by the veto of a single federal entity, including Kosovo or Vojvodina, the autonomous provinces of Serbia. The constitution did provide a means of breaking an impasse among the republics, but only for urgent measures, and only for the period of one year.[30]

Tito had understandable reasons for dividing power in this way. There was a real danger that one of the constituent nations – most likely Serbia – would get too much power and use it to dominate the others. Tito went to great lengths in order to guarantee the equality of these 'titular nations' and their respective sovereign republics. For example, the 1974 constitution specified that the presidency would consist of a group, not an individual. Each republic and each autonomous province (Kosovo and Vojvodina) would be represented, and the role of presiding over this eight-man presidency would rotate among the republics for a one-year term.[31]

If democracy was violated by the arrangement allowing each republic or autonomous province to veto the decisions of the whole, the rationale for the system would have been apparent to Alexis de Tocqueville, whose book *Democracy in America* constituted a cogent warning against the danger that rule by the majority would tend to become tyranny by the majority. Tito understood only too well the importance of institutionalizing protections for ethnic minorities; the consensual model was his method.[32] He sought to prevent any nationalist group from acquiring enough power to over-ride the concerns of minorities. His personal way of reaching decisions was by negotiating compromises to harmonize conflicting interests. But, though he usually succeeded with this approach himself, he worried about the prospect of nationalism after his own death, and introduced the constitutional provisions of 1974 with future dangers in mind. Unfortunately, his new safeguards would prove in the end to block the effectiveness of his successors altogether.

In fact, conflicts had been postponed and papered-over but not eliminated. As Tito's long life came to an end in 1980, there were very apparent conflicts everywhere and he left behind no democratic institutions to manage those antagonisms or to generate peaceful solutions.[33] There were calls for democracy, but a decade would pass before the League of Communists of

Yugoslavia (LCY) would become the first ruling communist party in the world to 'commit suicide' by dissolving itself.[34]

Throughout that initial post-Tito decade it was obvious that the federal constitution would have to be amended, but this could be accomplished only with the consensus of all components of the federal state. Instead of consensus, the positions of the various republics were becoming more contradictory, particularly when the topic of centralization was at stake. There was no agreement among them on any possible amendments. The more democratic citizens were especially favorable toward greater decentralization, on the theory that centralization equals totalitarianism. If that held true in other socialist countries, it was not true in Yugoslavia, where the main problem that destroyed the economy was the excessive decentralization created in the 1974 constitution. The weak central government no longer had sufficient power to make decisions that were required. The party tried twice in the early 1980s to develop a program of economic reform, but the parliament and cabinet were hamstrung by the constitutional requirement for a consensus among all the republics. Serbia wanted more central control, whereas Croatia and Slovenia insisted on further decentralization as a loose confederation of sovereign states. The Yugoslav People's Army (JNA) had a vote in the presidency along with the republics and it sided with Serbia. The great majority of its officers were Serbian.[35]

One reason for the dispute was the widening economic gap between the have- and have-not republics. Slovenia and Croatia were much more affluent than the other parts of Yugoslavia, partly because of their industrial base and partly because the spectacular scenery along the Adriatic attracted many tourists. Kosovo found itself at the opposite extreme as probably the poorest region in all of Europe. The resentment caused by the necessary transfer payments between republics worsened the conflict over the country's mounting foreign debt.

ECONOMICS AND NATIONALISM

Along with many other countries, Yugoslavia experienced a painful recession that began at about the same time as Tito's death in 1980. Everywhere the crisis was caused by the increasing oil prices of the 1970s. Because so many Yugoslavs were guest workers in Europe, sending remittances home periodically, the loss of much of this income due to recession, plus the world-wide rise in interest rates, added to Yugoslavia's already excessive foreign debt.[36] These loans had been incurred mostly by the republics, often without even informing the federal government about taking on these financial commitments.[37] By the mid-1980s most republics had already stopped paying their share of the federal budget, insisting instead that the federal government get by with only revenues from its own sources. Though Serbia favored a stronger central government, Kosovo and Vojvodina had enough autonomy to block economic reforms even in that republic. The central government was caught up in the irreconcilable conflict, with the decentralizing republican pressures on one side, and on the other the international financial institutions, which wanted centralization and majority rule. The smaller republics had a sound basis for fearing that majority rule would put them under the domination of the Serbs, who constituted the most numerous nation in the country.[38]

Ante Markovic became the new prime minister in March 1989. Economic reforms were initiated at the insistence of the International Monetary Fund (IMF), which demanded implementation of an austerity program to stop the inflation, which was running at 25,000% per year, the 14th highest rate of inflation in the history of the world.[39] The IMF demanded more accountability so that the federal government could enforce its economic decisions on the republics. This would have introduced majority rule instead of continuing to allow the bank governors of all republics to veto decisions that did not satisfy them. Slovenia and Croatia especially rejected these reforms, but all the republics resisted and continued to protect their own fiscal sovereignty.

Markovic accepted the painful requirements imposed by the IMF. He believed that the intense conflict over centralization would become manageable if the economic crisis could be resolved. Within a year he had succeeded in bringing inflation down to zero. He became extremely popular, yet he did not reap the political rewards that might have been accorded him for such an accomplishment. Some analysts maintain that it was the severity of the measures demanded by the West – especially the IMF's austerity program which Markovic adopted – that actually caused the ensuing collapse of the federal Yugoslav government. (See Michel Chossudovsky's article in this volume as an instance of this analysis.) On the other hand, the U.S. Ambassador to Yugoslavia, Warren Zimmermann, admired Markovic's reforms and pointed out that they became the model that was adopted by Poland and Czechoslovakia. Yet where reform in those countries succeeded, it failed in Yugoslavia because Markovic could not get a proper hearing in the media. The press was controlled by his political enemies, the leaders of the republics.[40] Zimmermann believed that Washington should have helped the prime minister, but no financial support came from that quarter. Markovic did, however, obtain some IMF support on the basis of beating back inflation and increasing his country's currency reserves.

To turn this economic triumph into a political breakthrough, Markovic knew he would have to hold federal elections immediately and win a restoration of legitimacy for the federal government. However, the Slovenian government had scheduled its own elections for April 1990 – well before elections could be held at the federal level. Since republican elections would probably capture the new legitimacy that he sought for the federal system, Markovic asked the Slovenians to delay their election. Slovenia's Prime Minister Kucan refused his request. Quitting the party, the communist politicians of both Slovenia and Croatia went to the voters before Markovic could do so. In this action they made it virtually certain that Yugoslavia would be dissolved. The republican elections were widely interpreted as a plebiscite on separatism, and the separatists won.[41]

By breaking with Stalin, Tito had become able to secure advantages for Yugoslavia by playing off the communist and capitalist blocs against each other. Both sides had courted him to maximize the advantages of his independence from the other bloc. In this way the Yugoslavs had received financial aid, military support, and cheap Soviet oil for about three decades.

With the ending of the Cold War, however, the Yugoslavs lost their position of power between the superpower blocs. Many politicians did not understand the changes in their situation and refused to adapt to it. They simply ignored the demands of the international financial institutions, assuming that the world would continue to coddle them as before.

At the same time it was becoming apparent that communism was either dead or likely to remain comatose for a long time. The political elite reacted to this situation, not by becoming democrats as one might expect, but by adopting the other ideology that socialism had forbidden: nationalism. This widespread response must be explained in terms of the similarity between communism and nationalism. Every communist had been taught a collectivist philosophy that constantly harped on struggles between enemies. Originally the struggle was to take place between various social classes, but almost any enemy was better than none at all.[42] Zarko Puhovski asserts that these people hardly knew how to act when they were deprived of an enemy to struggle against. After the Cold War the class struggle became politically unpopular, but the former communists needed a substitute enemy whose members could be identified almost at sight. As Puhovski explains, "Such group identification was most readily found in ethnicity, which can be demonstrated simply through language (or dialect) or even by name (or family name) in such a way that it cannot be easily stopped or controlled."[43]

Yugoslavs were not the only ex-communists who adopted nationalism as a substitute ideology; the same pattern can be seen throughout the former Soviet bloc. In Yugoslavia, there were plenty of ancient ethnic antagonisms that had been suppressed by the enforced outward display of 'brotherhood and unity', but which were suitable grounds for renewed struggle as soon as that suppression ceased. Chief among these rivalries was the antipathy between Serbs and Kosovars in the ancient 'cradle of the Serbian nation', Kosovo – or, to use the preferred Albanian spelling, Kosova.

In the interest of protecting and placating a minority of less than two million souls (who, in Kosovo itself, constituted a majority estimated at nearly 90%), Tito had taken the unusual step of granting an extraordinarily degree of autonomy to Kosovo in the 1974 constitution. The Serbs, however, never accepted his decision.[44] From a practical political point of view, that constitution actually may have conferred too much autonomy on both Kosovo and Vojvodina;[45] there are few other countries in the world in which the third tier of government is so powerful. By the late 1980s, Tito's influence was finally waning, after having survived his person by almost a decade. It became possible to take overt political measures to reverse the hated constitutional changes and give control back to Serbia proper. Slobodan Milosevic developed a scheme to induce Vojvodina to abandon its autonomy under rather minimal pressure, and this in turn would make the Kosovars more vulnerable. He organized mass rallies which proved successful, for the political leaders of Vojvodina did capitulate and allow themselves to be replaced by pro-Serbian nationalists. As predicted, this did leave the Kosovars in a more precarious position than before.[46]

At that time Milosevic was president of Serbia's League of Communists and a close friend of the President of Serbia, Ivan Stambolic, who sent him to Kosovo to cool the tensions there between Serbs and Kosovars. Milosevic, however, no longer considered himself a protégé of Stambolic. He had used his party position to gain control over the press and was prepared to grab power by challenging his former mentor.

Though they dominated Kosovo numerically, the Albanians of that republic were extremely disadvantaged in socio-economic terms. The Serbs, on the other hand, felt that the republic should belong to them because it had been the centre of their large medieval kingdom. Insofar as they could do so, the federal government had taken the Serbian side in this dispute and had jailed about 1600 Kosovar protesters as early as 1981, when martial law was declared and Prishtina University was closed. When Milosevic arrived on his assigned mission of conflict resolution, the Serbs claimed that they were being oppressed by the Albanians.[47] There were grounds for complaining; in 1986, some 60,000 Serbs in Kosovo petitioned for protection from the predominantly Albanian administration.[48]

Milosevic addressed a mass meeting of Serbians and Montenegrins at the legendary 'field of blackbirds' where the Turks had defeated the Serbian Prince Lazar 600 years before. A provocation was organized, which prompted the police (who were mostly Kosovars) to use force. Thereupon, Milosevic proclaimed that in the future, "nobody is going to beat these people." The crowd went wild with appreciation and at that moment he became the populist leader of Serb nationalism.

This was only the opening sortie in his new campaign. Thereafter, Milosevic kept up a steady mobilization in Serbia by holding mass rallies to whip up public emotions about the supposedly widespread abuse of Serbs in Kosovo. The ancient remains of Prince Lazar were placed in a coffin, which traveled from one Serbian village to another for a whole year. At each stop, the people reacted as if Lazar had been killed only the day before. They held a funeral ceremony that built up their anxieties and enabled them to consider themselves entitled to revenge against all Muslims in Yugoslavia, whom they portrayed as the same people as the Ottoman Muslims of 1389.[49]

This extremism gave Kosovar separatists an impetus to stage non-violent demonstrations. The majority of Albanians in Kosovo want their province to become an independent country. Every rally of theirs fueled Milosevic's own assertion that he was protecting Yugoslavia from a real threat of secession.

Probably the most serious limitation of democracy throughout Yugoslavia stemmed from the virtual lack of a free press. Urban residents had some access to unapproved information, but rural people had almost none. Governments controlled journalism to an extraordinary degree, not only limiting the flow of crucial information, but also whipping up chauvinistic emotions more than would have been possible, even in other Central or Eastern European countries. The impact of this can hardly be exaggerated when explaining the crude sentiments that rapidly supplanted civility in public discourse.[50] As George Urban has suggested, one special circumstance set Yugoslavia apart from the other socialist countries save Albania: its lack of access to broadcasts from Radio Free Europe. During the cold war, none of the Soviet bloc countries enjoyed freedom of the press, but millions of alert citizens managed to obtain information through 'surrogate broadcasting'. That is, dissidents presented their ideas to Western radio journalists, who beamed in this forbidden news via Voice of America, Radio Liberty, and other powerful foreign stations, stimulating discussions among the citizens and generally improving the popular level of knowledge concerning political affairs. Because Yugoslavia was courted as a potential ally, Western countries chose not to risk offending Tito and his successors by trans-mitting information to Yugoslavs that challenged the government's official statements. Indeed, it was only after the breakup of Yugoslavia that the first Radio Free Europe South Slavic Service was launched. As a result of this delay, the development of political discourse in that country had fallen behind, sorely hindering the spread of a democratic culture.[51]

ELECTIONS

Early in 1989 Serbia established a new constitution that revoked the autonomy of Kosovo and Vojvodina. This change not only gave Serbia political control over those two provinces, but it allowed Milosevic to pack the presidency with two additional of his political followers into the provincial seats, as well as control three delegations to the parliament. Since Montenegro was already in his pocket, Milosevic now controlled four of the eight positions in each federal institution. Multiparty elections were about to be held for the first time in Serbia, and by this political sleight-of-hand, Milosevic gained an advantage in the December 1989 Serbian presidential elections. He resorted openly to hate propaganda, which he disseminated through the press that he controlled, stimulating Serbs to demand the right to live in one state. Of course, those living in Yugoslavia already did live in one state, but what Milosevic wanted was to acquire for Serbia those regions of other republics, including Croatia, where Serbs were numerous. The popularity of this expansionist image of a 'Greater Serbia' enabled him to sweep to victory unopposed as president of Serbia. In other republics, however, most people felt entirely otherwise.

In 1990 there were multi-party elections in the republics for the first time. As noted above, the inability to hold the first democratic elections over the entire country as federal state sounded the death knell for Yugoslavia.[52] And although there were indeed many new parties, each republican election generally was an occasion for nationalists to beat their drums. In only one republic was a moderate elected: Kiro Gligorov became president of Macedonia. Warren Zimmermann had asked Milosevic what his strategy was for winning Albanian support in Kosovo, but discovered that the idea had never crossed his mind.[53]

Yet the outcome was not just of Milosevic's doing. The political leader of the Kosovars, a pacifist named Ibrahim Rugova, might very well have gained considerable power through those free elections, had he been willing to campaign against Milosevic. His Albanians vastly outnumbered the Serb voters in Kosovo. In the federal parliament, his delegation would have been a minority, to be sure, but he would have been able to hamstring the majority and win some of his objectives in that way. Yet Rugova refused to participate. His uncompromising demand was for independence and he had gained widespread support for that position. He insisted that the Kosovars would never again recognize Serbian authority and that he would not last a single day as their leader if he brought them into the electoral process.[54] Rugova's decision represented yet another failure of the democratic process.

In December 1990, Slovenia held a plebiscite asking whether voters wanted their republic to become an autonomous, independent state. It passed with 88.5%. This obliged the parliament to adopt within six months measures confirming this independence and enabling the republic to join a confederation of other Yugoslav peoples. No real negotiations took place; it was understood that a declaration of independence would simply be issued six months later, in June.[55] The JNA did not wait that long, but immediately went on combat readiness.

The federal presidency was headed by a president appointed in rotation by the various republics. In May 1991 the Croatian member Stipe Mesic was supposed to assume that role. However, he had earned the enmity of Slobodan Milosevic by openly supporting the dissolution of Yugoslavia. Since Milosevic now had control over the votes of Kosovo, Vojvodina, Montenegro, and his own republic, Serbia, he was in a position to prevent Mesic from taking office. This was so clearly a violation of democratic principles that within a few weeks the European Community's 'troika' of foreign ministers visited Belgrade and prevailed upon Milosevic to let Mesic be appointed to the presidency. By that time, insiders knew that it was impossible to keep Yugoslavia from breaking up. War was inevitable. By agreeing to seat Mesic, Milosevic was simply stalling in the knowledge that his Serbia, as well as Slovenia and Croatia, were rushing their preparations for the violent breakup that was imminent.[56] The Slovenian and Croatian national guards were beginning to be transformed from police forces into armies. Slovene and Croat soldiers deserted from the JNA to join the them, leaving the JNA high command increasingly Serb, though it already had been disproportionately so since the 1980s.

Milosevic may actually have wanted the Slovenes to leave so that Croatia would lack allies to help fight against him in the upcoming war.[57] President Kucan said that Milosevic offered him a deal: He would accept secession on condition that Kucan agree to re-write the constitution of Yugoslavia and extend the right to secede not just to all republics but also to all ethnic groups.[58] For their part, the Slovenes simply wanted independence and were unconcerned about the consequences of their secession for the rest of Yugoslavia.

Only in Bosnia-Herzegovina and Macedonia did politicians cherish any hope that Yugoslavia would hold together. Presidents Gligorov of Macedonia and Izetbegovic of Bosnia-Herzegovina believed that their republics could not sur-vive without the entirety of Yugoslavia remaining intact. They tried to promote a new structure for Yugoslavia as a loose confederation, but Presidents Milosevic of Serbia, Tudjman of Croatia, and Kucan of Slovenia dismissed the idea as con-trary to the interests of their republics.[59] The breakup was now inevitable.

Sporadic ethnic confrontations had turned violent beginning in March 1991 in western Slavonia, where the JNA intervened. The same month, Croatian policemen and Serb paramilitary forces engaged in a battle in a national park near Knin, where two were killed and 20 injured. Again, the JNA intervened.[60]

INTERNATIONAL INTERVENTION

Only when Croatia and Slovenia were on the verge of declaring their inde-pendence did the international community become seriously engaged in trying to prevent the looming disaster. The Conference on Security and Cooperation in Europe (later the OSCE) opposed the breakup of Yugoslavia and in this position were supported by the European Community's foreign ministers (including Germany), who warned that they would not recognize Slovenia and Croatia if they seceded unilaterally.

U.S. Secretary of State James Baker made a quick trip to Belgrade on June 21, 1991 and took the same position as the European organizations, adding that the United States also objected to any use of force to prevent the breakup. His visit seems to have left everyone confused, believing that Baker had supported their own position. Serbians believed that his message was: Stay united. The Slovenians and Croatians understood him to emphasize only that, if they decided to secede, no one should forcibly prevent that.[61] When Baker flatly told Tudjman and Kucan that the U.S. would not recognize any unilateral secession on their part, Tudjman replied by simply expressing certainty that the JNA would not fight to stop a Croatian declaration of independence.[62]

Four days after Baker's visit, Slovenia and Croatia declared independence. Two days later war started between Slovenia and the JNA. Croatia would not get off so lightly, but Slovenia's war of independence was extremely brief and cost the lives of only nine of its soldiers. The Slovenes shot down a helicopter that was carrying bread; its pilot was a Slovene soldier in the JNA, which lost 37 men in all.

On June 30 the EC sent the troika of foreign ministers to halt the bloodshed. EC representatives attended a presidency meeting the next week and persuaded Slovenia and Croatia to delay independence for three months. This, the Brioni Accord, did bring the war in Slovenia to a close, but it did not prevent the Croatian war. Though negotiations were supposedly going on during this period, everyone knew that these were not serious, and it came as no surprise that Serb irregulars and the JNA were engaged in heavy fighting against Croats by the end of June.[63] Milosevic told the Italian Foreign Minister Gianni De Michelis that Croatia would not be allowed to leave Yugoslavia because 600,000 Serbs lived there.[64]

THE CROATIAN-SERBIAN WAR

The real war began in August 1991 in Eastern Croatia around Vukovar, where Serb nationalists had seized power and were trying to expel their Croat neighbors and bring the region into a 'greater Serbia'. EC observers were reporting ethnic cleansing and atrocities on the part of the JNA.[65] Extremist allies of President Milosevic were inciting conflict between the Serbs and Croats.

Between September 7, and December 15, 1991 the EC convened a Peace Conference on Yugoslavia in The Hague, with Lord Carrington as its chairman.

It did not, however, find any basis for compromise.[66] Carrington asked Milosevic whether he would accept the independence of Croatia if protection could be assured to the Serbs living there. Milosevic agreed, and Carrington tried to turn this agreement into a substantial plan to end the conflict throughout Yugoslavia. It would give all the republics sovereignty and independence. In fact, Milosevic had accepted Carrington's plan only because he assumed he would get what he wanted – that the rest of the republics besides Croatia and Slovenia would reject independence and stay in Yugoslavia. In order to defeat the Carrington plan, at least one president of a republic would have to agree to stay in Yugoslavia. Milosevic expected Montenegro's Momir Bulatovic to do so, but in fact Bulatovic announced the intention of seceding as well, noting that the Carrington plan was an excellent means of stopping the war. It would have dissolved Yugoslavia. Shocked, Milosevic walked out of the meeting and summoned his supporters to confer about the new situation. These people took Bulatovic to task for his 'treachery', accusing him of accepting money in exchange for his vote. In fact, Bulatovic could not deny that the Italian foreign minister De Michelis had held out a powerful incentive for Montenegrins: aid from, and speedy entry into, Europe. The intense pressure from Milosevic's team convinced Bulatovic to sign a retraction of his vote. The Carrington plan was dead.[67]

With the way cleared for further fighting, the Serbs went about completing their battle against Vukovar, now using heavy artillery instead of fighting from house to house. Although Tudjman realized that Vukovar could not be saved, he chose to use the defeat of the city as a means of winning sympathy from the West. He promised the supplies and artillery pieces that his Croat general requested, but these never arrived. The commander openly accused Tudjman of sacrificing the city to gain international sympathy. If that was the leader's plan, it succeeded. Vukovar fell, but Hans Dietrich Genscher, the experienced but sympathetic foreign minister of Germany, argued in favor of recognizing Croatia and Slovenia. He tried hard to persuade other countries to do so.

However, most other countries, including the United States, regarded recognition as a serious mistake. Both Cyrus Vance and Lord Carrington opposed recognizing any Yugoslav republic until they had all agreed on their relationships. Vance, the former U.S. Secretary of State, now the personal envoy of the United Nations Secretary-General Javier Perez de Cuellar, was making some progress in negotiating a truce between Croatia and the Yugoslav army. He told U.S. Ambassador Warren Zimmermann, "My friend Genscher is out of control on this. What he's doing is madness."

Just then, in late November, Bosnian President Alija Izetbegovic went to visit Genscher in Bonn, hoping to convince him that EC recognition of Croatia and

Slovenia would bring war to Bosnia. When they met, however, Izetbegovic evidently forgot to bring the subject up, perhaps allowing Genscher to suppose that he had no objections.[68]

INDEPENDENCE FOR WHOM?

The EC had decided on August 27 against recognizing any Yugoslav republic until its claim had been investigated by a commission which it established under the judicial supervision of Robert Badinter. It had also proposed a peace conference – the one over which Lord Carrington unsuccessfully presided. However, during that same period other diplomats were also at work on the Balkan problems. Cyrus Vance, who was working separately for the United Nations, succeeded in brokering a cease-fire between Croatia and the JNA on November 23, 1991. It was not finalized until January 2, 1992, but it proved to be lasting. The UN Security Council also sent 14,000 peacekeeping troops to Croatia in early 1992. The cease-fire also called for UN peacekeepers to be based at Sarajevo, and to patrol it in an attempt to prevent war in Bosnia.

Meanwhile, the EC-sponsored Badinter Commission was investigating the applications for sovereignty of all the republics, including Macedonia, Bosnia-Herzegovina, and Kosovo, which had followed the separatist example of Slovenia and Croatia. Badinter gave favorable recommendations with respect to sovereignty to them all except Bosnia-Herzegovina. It did not issue any ruling on Kosovo. Despite its favorable recommendation, Macedonia was not immediately recognized as an independent state, mainly because Greece objected. Badinter demanded that Croatia improve guarantees for human rights and protection for minorities and that Bosnia-Herzegovina hold a referendum before attempting to secede.

Before an official referendum could be held, the Bosnian Serbs held a referendum of their own in which 99% took the position that if the republic tried to secede from Yugoslavia, they would demand the right to create a Serb republic within Bosnia-Herzegovina. Since the Serbs constituted only one-third of the Bosnian population, they were outnumbered by Muslims and Croats and would surely be outvoted by people who did want secession from Yugoslavia. Therefore the Bosnian Serb leaders proclaimed their own new republic, Republika Srpska, with its own government and currency. The EC ignored their referendum, as they had also ignored a referendum that the Serbs in Croatia had held, showing their desire to stay in Yugoslavia.

In retaliation, the Serbs boycotted the referendum which the Badinter Commission had requested, which occurred at the end of February, 1992. In an extraordinary polarization of opinion, some 99% of those voting in that official referendum supported independence. Accordingly, Bosnia did obtain speedy diplomatic recognition and a seat at the United Nations.

Whereas the United States (in contrast to its earlier position vis à vis Slovenia's and Croatia's status) promoted the recognition of Bosnia-Herzegovina and Macedonia as sovereign states, the European Community remained hesitant, on the grounds that Bosnia had become so weakened politically that it could not function as a sovereign country. The EC negotiators, now led by the Portuguese mediator José Cutileiro, convened meetings with a view to keeping all three of the Bosnian party leaders committed to existing Bosnian borders and thereby (they hoped) preventing war, without recognizing Bosnian sovereignty. In Lisbon in March 18, 1992 the three parties did meet and signed an agreement to form a confederation of three constituent nations, each of which would form an ethnic canton. However, this agreement did not hold up. Both Izetbegovic and Bosnia's Croatian leader, Mate Boban, reversed themselves within days and withdrew their consent from the Lisbon Accord.

Now there was certain to be a war in Bosnia-Herzegovina. President Izetbegovic, who foresaw this outcome, had been open to almost anything that would have held Yugoslavia together. He knew, nevertheless, that if Slovenia and Croatia were to secede, Bosnia would not accept its new position as a minority within a Serbian-dominated Yugoslavia. His government would have to secede too, though that would mean war.

The logic of the Bosnian Serb position was similar to Izetbegovic's. They too wanted to hold Yugoslavia together at all costs. If, nevertheless, Bosnia were to secede, they were unwilling to accept a new position as a minority within a Bosnia dominated by Muslims and Croats. They would have to secede too.

Moreover, both of these positions were logically identical to the recent position of the Serbs in Croatia. They too had refused to be forced against their will to separate from Yugoslavia. And, had Tito's practices been followed, none of these groups could have been forced to accept the majority's decision in their republics. Tito had treated each constituent nationality as if it had veto power, whether or not this was absolutely guaranteed constitutionally. His system of consensus would have required a majority of voters of each nation and each significant national minority to accept the proposal before it could be adopted. Had this practice been maintained during the early 1990s, a stalemate might have occurred, but not the wars that actually took place.

The liberal democracies of Europe and North America were not in the habit of using consensual (consociational) forms of democracy, but believed that decisions should be made on the basis of the majority of voters in the whole state. Their procedures were imposed. The referendum was decisively won by the separatists, which legitimized Bosnia's right to secede.

There were, of course, cogent reasons for some groups in Yugoslavia to fear for their safety under a simple system of majority rule. Minorities recalled their experiences during World War II but these fears were not taken particularly seriously. Warren Zimmermann reports that he kept trying to convince the Bosnian Serbs to take part in the Bosnian referendum and accept its outcome. Had they done so, they would have become a minority within an independent Bosnia.[69] As Susan Woodward notes,

> It was a matter of unresolved constitutional interpretation whether republics had the right to secede and, if so, whether individuals who identified with another constituent nation within these republics had to give their consent. In choosing [to define sovereignty in terms of] the republican borders and the claims of the majority nation for an independent state, the EC politicians made no accommodation to this second, constitutionally equal category of rights to self-determination . . .
>
> Yet the EC] had no leverage with which to persuade the Serbs in Croatia and Albanians in Serbia to be satisfied with minority rights when their rebellions were motivated, in part, by real discrimination by the governments that would be expected to guarantee them protections.[70]

THE WARS IN BOSNIA-HERZEGOVINA

In April 1992 Sarajevo, the famously tolerant multi-ethnic capital of Bosnia-Herzegovina, was about to become a besieged city. On March 3, Bosnia had declared itself an independent nation. The Bosnian Serbs, rejecting this declaration, had begun to fight. Across Bosnia there were local hostilities in February and March 1992.

Soon the Muslims would be engaged in two wars: against both the Serbians and, sporadically, the Bosnian Croats. Despite their own fight over the Serb areas of Croatia, Tudjman and Milosevic had a common vision of the future: partition Bosnia so that each of its three constituent peoples would have a country of its own. (Presumably, the Bosnian Croats and Bosnian Serbs would link their own territories to Croatia and Serbia respectively.) To this end, Tudjman initiated secret meetings between the Bosnian Serbs and his clients, the Bosnian Croats, after the cease-fire of January 1992 in Croatia. According to Stipe Mesic, they reached an agreement to divide Bosnia along the Neretva River between Croatia and Serbia.[71] The plan would have left the third constituent people, the Muslims, essentially stateless. In February and March, Croatians – who had been allied with the Muslims against the Serbs – started their own 'ethnic cleansing' program against the Muslims to create an ethnically 'pure' region. There were reports of concentration camps. More serious, however, were the Bosniaks' fights against Serbs.

There were some 100,000 JNA troops in Bosnia. As soon as Bosnia was recognized as an independent state in April, their presence would mean that

Serbia would be seen as an aggressor state. To prevent this, Milosevic transferred every Bosnian Serb in the JNA to a unit in Bosnia, along with abundant JNA weaponry. The Bosnian Serb leader Dr. Radovan Karadzic thereby acquired a well-trained and equipped army of 80,000 men, still on Belgrade's payroll. Milosevic also sent Serb paramilitaries who were nationalists and specialists in terror, led by the self-proclaimed Chetnik Vojislav Seselj.[72] These fighters quickly captured the city of Bijeljina in northwest Bosnia, then rounded up activist Bosniak civilians and killed them.

Izetbegovic called upon the police and militia to defend his new state. Instead, thousands of Sarajevo's citizens marched to the parliament chamber, shouting their opposition to the nationalism of both the Bosniak government and the Serb rebels. Then they attempted to march on the Serb headquarters in the downtown Holiday Inn, but Karadzic told his guards to fire on the crowd. Six demonstrators were killed. Karadzic and his men took to the hills above the city and began mortaring and sniping at the downtown area. Izetbegovic had the option of partitioning the city and thereby stopping the fighting but he refused. The Serbs would hold the city under siege for 40 months, and some 10,000 of its inhabitants would be killed.

The Serbs' main goal was not to seize Sarajevo but to take control of Serb areas in the region adjoining Serbia itself. However, there were some towns there whose inhabitants were mostly Muslim. Most of them were expelled as displaced persons.[73]

In an effort to forestall the fighting, the French government had proposed to President George Bush that preventive peacekeepers be sent to help the Bosnian government. Still smarting from the defeat in Vietnam, the U.S. administration declined, saying that the proposal did not provide for any 'exit strategy'.[74] Indeed, the State Department officials [notably Lawrence Eagleburger] made sure *not to threaten any use of force*. As Wayne Bert notes,

> Eagleburger seemingly had no misgivings about the value of American credibility unless some overt threat was made for which there was no follow-through. Complete inaction, in his view, did not compromise U.S. credibility.[75]

President Clinton would change this policy only after three years had elapsed and thousands of uprooted persons had been massacred while under the protection of peacekeepers.[76] However, the international community was not entirely passive during this time. In June and July 1992, peacekeepers took charge of the Sarajevo airport and maintained it for an international airlift. In August there was a major international conference on Yugoslavia in London. The agreements reached there dealt with aid and a cease-fire, but were never implemented. In September, the UN Security Council suspended Yugoslavia from its membership in the General

Assembly. In November, the Security Council authorized a naval blockade of the new, smaller Yugoslavia – now consisting only of Serbia and Montenegro.

None of these measures had much effect. Throughout the winter and spring of 1992–93, the beseiged Sarajevans lost most access to gas, water, electricity, food, fuel, and medicine. The humanitarian air-lift met only a small fraction of the needs. However, the United Nations defined enclaves – so-called 'safe areas' – where displaced persons could expect some measure of security during that terrible winter. The first declaration of such a safe area came about unexpectedly. General Morillon, the French peacekeeping commander, went to Srebrenica with a convoy of food which the surrounding Serbs would not allow through. Morillon himself proceeded to the town through a minefield and saw there the ghastly human consequences of ethnic cleansing. When he prepared to leave, the angry people would not permit him to go. Immediately he decided to stay there and proclaimed that his U.N. peacekeepers would protect the enclave. Morillon had no authority to promise to protect the Muslim side in the war. The astonished Security Council had to react. One resolution called for Srebrenica to become a safe haven, which would have required the United Nations to defend it. Eventually, however, the European view prevailed. The final resolution, adpted on April 16, 1993, did not call Srebrenica a 'safe haven,' but merely a 'safe area' – a term that lacked any official meaning. The United Nations sent peacekeepers – the United Nations Protection Force, or UNPROFOR – but they were not supposed to protect anyone. The belligerents of all sides resented or even hated them for their impotence. A few weeks later the Security Council set up five other safe areas for Sarajevo and the Muslim towns of Tuzla, Bihac, Zepa, and Gorazde.[77] According to Mark Danner, the Bosniak government sought to "make use of the misery of the enclaves to force action by the United Nations and Western countries... [They] were simply trying to make use of the only weapon the peculiar and hypocritical international involvement in their country seemed to offer them."[78]

Most Western reporters allege,[79] on the other hand, that the Serbs were disproportionately responsible for war crimes in Croatia, Bosnia-Herzegovina, and later in Kosovo. Ironically, the Serbs' initial arguments against the breakup of Yugoslavia were as logical as the demands of the Slovenians, Croats, Bosniaks, and Kosovars, but their claims were rejected in the public mind abroad because of their military excesses. Reading the numerous reports of Serb atrocities, Westerners necessarily placed most of the blame on the Serbs and criticized UNPROFOR for doing too little to protect their victims.[80]

The Geneva conference – International Conference on the Former Yugoslavia (ICFY) – was a joint project of the EC and the United Nations. It was headed by Cyrus Vance and Lord David Owen, who carried on negotiations during the winter of 1992. On January 2, 1993 they unveiled their proposed solution. The Vance-Owen plan would have divided Bosnia into ten cantons on ethnic lines and would have limited the Serb portion to 43% of the territory but – at least nominally – Bosnia would have remained a single sovereign country. The Croats welcomed the plan, while the Bosniaks were slower to accept it.[81] Under considerable pressure, Karadzic signed the document, knowing that it would have to be ratified by the Bosnian Serb assembly. After the assembly in Pale had listened to all the arguments, they went into a closed session, excluding Milosevic and Montenegro's President Bulatovic from the room. Milosevic left. The Assembly did not directly reject the plan, but called for a referendum, which was held in mid-May 1993; the Bosnian Serbs decisively rejected Vance-Owen plan.

Still, the mediators continued. Cyrus Vance was replaced by Norwegian foreign minister Thorwald Stoltenberg, who with Lord Owen worked out a new peace plan, sometimes referred to as the 'Invincible' plan after the British battleship on which some of the negotiations were held. In June Milosevic, Izetbegovic, Tudjman, and other leaders met in Geneva. However, seeing a plan emerge to split Bosnia into three parts, Izetbegovic walked out of the talks. On August 30, 1993 the Bosnian government rejected the Owen-Stoltenberg plan, as it had rejected earlier proposals.

Now the Croats and Muslims were seriously fighting. The Bosnian government sought to capture Croat areas of Bosnia. Tudjman's army fought aggressively to keep that from happening, but his army contained thousands of Muslims, so the officers set up camps and imprisoned the Muslims soldiers who had been their comrades only days before. The appalling conditions of these camps resembled those of the concentration camps of World War II; many died. The responsibility for these conditions went right to the top of the Croat government. Stipe Mesic said that Tudjman himself had said of these camps, "We shouldn't be ashamed of ourselves."[82] However, throughout the fall of 1993 the Bosnian army and Muslim militia were winning battles against the Bosnian Croat Army, which had the active support of the Croatian Army. Yet the Bosniaks' war against the Serbs was not going well; Bosnian Serbs (sometimes with the support of Yugoslav regulars) continued to prevail in most of their fights against the Bosnian Army in Sarajevo, the enclaves, and Mostar, and they were mobilizing an even stronger army from the Serb refugees.

The Serbs were still in the hills around Sarajevo and the war continued. On February 4, 1994, the Markela Marketplace in Sarajevo was shelled, killing 68 people and wounding over 200. This event induced Bill Clinton to issued an ultimatum to the Serbs. NATO was preparing to bomb them unless they handed over their heavy weapons. Karadzic agreed, on condition that the Russians join the peacekeeping forces; that demand was met.

American diplomats decided to pursue the solution to one of the two wars which the Bosnians were then fighting. They would propose a Muslim-Croat Federation supported by Croatia. There was one incentive for Tudjman to accept this: He needed American support for his plan to recapture those parts of Croatia that had been taken by Serbia. Seeing certain possibilities in this deal, Tudjman agreed. On March 18, 1994 he and President Izetbegovic met in Washington and signed an alliance to end their war.

But the other war continued. NATO threatened to strike the Serbs if they did not stop firing and pull back from Sarajevo. By April 27, they had essentially complied. In May the major powers formed the 'Contact Group' – representatives from Britain, France, Germany, Russia, and the United States – offered yet another peace plan, starting with a four-month cease-fire and moving eventually toward the partition of Bosnia. The Croats accepted the plan willingly, whereas the Bosniaks were more reluctant. The Bosnian Serbs rejected the proposal, against the wishes of Milosevic, who refused to give them any further help.

During the summer and fall of 1994 the Bosnian Army became more successful in the campaigns against the Serb separatists. The cease-fire around Sarajevo held, and the Bosnian troops won a major victory at Bihac. A week later, however, the Serbs retaliated against NATO, whose planes had begun bombing Serbian forces, by taking peacekeepers and holding them hostage. Former U.S. President Jimmy Carter undertook to mediate a solution to the conflict and on December 20 announced a Bosnian cease-fire, which (except in Bihac) lasted four months, ending in May 1, 1995.

Now the war in Croatia flared up again. The Croatian government undertook a new offensive against the Croatian Serbs, seeking to recapture land. In retaliation, the Serbs shelled Zagreb. By May 24, 1995 the United Nations ordered Serbs to remove heavy weapons from the Sarajevo area. When this order was not obeyed, NATO attacked. In response, the Bosnian Serbs shelled safe areas, including Tuzla, and used some 370 U.N. peacekeepers as human shields against NATO airstrikes. By mid-June, however, almost all the peacekeepers had been released. The Bosnian government launched an offensive to break the siege of Sarajevo, which ended in May 1995.

Still, the Muslim enclaves in eastern Bosnia were extremely vulnerable. For three years, tens of thousands of displaced Muslims had been in Srebrenica. After Bosniak fighters in their midst launched an attack against the Serbs, the Serbs retaliated on July 6, 1995 by shelling Srebrenica and clearing its inhabitants. People were herded into a soccer field and killed in large numbers.[83] The United Nations tried to evacuate some of them but General Mladic, the Serb commander, shoved them aside and put the people onto waiting buses. When the buses reached Muslim-controlled territory the only passengers were women and children; all the men had been taken off, shot, and buried. The death toll of Muslims there between July 12–16 was 7,079.[84]

It was the massacres at Srebrenica and later at other 'safe areas' such as Zepa, that finally forced the Western states to consider stronger responses. President Chirac of France called for the United Nations to retake Srebrenica, but the United States did not accept that idea. Instead, they agreed to shift populations around inside Bosnia so as to create a stable balance of power. Earlier such a proposal would have been considered ethnic cleansing, but now it seemed necessary.

Another aspect of the new plan was to give Croatia a green light to retake the one-fourth of its territory that the Serbs had previously captured – Krajina. Now holding a clear military advantage, Tudjman's troops began shelling the rebel capital Knin. The Serb soldiers had fled and now also fleeing were 100,000 Serbs who had long lived in Croatia. Tudjman's army burned their villages and murdered hundreds of old people. This offensive, 'Operation Storm,' resembled the U.S. military doctrine called 'AirLand Battle 2000' because the Croats were being trained by American officers based in Germany.[85] Though some of the American negotiators wanted Tudjman to call off the offensive, Ambassador Richard Holbrooke and others were pleased to see the Serb wave reversed.[86] However, they were instructed not to say publicly that they encouraged more bloodshed and fighting.[87]

While this military campaign was going on, another shell fell in Sarajevo near the marketplace, dismembering 37 people. This finally pushed the United States to undertake serious bombing of the Serbs. Milosevic summoned the Bosnian Serb leaders to Belgrade and demanded that they give him full authority to negotiate a peace on their behalf. Their acceptance, it seemed, would amount to political suicide, for the Serb Bosnian population would not have approved. However, there was one person who could provide an alibi that would be accepted by the Bosnian Serb population – the patriarch of the Orthodox Church. Following his 'suggestion,' Karadzic capitulated, and serious negotiations would begin, with Milosevic representing his side.

This time, however, the prospect of a political agreement did not suffice to halt the military actions. Within a fortnight NATO pilots flew 3400 sorties, pounding especially hard the Bosnian Serbs' communications network. Within a month the Serbs had lost much of the territory they had won in previous battles, so that they held approximately half of Bosnia's land.[88] At the same time, Holbrooke was carrying on negotiations in Belgrade, insisting that the Serbs must stop bombing Sarajevo and withdraw their weapons from its hills. Only pressure from Milosevic and Bulatovic forced Karadzic and Mladic to agree. They withdrew their heavy weapons and, in response, the Americans called off their air strikes.

Seeing that the war would soon be over, the Muslims and Croats, who were already winning their battles, pushed through to seize as much territory as possible while the opportunity still existed. Holbrooke realized that the more land they grabbed, the more they might be able to keep in a peace agreement. However, Holbrooke worried that the offensive might not succeed much longer, so he warned Izetbegovic that his lines were so thin that the Serbs might be able to punch holes in them. He proposed a cease-fire.

Izetbegovic agreed, but only on condition that the water, gas, and electricity be turned on in Sarajevo now and that the road to Gorazde be opened before the start of a peace conference. He was buying time, but he accepted the cease-fire after five days.[89] Holbrooke told Tudjman that he had five days left to take Prijedor and Sanski Most, two towns in northwestern Bosnia: "What you don't win on the battlefield will be hard to gain at the peace talks. Don't waste these last days."[90]

Holbrooke was working hard at establishing the terms of the peace. He knew that the peacekeepers, IFOR, had to be headquartered in Sarajevo and would also have to be deployed in the Serb portion of Bosnia, which would be called Republika Srpska. There was some debate about the duties of the military. They did not want to take on many responsibilities and, although Holbrooke believed they should have the duty to arrest war criminals, they accepted the authority but not the *obligation* to undertake such tasks.[91]

DAYTON AND POST-DAYTON

The leaders of all the belligerents came to Wright-Patterson Air Force Base in Dayton, Ohio in November 1995 with a team of negotiators that included Secretary of State Warren Christopher and Richard Holbrooke. Christopher laid down four conditions for a settlement: (1) Bosnia must remain a single state; (2) the settlement must take into account the special history and significance of multiethnic Sarajevo; (3) human rights must be respected and those responsible for atrocities must be brought to account; and (4) the status of eastern Slavonia must be resolved.[92]

The process was scheduled to last 17 days. They first addressed the reaffirmation of the Federation of Muslims and Croats, which had been established in September 1993 by the Washington Agreement. Then there was a solution to the timing of a reversion of eastern Slavonia to Croatia.[93] It was decided that each entity within Bosnia might have its own military force; this was recognized as a flaw, but an inevitable one.[94]

Next Holbrooke brought together Milosevic and Haris Silajdzic, the prime minister of Bosnia. One issue confronting them was the security of Gorazde, a

Muslim enclave in eastern Bosnia that had to be connected to the Muslim-Croat Federation. NATO would build the road, but Milosevic had to give up a corridor through which it would pass. Any corridor that was sufficiently wide to be secure would significantly reduce the Serbs' proportion of Bosnian territory. There was a complicated process of shaving the lines on the map to make each side's territory conform to the agreed proportions.

Also, the displaced persons voting issue was resolved. The compromise allowed people to vote in the area where they had lived in 1991, but apply to an electoral commission for the right to vote elsewhere.[95] One important issue was left unresolved: the fate of Brcko, a town in vulnerable corridor of Republika Srpska, was to be decided by arbitration at a later time. (Not until March 6, 1999, did the arbitrators decide to leave the city permanently under international control rather than return it to the Serbs.)

These, then, were the terms of the 'Dayton Accords,' which were signed a month later in Europe, where it is called the 'Paris Peace Plan'. It stopped the war without bringing peace. The document called for the presence in Bosnia-Herzegovina of an international group of peacekeepers under NATO command, to be called SFOR (Stabilization Force), plus a High Representative named by Europe to command the civilian operations. This post was initially held by Carl Bildt, former prime minister of Sweden, followed by Carlos Westendorp of Spain, and then by Wolfgang Petritsch of Austria.

Bosnia's infrastructure and economy remained terribly damaged for years before significant reconstruction began, nor was much progress made in creating joint Bosnian institutions. Everywhere the police were mainly strongly partisan and could not be counted on to impose justice. Refugees were supposedly free to go back to their former homes but usually they were justifiably afraid to return. Rebuilding was slow because foreign sources of capital were dubious, for good reasons, about the prospects of an enduring peace, especially one that would require the cooperation of all the nationalities of Bosnia.

In September 1996, probably too quickly after the war, a national government was elected, though many Serbs boycotted the collective presidency election. The winners were mainly the same people whose hateful decisions had led the country into war. The OSCE certified the elections as 'free and fair' over numerous objections, and local elections were held a year later. The High Representative accused officials of the Muslim-Croat Federation of diverting some $30 million in aid. In Republika Srpska, there was a bitter conflict between the hard-line Serb nationalists based in Pale, followers of the indicted war criminal Radovan Karadzic, and a group located in Banja Luka headed by the president of Republika Srpska, Mrs. Biljana Plavsic, who agreed to honor the Dayton agreement. A struggle developed over control of the media.[96]

President Plavsic won the first rounds of this dispute, but in September 1998 elections she was defeated by an ultra-nationalist supporter of Karadzic, Nikola Poplasen, who was later dismissed by the High Representative for consistent obstruction of Dayton implementation. On the other hand, moderates generally fared well in the parliamentary races in an election that was declared even more 'free and fair' than the previous one. The new three-member presidency now consisted of Alija Izetbegovic (overwhelmingly re-elected by the Bosniak community), the moderate Serb Zivko Radisic (who narrowly defeated a hard-liner) and Ante Jelavic, a hard-line Croat.[97]

Although in February 1993 an international war crimes tribunal had been established in The Hague and 21 Bosnian Serb commanders had been charged with genocide and crimes against humanity, few of those who were indicted were actually captured and brought to trial. Neither Mladic nor Karadzic had been allowed to run for office, but they remained free and were supported by the dominant political factions among the Serbs. The SFOR officers refused to assume the risks connected with any initiative to capture them until 1999 and 2000, when the previously reluctant French peacekeepers arrested several high-ranking indicted persons, but not yet either Karadzic or Mladic, who were believed to be in their sector in eastern Bosnia-Herzegovina.[98]

Still, Freedom House ranked Bosnia as 'partly free' in 1997–98, and in spring elections of 2000, the voters did make some important changes in the Federation by supporting candidates of the Social Democratic Party – formerly the Communists – the least nationalistic of all the parties in Bosnia and Hercegovina.

In the aftermath of the Dayton settlement, the various states had quite different experiences. Of the former Yugoslav Republics in the first few years of the post-Dayton period, *Slovenia* came closest to meeting its goals, both economically and politically. It is the only one of the states that had been part of Yugoslavia rated as fully free (i.e. democratic) by Freedom House.[99]

Croatia – or at least Franjo Tudjman – was the main winner from the Dayton settlement. The Croatian entity within Bosnia was merged into the Muslim-Croat Federation, but nevertheless it continued to function almost as an integral part of Croatia.[100] According to Freedom House, Croatia soon achieved a rating of 'partly free'. In December 1999 Tudjman, who was being investigated for war crimes by the tribunal,[101] died of cancer. A national election was held soon and he was succeeded by the same Stipe Mesic whose election to the presidency of the former Yugoslavia had been blocked by Milosevic in the final months before the country broke up. Mesic immediately displayed a new commitment to democratization and to reforming the conspicuous corruption which had characterized Tudjman's government.

Macedonia, with its two million inhabitants, was the only republic that managed to secede peacefully from Yugoslavia, almost by default. There had been no strong nationalistic movement there until 1992. When the new nation adopted the Macedonian name and symbols, Greece objected strongly, claiming the right to reserve these symbols for use by the region of its own country that is also called Macedonia. This Greek response largely created Macedonian nationalism. Besides these symbolic conflicts, the new state confronted some realistic threats. It was subjected to a double blockade for two years: Greece embargoed trade on the one hand and there were international sanctions against Serbia, on which it was dependent for trade routes as well as goods. Nevertheless, Macedonia came through that period with considerable stability, primarily thanks to the U.N. preventive force (UNPREDEP) sent by the Security Council. Indeed, in 1999 it found ways to accept an enormous influx of Kosovar refugees and continued to avoid the strife typical of the other former republics of Yugoslavia.[102] It was rated as 'partly free' by Freedom House in 1997–98.[103]

Serbia remained under the control of Milosevic, who continued to function as a dictator, occasionally curtailing the freedom of the few radio stations and newspapers that dared criticize his regime. In the winter of 1996–97 there arose a popular opposition movement called *Zajedno* ('Together') in Belgrade and a few other cities.[104] This occurred because Milosevic openly prevented elected candidates from taking their seats if they were critical of him. A coalition of independent political factions came together and marched daily in the streets for three months, banging on pots and pans to make noise, especially when the state-run newscasts portrayed events through its own self-serving lens. Eventually Milosevic relented somewhat, but the movement did not hold together long enough to reach its objectives. It was composed of three very dissimilar factions that could not cooperate long. At one side there was a genuinely democratic movement led by sociologist Vesna Pesic; in the centre there was a movement of mixed policies led by Zoran Djindjic, and on the other side was a Serb nationalist movement led by Vuk Draskovic, who had called for the expulsion of Kosovars from Kosovo. Later Milosevic coopted Draskovic to serve in his government, but this connection did not survive NATO's military intervention over the human rights violations in Kosovo in 1998–99.

Montenegro, the only republic to remain part of Yugoslavia wtih Serbia, soon became less satisfied with its status. By early 2000, democratically oriented President Milo Djukanovic was engaging in a sustained conflict with his counterpart, President Milosevic and had proposed a referendum on the independence question. Belgrade then countered with a media campaign and threats of military intervention to prevent secession. Djukanovic's Democratic Party of Socialists then accused Yugoslav generals of forming a paramilitary unit in Montenegro with loyalties to Milosevic. Many informed political observers were predicting war.[105] The leader of the Social Democratic Party, along with some officials of the OSCE, urged that international observers be deployed in Montenegro early, to prevent the fighting.[106] This may not happen, since previous bad experiences have led the international community to try to prevent secession in any of the countries in the region, even including Serbia. There are certain examples of such missions that have worked – notably the presence of observers along Macedonia's border with Kosovo and Serbia, which is one reason why Macedonia avoided war. On the other hand, UN peacekeepers have sometimes been in place without being able to prevent violence – as for example in Croatia in 1991 and most conspicuously in Srebrenica in 1995.

According to Freedom House, the rump Yugoslavia (*Serbia* and *Montenegro*) ranked as 'not free' in 1997–98. And, not surprisingly, *Kosovo,* which was rated separately, was also designated as 'not free'.[107] This reflected the plight of the Albanians in that province during the period preceding NATO's decision to resort to bombing to prevent the ongoing violations of their human rights.

THE WAR IN KOSOVO

Oddly, it was probably Kosovo that suffered most (albeit indirectly) from the Dayton Accords. It was in Kosovo in 1987 that Slobodan Milosevic had launched the nationalistic chauvinism that led to secession and wars throughout Yugoslavia, yet for several years the outside world hardly heard of Kosovo, while the news was full of stories about the troubles in Bosnia, Croatia, and Serbia. This quiet was misleading, for human rights violations were continuing in Kosovo throughout that period, and the Albanian population was intensely engaged in a plan to rescue its culture. Had the international community paid more attention to the Kosovars and their non-violent campaign for independence, the transition to democracy there might have been uneventful. This did not happen.

When President Clinton summoned the belligerents of the former Yugoslavia to Dayton, he did not include any representatives from Kosovo. Milosevic would not have come to Dayton had Kosovo been on the agenda and had Kosovars been invited. The United States' priority was stopping the war in Bosnia-Herzegovina. Nevertheless, the omission proved to many Kosovar activists that their non-violent campaign was completely failing, and that they would have to take up arms in pursuit of their goals. That realization became the impetus that led to the rise of the Kosova Liberation Army (KLA) and its armed struggle against the Serbian army and government. The conflict would lead the NATO countries to bomb throughout Serbia on behalf of the Kosovars.

According to Human Rights Watch, during the decade after Milosevic removed the autonomy of Kosovo, the province underwent the worst record of human rights violations in Europe. There had been 'creeping ethnic cleansing',[108] as every possible incentive was created to induce the Albanian population to emigrate and Serbs to settle in the region.

From 1974 to 1987, when Kosovo enjoyed considerable provincial autonomy, the mainly Albanian leadership of the local League of Communists reversed much of the official discrimination against Albanians. This change led to allegations that the Serbs had become a disadvantaged group. In the intense period of the mid-1980s, this reverse discrimination was escalated into a charge of 'cultural genocide' of Kosovo's Serbs. In fact, radical Albanians also opposed the ethnic Albanian provincial leadership, which they identified with the 'repressive' bureaucratic federal system and Party.

The Albanians of Kosovo had been in control of their own administration since 1974, when they began introducing bilingualism into the government (instead of using only Serbo-Croatian) and equalizing opportunity for employment. (Though outnumbered, the Serbs had previously dominated in language, status, and occupational ranks.) The miners and other Kosovar workers went on strike after Milosevic stripped their province of its autonomy in 1989–90. In retaliation, at least 70% of all employed Albanians were dismissed from their jobs. The schools were forbidden to instruct in Albanian, but the teachers defied this order and were in turn dismissed too. Albanians were evicted from the University and from all except two high school buildings.[109] At that time, though Kosovo had the highest birthrate in Yugoslavia and the highest illiteracy rate, it also had the highest proportion of youths in higher education,[110] and was fully committed to their training.

The League for a Democratic Kosova (LDK), was led by Ibrahim Rugova, who was chosen in clandestine elections as president of a separatist republic in 1992. Rugova adopted methods that he had learned from reading Gandhi and Martin Luther King, Jr. – especially non-cooperation tactics and the maintenance of parallel institutions.

Therefore, though most Albanian physicians had been fired from the state hospitals and Serbian was the only language allowed, the movement set up private clinics to provide free treatment. Wherever workers were fired, the movement encouraged them to set up small businesses. Belgrade fired 70% of all employed Albanians and imposed a Serbian curriculum of education. The teachers had insisting on teaching the Albanian curriculum, so the government stopped paying them, yet they carried on anyway for nine months – until they were bodily kept out of the schools. At that point, the teachers set up a parallel school system in private places such as empty houses and garages, serving their children of all ages, including university students. The people began a system of voluntary taxation that raised about 70% of the money required for their schools. The rest was raised by their government-in-exile in Germany. The Kosovars living in exile – who numbered some 300,000 – also sent remittances home.[111]

Rugova and his party had concluded that the Serbian reprisals were so harsh that public demonstrations were not worth the penalties they incurred. Instead, they began to emphasize lobbying foreign governments, but they received little overt international support. Popular support for the non-violent campaign waned, but the mounting dissatisfaction did not immediately translate into a desire to take up arms, but rather into a sense of urgency about finding other, more effective, non-violent methods of struggle. University students took to the streets, ignoring the warnings of Rugova. Police used tear gas to disperse their march.

At about that time the first public actions were undertaken by the Kosova Liberation Army (KLA, or, to use the Albanian acronym, UCK), a guerrilla army that had begun forming in the spring of 1996, shortly after the Dayton agreement. Its political base was a political party that had existed since 1982: the Levizja Popullore e Kosoves (LPK), or Popular Movement for Kosovo.[112] According to Adem Demaci, the KLA's political representative, the disaffected Albanians had drawn two lessons from Dayton. Lesson one: non-violence does not work. Though they had pursued a five-year strategy of exemplary non-violence, the United States had ignored their concerns and had made a deal with the warrior Milosevic. Hence lesson two: violence pays.

In early 1997 the state collapsed in Albania and widespread looting of military bases took place. One could buy a Kalashnikov for $15, and soon these weapons were being put to use across the border in Kosovo. The non-violent Kosovar activists were turning to violence. The KLA ranks swelled, predominantly by rural people who had been political prisoners in the former Yugoslavia and who were dissatisfied with Rugova's moderate methods. That summer the KLA took over large towns, which they could not defend. They fled when the Serbs came back and wreaked vengeance on the civilian inhabitants, yet those same civilians considered them as heroes.[113]

By the spring of 1998, Rugova was still the elected president of the Kosovars, yet the KLA had stolen his initiative. Fighting was increasingly widespread, but outside observers still believed that compromise was possible. Rugova had never wavered from his commitment to independence for Kosovo, but there was some basis for hoping that he might accept virtual statehood within a new Yugoslavia comprising three republics: Serbia, Montenegro, and Kosovo. This was the solution with the widest appeal, since the rest of the world was coming to oppose secession, which seemed to be a formula for additional trouble everywhere it was tried. Yet this three-state solution did not prevail because the KLA offensive, having seized control of 40% of the province, was on a roll and the Kosovars were euphoric about their prospects.[114]

They should not have been so optimistic. The Serbian army had recovered some of its strength after the cease-fires preceding Dayton. Yugoslavia's replenished military forces now included about 15,000 members of the air force; 85,000 army troops (mostly poorly trained conscripts) plus possibly 200,000 reservists; and 15,000 personnel in the navy. They possessed 15 of the best MiGs and about 50 obsolete MiG interceptors, plus about 90 light ground-attack jets, in addition to various missiles. The army had about 1,270 battle tanks, mostly obsolete, plus 300 modern Soviet ones (which had fared poorly against the Western tanks in the Gulf War); and numerous armored fighting vehicles and guns. However, morale was low, and in some areas only one in ten reservists who were called up actually reported for duty. The country was drifting toward civil war, although no commanding officer had yet emerged to give the KLA a clear direction.

In September 1998 Serb forces attacked central Kosovo. Now the international community became more seriously involved. The U.N. Security Council adopted a resolution calling for an immediate cease-fire. Some diplomatic monitors had been accepted in Kosovo, beginning in July. Holbrooke met with Milosevic in Belgrade in October and threatened that NATO would bomb if the Serbian army's offensive did not cease. A cease-fire was agreed and Milosevic agreed to withdraw troops. The deal also permitted some 2,000 unarmed verifiers to be sent by OSCE to join the other monitors and try to verify compliance with the cease-fire. However, Holbrooke agreed that Milosevic might retain nearly 20,000 armed personnel in Kosovo, and in fact Milosevic did not keep his commitment to reduce Serbian forces. The failure of this deal set the stage for a renewal of fighting in December and January.[115]

U.S. Ambassador Christopher Hill, though based in Macedonia, undertook shuttle diplomacy, negotiating with the KLA and Milosevic.[116] Yet the responses in Serbia were not encouraging. Most Belgrade politicians – including those who opposed Milosevic and wanted democracy – promised to fight to keep Kosovo. Even if an opposition politician could oust Milosevic, the fact remains that he had systematically closed down freedom of press, so the public knew little about the Serb atrocities in Kosovo. Therefore, Milosevic's successor – had he been chosen in a free election at that time – might have been a nationalist of truculent disposition, such as Vojislav Seselj.[117] Between October and December 1998, there was scattered violence in Kosovo every day.

In early February 1999 a first round of peace talks began in Rambouillet, France between Kosovo Albanian rebels and Serbs, even as intense fighting occurred on the ground. The Contact Group (France, Germany, Italy, Russian Federation, United Kingdom, and the United States) had issued ultimatums to Serbs to get the process underway: Come to Rambouillet and sign the documents there, or face air-strikes.

The Serb delegation consisted of lawyers and party members from Belgrade, as well as unknown representatives from minorities throughout the region. The delegation excluded academics and Serbian Orthodox priests.

The Albanian delegation was headed by the KLA, though that army had no legitimate legal status in Kosovo and was never endorsed by President Rugova, the elected leader of the republic.

The delegations arrived in Rambouillet to find a document already on the table. It had been prepared in advance by Ambassador Hill. There were two co-chairs, France and England, with a Russian mediator. During the three weeks of meetings at Rambouillet, there were no meetings between the two delegations. The draft text indicated that Kosovo would be supervised by a peacekeeping force, but would have a fully-functioning indigenous government after 18 months. The terms that were most problematic to Serbia were those suggesting that the Kosovars might be permitted to secede three years in the future.[118] This condition was demanded by the KLA and adopted only reluctantly by NATO when it became clear that the Kosovars would not otherwise sign the Rambouillet document. According to some observers, the Americans had seriously bungled the negotiations, first by weakening the status of the non-violent elected Kosovar government, then by taking for granted the KLA's acceptance of the text without bothering to address their security concerns by including NATO officials in the negotiations.[119]

Though ultimately the Kosovars proved to be conditionally willing to sign it, Milosevic was not. The clause promising a future meeting to reach the final settlement three years later implied to him that the Kosovars would be permitted to secede at that point. This was the main issue of contention between his government and the Kosovars; to accept that would be tantamount to capitulation. For their own internal political reasons, the Russians did not put pressure on Milosevic to sign the text and only later, during the bombing, began to seriously work on brokering an agreement.

NATO BOMBS SERBIA AND KOSOVO

The issue was not brought before the U.N. Security Council, since the Russians would certainly have vetoed any resolution calling for military action to enforce adoption of the Rambouillet text or any other similar document. President Clinton and the other NATO leaders therefore took the initiative without authorization and launched bombing attacks on March 24, 1999 against military targets throughout Serbia and Kosovo.

It was a draconian action ostensibly intended to save Albanians from atrocities, but it caused more immediate tragedies than it prevented. The Serbian forces had been assembling in preparation for a major assault against Kosovo, and when the bombing began, Milosevic ordered that invasion. Hordes of refugees fled or were expelled forcibly from the country, and mass killings were carried out on the ground in Kosovo. According to estimates by a Swedish peace group, the Transnational Foundation (TFF) on April 16, 1999, there had been 13 months of warfare in Kosovo before the NATO bombings began, with an average of about five persons being killed per day and perhaps 632 persons fleeing from their homes. After the NATO bombing began, TFF asserts that the international community estimates that 15 persons were day were being killed and as many as 20,833 refugees per day running or being expelled from their homes. "Even if the figures above are estimates, there must be something fundamentally wrong with a peace policy that seems to kill 3 times more civilian people and produce 33 times more refugees per day than did the war it aims to stop," said TFF director Jan Oberg.[120]

The Russian prime minister met with Milosevic in March, but failed to reach a breakthrough that would halt the bombing, which included strikes in the heart of Belgrade, where the headquarters of the Yugoslav First Army was destroyed, along with civilian infrastructure such as bridges, and even (by accident) the Chinese Embassy. In fact, the attacks consolidated the morale of the Serbian population, including persons who had long been strong opponents of the regime. It was not that they liked Milosevic any better, but that they were angry toward NATO about the casualties and destruction of the infrastructure.

Milosevic may have actually wanted the bombing because it gave a cover for the ethnic cleansing that he had planned in advance.[121] Refugees who succeeded in escaping from Serbia reported that the 'cleaners' had told them they had orders to 'clean' Kosovo within a week. By May, NATO claimed to have destroyed about one-quarter of the Serbian military capacity, but instead of halting the ethnic cleansing, it had intensified it. Some 1.2 million Kosovars had fled, leaving only about half a million persons of their community in Kosovo, mostly hiding in woods or being moved from place to place. About 100,000 men were then missing, thousands others had been executed; women had been raped, and large sums of money had been exacted before refugees were permitted to go to the border.[122] Milosevic was indicted for war crimes by the international tribunal.

The bombing continued until June 22 – much longer than the NATO leaders had expected would be necessary. In May, after a series of mistakes, including the bombing of a refugee convoy and a strike against a Serbian old people's home, the NATO leaders were feeling desperate, as opinion polls showed support for the bombing was dropping sharply. The commander, General Wesley Clark, succeeded in obtaining permission to hit the Yugoslav power grid. This was the most effective military strike of the campaign, knocking out computers and civilian amenities throughout the country. Although this move had been avoided until that time, some analysts argue that it was more merciful than other targets, causing few casualties but immediately making it clear to the population that the regime's days were numbered.[123] Shortly thereafter, the Serbs began to negotiate seriously.

The Russian envoy, Victor Chernomyrdin, was representing the Serbian government in negotiations with the American Deputy Secretary of State, Strobe Talbott, and Finnish President Martti Ahtisaari. When they reached an agreement on June 4, Milosevic accepted it and agreed to withdraw all Yugoslav military and police forces from Kosovo within seven days and allow more than 50,000 foreign troops under a United Nations flag – many of them from NATO and under NATO command – to police the province.[124] Even so, the insistence by Milosevic and the Russians that the U.N. and not NATO should govern Kosovo meant that the United Nations was forced to undertake an operation for which it was not prepared and had no funds.

By June 13, about 200 Russian troops entered the Kosovo capital and took up positions at the military airport, much to the consternation of NATO. A few hours later, the first NATO troops arrived. An agreement was established for joint peacekeeping operations between the Russians and Westerners, though the relations between these two sides had become more acrimonious during the war than at any other time since the ending of the Cold War.

The terms of the peace agreement were basically the same as those that the G8 industrial nations had laid down a month before. They permitted a few hundred uniformed Serbs to return to Kosovo and guard border posts and Serbian holy sites as a symbol of Yugoslav sovereignty over the province, but there was little sovereignty left otherwise for Belgrade. Kosovo was to have 'substantial autonomy' and was to be under the supervision of an international authority that would establish new democratic institutions and elections, as well as seeing to the return of refugees.

Nevertheless, it can be argued that Belgrade made major gains from the conflict with NATO, for the terms of the peace agreement are more to its liking than the one produced at Rambouillet. Nothing was said in the new text about any referendum or re-examination of Kosovo's sovereignty after three years, as Rambouillet did. Indeed, the final peace agreement does not differ greatly from the one that the Serbs would have accepted before the bombing began.

However, they had refused to allow armed foreign peacekeeping forces into the country, even though the Russians had tried to persuade them to do so. A primary goal of the bombing was to force Serbia to admit a peacekeeping force, and it achieved this objective – although at an extraordinary cost and within the context of no clear overall plan for the region.

The war also produced an unexpected result: It strengthened the KLA and made peacekeeping in autonomous Kosovo much more difficult than SFOR's operation in Bosnia. Since the KLA served essentially as NATO's ground force during the war in Kosovo, it was able to wrest concessions that delayed its dissolution after the war. Its continued armed presence encouraged Albanian atrocities against Serbs and contributed to a massive flight of this minority from Kosovo. Thus, critics of the war have argued that the bombing had been unnecessary and, apart from forcing Serbia to admit the peacekeeping forces (KFOR) to Kosovo, did not accomplish what had been intended.

KFOR's mission – to ensure that a multiethnic, autonomous Kosovo operates within an impartial rule of law – could not be accomplished. Instead, the recriminations of those Albanians returning to Kosovo meant that it was the turn of the Serbs to flee in terror – most of them into Serbia. Two-thirds of the pre-war Serb population (200,000 people) fled, along with more than 50,000 Romas, Slav Muslims, and Croat Catholics. Though officially disbanded, the KLA was actually still active, their secret police carrying out executions and intimidations. More than 80 Orthodox churches were severely damaged after the war, often in the presence of KFOR troops. Almost all Serb shops were acquired by Albanians.[125] In March US troops had to rush into Mitrovica in the northeast part of Kosovo to reinforce French troops that were trying to collect arms and separate the Kosovars and Serbs. Mitrovica was almost the only place in Kosovo where significant numbers of Serbs remained. After the peacekeepers intervened to keep the battling groups apart, the townspeople attacked them and the NATO troops were sustaining casualties from both sides. General Clark told the Senate Armed Services Committee that "the hardest part of securing peace in Kosovo lies ahead."

Elsewhere, the supposedly demobilized KLA leaders were involved in several new political parties, especially the Party of Democratic Prosperity of Kosovo, led by Hashim Thaci. Nevertheless, polls in early 2000 indicated overwhelming popular support for Rugova and his LDK party, as they had done before the war. Had the NATO countries supported Rugova or any other non-violent leader, the war might have been avoided.

But there are countless additional ways in which, in retrospect, one can see that Kosovo's tragedy might have been avoided. Yugoslavia could have prevented the violence, had Milosevic wanted to do so, by accepting international peacekeeping forces early on, thereby making the Rambouillet meeting and the subsequent

NATO bombings unnecessary. Before Rambouillet, both sides could have reached the same agreement (apart from the presence of peacekeepers) as the one they eventually adopted on June 4. Great suffering might have been prevented and the final peace agreement would have been acceptable to the liberal elements of Serbia's democratic opposition and to the Orthodox Church.

TOWARD RECONSTRUCTION: THE STABILITY PACT FOR SOUTHEASTERN EUROPE

In July 1999 some forty heads of state attended a summit in Sarajevo to inaugurate a new institution, the Stability Pact for Southeastern Europe. This ambitious plan is designed to foster democracy, human rights, and market-oriented economic reforms throughout the Balkan peninsula, and to offer financial assistance and, eventually, membership in Western Europe's political and economic institutions. All the Balkan states will be brought into the scheme – from Albania to Slovenia, and from Croatia to Bulgaria.[126] Its premise is that the region's problems are inter-dependent and can be handled only by viewing the Balkans as a single political and economic zone. For example, the high levels of organized crime and corruption in the area transcend borders and must be addressed comprehensively. Similarly, there are displaced persons throughout the region who cannot return to their homes unless there is a new level of protection offered to minority persons. The new organization is promoting cross-border cooperation and infrastructure development.

Among the many initiatives taking shape within the Stability Pact framework are a scheme to reduce the flow of small arms; a pledge by Bosnia to reduce its military budget; and a regional task force on gender issues, already operating in Bucharest. Unfortunately, the Stability Pact has no resources of its own. It depends largely on the EU, which is negotiating agreements with almost all the states in the region, applying 'conditionality' to the process. If the politicians fail to deliver on their promises, the promised integration will halt from the Western side in response.[127]

CONCLUSION

I have offered almost no theoretical framework here for addressing the big question about the collapse of Yugoslavia. However, it may be useful to name the three most common frameworks that underpin most discussions of the former Yugoslavia's demise. There may be some validity in all three models, though perhaps more in one than another.

First, one explanation for the tragic and violent breakup of the country can be called the theory of a 'democratic deficit'. This model focuses attention on Yugoslavia's political history before the end of the Cold War. It portrays Yugoslavs as inexperienced with democratic institutions, unfamiliar with the democratic habits of civil society, and as having inherited an unworkable constitution that could not yield solutions to the challenges of the 1990s.[128] Had democracy been institutionalized earlier, this theory suggests that the collapse might have been avoided. The model criticizes the Serbian leader Slobodan Milosevic as particularly anti-democratic and suggests that a different leader with a proper grasp of pluralism might have facilitated a smooth transition to democracy.

Second, there is the 'nationalist theory' – an explanation for Yugoslavia's troubles that refers to the long history of ethnic hatred in the Balkans. According to this model, democracy could not have prevented the crisis. Instead, the coming of greater freedom of expression actually allowed the Yugoslav citizens to demonstrate overtly their centuries-old hatred of other nationalities, in contrast to the communist period when animosities had been repressed. This model, which takes nationalistic hatred as immutable, tends to assume that the solution to the crisis must involve a partition of the state to separate populations who will never learn to live together peaceably because of their long history of violence.

Finally, the 'economic model' is the third explanation for the collapse of Yugoslavia. It points to economic factors, especially the foreign debt that had accumulated during the Cold War and the stringent measures imposed by the International Monetary Fund to curtail inflation. Yugoslavia was never a rich country, and it is true that it underwent a painful belt-tightening shortly before the country broke apart. Some observers blame the crisis on the poverty that accompanied the Western-imposed economic austerity program of the late 1980s and early 1990s.[129] One the other hand, many other analysts, particularly those of Yugoslav background, believe that the harsh economic measures were necessary, successful, and even accepted by the citizenry.

It is unlikely that the readers will adopt all the same conclusions about these interpretations. However, it may be useful to read the pages ahead with these three explanatory models in mind. I shall not argue in favor of any of them, though I may say that the economic explanation seems least likely to be sustained except perhaps in combination with one of the other factors. The nationalist model was rejected by almost all the contributors to the conference and is equally unpopular among the other contributors to this volume, who, I think, predominantly hold to the democratic deficit model. I shall not pursue this question further here, for the authors of the papers that constitute this book have interesting analyses to offer the reader, and it will be more valuable to move directly into an exploration of their ideas.

NOTES

1. Constantine A. Chekrezi, *Albania: Past and Present* (New York: Macmillan 1919).
2. This is a common estimate but it may be wide of the mark. It is not based on any census, for the Albanians of Kosovo boycotted the 1991 census, as well as the elections, as a means of political protest.
3. Mihailo Crnobrnja, *The Yugoslav Drama* (2nd ed.) (Montreal and Kingston: McGill-Queen's University Press, 1996) pp. 16–20.
4. Crnobrnja, p. 23.
5. Christopher Bennett, *Yugoslavia's Bloody Collapse: Causes, Course and Consequences* (New York: New York University Press, 1995) pp. 22–23.
6. Bennett, p.29.
7. Crnobrnja, p. 51.
8. Bennett, p. 35.
9. Dusan Necak, 'Historical Elements for Understanding the "Yugoslav Question"' in Payam Akhavan and Robert Howse, eds. *Yugoslavia: The Former and Future* (Washington: The Brookings Institute, 1995) p. 23.
10. Bennett, p. 40.
11. Crnobrnja, p. 65.
12. Bennett, pp. 43–44. See also Michael Ignatieff, *Blood and Belonging: Journeys into the New Nationalism* (New York and Toronto: Viking, 1993) pp. 22–24.
13. For an evaluation of the various estimates, see Bennett, pp. 45–56.
14. Crnobrnja p. 66.
15. Warren Zimmermann, *Origins of a Catastrophe* (New York: Random, 1996) p. 39.
16. Bennett, p. 47.
17. Crnobrnja, p. 66.
18. Ignatieff, p. 16.
19. Crnobrnja, p. 69
20. Susan L. Woodward, *Balkan Tragedy: Chaos and Dissolution After the Cold War* (Washington, D.C.: The Brookings Institution, 1995) p. 31.
21. As Warren Zimmermann, who was U.S. Ambassador to Yugoslavia at the time of its breakup, wrote, "The government of Yugoslavia was constitutionally the weakest in Europe – a condition that contributed critically to the rise of the republican nationalism that killed it. I cautioned Washington not to equate decentralization with democracy or centralism with authoritarianism. Those equations might have described the Soviet Union, a ruthless dictatorship from the center, but they didn't describe Yugoslavia." *Origins of a Catastrophe*, p. 17.
22. Robert Howse, 'A Horizon Beyond Hatred?' in Akhavan and Howse, p 4.
23. Zoran Pajic, 'Bosnia-Herzegovina: From Multi-ethnic Coexistence to "Apartheid" and Back,' in Akhavan and Howse, p. 154.
24. Woodward, pp. 25–26.
25. Woodward, p. 37.
26. Vojin Dimitrijevic, 'The 1974 Constitution and Constitutional Process as a Factor in the Collapse of Yugoslavia,' in Akhavan and Howse, p. 45.
27. Mitja Zagar, 'Yugoslavia: What Went Wrong? Constitutional Aspects of the Yugoslav Crisis from the Perspective of Ethnic Conflict', in this volume.
28. Dimitrijevic, p. 55.

29. Dimitrijevic, p. 60–61.

30. Zagar, ibid. and Dimitrijevic, p. 61.

31. Zimmermann, p. 9.

32. Compare this to the 'consociational model of democracy' frequently analyzed by Arend Lijphart as mechanisms for balancing interests in multi-ethnic states. See Arend Lijphart, Bernard Grofman, *Choosing an Electoral System : Issues and Alternatives* (New York : Praeger, 1984)

33. Pajic, p. 154.

34. Zimmermann, p. 54.

35. Zimmermann, p. 86. Also, see Savka Dabcevic-Kucar: '71, *Croatian Dreams and Reality* (1997) (in Croatian) for the ethnic composition of the fifth (Zagreb) Army in 1969. Yugoslavia was militarily divided in seven army areas: Ljubljana, Zagreb, Split (the navy), Sarajevo, Belgrade, Nis, and Skopje.

36. Woodward, pp. 49, 51.

37. Bennett, pp. 69–70.

38. Woodward, pp. 74, 78, 80.

39. Zimmermann, pp 42, 46. Crnobrnja's account of the inflation rate is lower, but he agrees that Markovic brought it down in a surprisingly short time to a level of 60% annually and, by the spring and summer of 1990, to zero. See Crnobrnja, p. 90.

40. Zimmermann, pp. 48–49.

41. Bennett, pp. 117–19.

42. See Vamik D. Volkan, *The Need to Have Enemies and Allies: From Clinical Practice to International Relationships.* (Northvale, N.J.: Jason Aronson, 1988).

43. Zarko Puhovski, 'The Bleak Prospects for Civil Society,' in Akhavan and Howse, eds. p. 122.

44. Bennett, pp. 86–87.

45. Crnobrnja, p. 95.

46. Crnobrnja, p. 103.

47. Bennett, pp. 86–91.

48. Bob Allen, 'Why Kosovo: Anatomy of a Needless War', Canadian Centre for Policy Alternatives, July 1999, p. 10. See also reports of persecutions by ethnic Albanians in the *New York Times*, Nov. 1, 1987, Sunday, Late City edition, Section 1, Part 1, p. 14, Col. 1.

49. Vamik D. Volkan, 'Post-Traumatic States: Beyond Individual PTSD in Societies Ravaged by Ethnic Conflict', a lecture presented at the United Nations in 1998.

50. Mark Thompson, A Paper House: The Ending of Yugoslavia (London: Hutchinson/Radius, 1992).

51. George R. Urban, Radio Free Europe and the Pursuit of Democracy (New Haven: Yale University Press, 1997).

52. International Commission on the Balkans, Unfinished Peace (Berlin: Aspen Institute and Carnegie Endowment for International Peace, 1996), p. 27.

53. Zimmermann, p. 57.

54. Zimmermann, pp. 80–81.

55. Woodward, p. 138.

56. Darko Silovic, 'The International Response to the Crisis in Yugoslavia,' in this volume.

57. This was the opinion of Warren Zimmermann as expressed in Origins of a Catastrophe, p.125.

58. The BBC Documentary, The Death of Yugoslavia shows portions of interviews with both Kucan and Milosevic, who acknowledge having discussed the possibility of Slovenian secession.

59. Dusan Janjic, 'Resurgence of Ethnic Conflict in Yugoslavia: the Demise of Communism and the Rise of the "New Elites" of Nationalism,' in Akhavan and Howse, pp. 31–32.

60. Crnobrnja, p. 157.

61. Silovic, ibid.

62. Zimmermann, pp. 132–37 describes Baker's visit, portraying it as an admirable performance.

63. Zimmermann, pp. 148–49; Bennett, p. 159.

64. *The Death of Yugoslavia.* BBC Documentary. First episode, 'War of Independence'.

65. Bennett, pp. 160–66.

66. Crnobrnja, p. 194.

67. BBC Documentary, *The Death of Yugoslavia*, 'Independence'.

68. Zimmermann, p. 176–77.

69. Zimmermann, p. 187.

70. Woodward, p. 210.

71. BBC Documentary, *The Death of Yugoslavia*, 'The Gates of Hell'. See also Laura Silber, and Allan Little, *The Death of Yugoslavia.* (London: Penguin, 1995) p. 213.

72. In 'The Gates of Hell' Seselj explicitly states that Milosevic personally asked him to send his fighters to Bosnia.

73. In 'The Gates of Hell' this official, a Mr. Mendiluce, recounts his experience.

74. Mark Danner, 'America and the Bosnia Genocide', in *The New York Review of Books*, Dec. 4, 1997, pp. 60–61.

75. Wayne Bert, as quoted by Danner, 'America and the Bosnia Genocide', p. 63.

76. Danner, 'America and the Bosnia Genocide', p. 57.

77. Woodward, p. 307.

78. Mark Danner, 'Clinton, the U.N. and the Bosnian Disaster,' *New York Review of Books*, Dec. 18. 1997, p. 74.

79. Jan Willem Honig and Norbert Both, *Srebrenica: Record of a War Crime* (New York: Penguin, 1996).

80. James Gow, *Triumph of the Lack of Will: International Diplomacy and the Yugoslav War* (New York: Columbia University Press, 1996).

81. Zimmermann, p. 222, says that he knew that the Bosnian government was actually enthusiastic about the plan but that their seeming reluctance was only a tactical move. Other observers, on the other hand, believe that Izetbegovic held serious objections. For example, Paul Szasz recounts being present when a crucial ultimatum from the Croatian nationalist leader Mate Boban was delivered to Izetbegovic demanding that it be accepted.

82. 'Gates of Hell'

83. Richard Holbrooke, *To End a War* (New York: Random, 1998).pp. 69–70.

84. According to Mark Danner, Bosnian Foreign Minister Sacirbey had telephoned Madeline Albright to discuss the massacre in Srebrenica on July 13. She asked the intelligence officers for corroborating evidence, but they ignored her request. See Mark Danner, 'The Killing Fields of Bosnia', *New York Review of Books*, Sept. 24, 1998, pp. 63–77.

85. Mark Danner, 'Operation Storm' in *New York Review of Books*, Oct. 22, 1998, pp. 74.

86. Holbrooke, pp 73, 160.

87. Holbrooke, p. 172.

88. Mark Danner, 'Operation Storm', pp. 73–79.

89. Much of this account comes from 'Gates of Hell' by BBC.

90. Holbrooke, p. 199.

91. Holbrooke, pp 221–22.

92. Holbrooke, p. 237.

93. Holbrooke, p. 264.

94. Holbrooke, p. 277.

95. Holbrooke, p. 309.

96. *Freedom in the World*, 1997–98, p. 156.

97. RFE/RE *Balkan Report*, 7 October 1998.

98. Institute for War and Peace Reporting, *Balkan Crisis Reports*, Feb 29, 2000.

99. Freedom House, *Freedom in the World–The Annual Survey of Political Rights & Civil Liberties 1997–98* (New York, NY: Freedom House.) This organization is devoted to measuring democracy by a single standard on an annual basis. Its summary ratings ('Free', 'partly free', and 'not free') are based on the manifestations of civil liberties and political rights. See its web site for further methodological details: <www.ciesin.org/docs/>

100. Zimmermann, p. 233.

101. The files which might have incriminated him went missing after his death, allegedly because they were controlled by his son, then the head of Croatia's intelligence service. See Institute for War and Peace Reporting, *Balkan Crisis Report*, Feb. 26, 2000.

102. See Alice Ackermann, 'Conflict Prevention in Macedonia', *Peace Magazine* May, 1998, p. 26.

103. *Freedom in the World*, 1997–98, p. 340.

104. Ken Simons, 'Islands of Civility and Resistance', *Peace Magazine*, May/June 1997, pp. 16–19.

105. Gordana Igric in *Balkan Crisis Report*, Feb. 29, 2000.

106. Radio Free Europe/Radio Liberty Balkan Report, Vol. 4, No. 1;8, March 3, 2000.

107. *Freedom in the World*, 1997–98, pp. 546–61.

108. Mient Jan Faber and Mary Kaldor, 'What is Humanitarian Intervention?' a paper distributed through the network of Helsinki Citizens Assembly, May 1999.

109. Howard Clark, 'Prospects for Peace in Kosova', *Non-violent Action*, a publication of War Resisters League, also available on the Nonviolence Web.

110. Howard Clark, 'Kosova's Nonviolent Struggle', *Peace Magazine*, January 1998, p. 10.

111. Clark, 'Kosova's Nonviolent Struggle'

112. Tim Judah, 'Impasse in Kosovo,' *New York Review of Books*, Oct. 8, 1998, pp. 4–6.

113. Timothy Garton Ash, 'Cry, the Dismembered Country', *New York Review of Books*, Jan. 14, 1999, pp. 29–33.

114. Judah, 'Impasse in Kosovo.'

115. James Hooper, 'Kosovo: America's Balkan Problem', *Current History* April 1999, p. 180

116. Ash, pp. 30.

117. Ash, p. 32.

118. The clause read: "Three years after the entry into force of this Agreement, an international meeting shall be convened to determine a mechanism for a final settlement for Kosovo, on the basis of the will of the people, opinions of relevant authorities, each Party's efforts regarding the implementation of this Agreement, and the Helsinki Final Act, and to undertake a comprehensive assessment of the implementation of this Agreement and to consider proposals by any Party for additional measures." Apparently the 'will of the people' referred to Kosovars.

It should be mentioned that there were other clauses in the document (especially Chapter 4, Art. 1) to which a number of Western critics took exception. These clauses seem to have stipulated that the peacekeeping force would have the right to occupy the entire Federal Republic of Yugoslavia, not just Kosovo, and that the economy of Kosovo must function in 'accordance with free market principles'. That final clause is indeed strange, but the one about occupying the entire country is evidently a standard clause in such documents, basically intended to guarantee free passage through the country, and did not meet with objections from the Serbian delegation.

119. Hooper, p. 163.

120. TFF Press information #63. 'NATO Mistakes Take More Lives Than The Serb-Albanian War Did', April 16, 1999.

121. Faber and Kaldor, 'What is Humanitarian Intervention?'

122. Faber and Kaldor, op cit.

123. Michael Ignatieff, *Virtual War: Kosovo and Beyond* (Toronto: Viking, 2000), p. 107.

124. 'Milosevic Yields on NATO's Key Terms', *New York Times*, June 4, 1999.

125. Bishop Artemije, in a speech to the Helsinki Commission Hearing of the U.S. Congress, Feb. 28, 2000.

126. Tim Donais, 'Steering a New Course in the Balkans', *Peace Magazine* Vol XVI, Issue 1, Jan 2000, pp 26–28.

127. Fabian Schmidt, 'Is Southeastern Europe on its Way Towards European Integration?' *Radio Free Europe/Radio Liberty, Balkan Report,* Vol. 4, No. 28, 18 April, 2000.

128. Although they do not discuss Yugoslavia much, the democratic deficit model is especially well explained by Robert Putnam and Robert Leonardi in *Making Democracy Work: Civic Traditions in Modern Italy* (Princeton, NJ: Princeton University Press, 1993).

129. Allen, pp.. 8–13.

2. DEMOCRATIZATION, DIVISION AND WAR IN YUGOSLAVIA: A COMPARATIVE PERSPECTIVE

Robert K. Schaeffer

Between 1974 and 1994, economic and political crises forced dictators to surrender power in more than 30 countries around the world.[1] These economic crises had different regional origins: oil embargo in Southern Europe, debt in Latin America, slowed growth in East Asia, stagnation in the Soviet Union and Eastern Europe, and trade embargo in South Africa. For most dictators, capitalist and communist alike, economic crisis was compounded by the political turmoil associated with the death of a dictator, defeat in war, popular protest, withdrawal of superpower support, or electoral defeat. When economic and political crises joined, dictators transferred power to civilian successors, who then rewrote constitutions, held multi-party elections, restored civil society, and moved to address the economic problems that precipitated crisis. Remarkably, the dramatic political and economic change that accompanied democratization was, for the most part, peacefully accomplished.

Like many other states in recent years, Yugoslavia 'democratized,' meaning that political power was transferred to or shared with people who had not previously made decisions or participated in the political life of the republic. But even though political power was enlarged, it was simultaneously narrowed because Yugoslavia was also divided and the political powers of the unitary state distributed to successor republics, some more 'democratic' (Slovenia) than others (Serbia). In Yugoslavia, as elsewhere, democratization did not necessarily result in 'Democracy', with fully functioning constitutional institutions and a vigorous civil society. Indeed, few democratizing states in recent years

Research on Russia and Eastern Europe, Volume 3, pages 47–63.
2000 by Elsevier Science Inc.
ISBN: 0-7623-0280-1

have met western standards. But wherever dictators fell and civilians came to power, democratization was still a significant achievement. The partial democratization of parts of Yugoslavia occurred for many of the same reasons that it did elsewhere: economic and political crises combined to create political change. But unlike most other states, democratization in Yugoslavia was accompanied by division, which was uncommon, and by war, which was rare. From a global and comparative perspective, events in Yugoslavia illustrate both the rule (democratization) and the exception (division and war).

Many scholars have argued that Yugoslavia was exceptional because indigenous conflict was too strong and foreign diplomacy too weak to ensure peaceful conformity with democratizing norms.[2]

I take a different view, arguing first that Yugoslavia, like other states, democratized in response to common economic and political problems, and second, that the unconventional results of democratization were largely a product of the regime's postwar principles and practices, which facilitated division and legitimized war. To explain why democratization in Yugoslavia was accompanied by division and war, it is necessary to examine economic, political, and military developments since 1941, using comparisons with other states to bring the common and exceptional character of Yugoslavia into relief.

ECONOMIC DEVELOPMENT AND DIFFERENTIATION

Although Yugoslavia's postwar economic policies are regarded as unique among socialist countries, they were actually quite similar to those of other states in Southern Europe. In economic terms, the policies adopted by Tito's regime closely resembled those adopted by regimes in Spain, Portugal, and Greece between 1945 and 1974, and those adopted by dictatorships in Latin America and Eastern Europe after 1974.

Throughout the postwar period, Yugoslavia, like its counterparts in Spain, Portugal and Greece, ran persistent trade deficits because its export industry was too weak to pay for the imports the country needed.[3] Trade deficits are typically a problem because they deplete monetary reserves, force currency devaluations, reduce imports, and slow the industries that rely on imports. This was a particular problem for Southern European countries whose industries needed imported energy to survive. Without some way to pay for imports, Southern European states would have been unable to recover from civil war or wartime destruction and spur economic growth, which would have undermined their political authority.[4] As it happened, they were all able to mitigate their persistent trade deficits because they received large infusions of U.S. economic and military aid in the 1940s and 1950s, and then adopted policies that produced income from the import and export of workers in the 1960s.

In the late 1940s, U.S. and European aid enabled Yugoslavia to close trade deficits and promote economic growth. U.S. officials had been unwilling or reluctant to provide aid either to fascist or to communist dictatorships after the war, but as U.S.-Soviet conflict sharpened, U.S. officials reconsidered, and began providing aid to dictatorships on its side of the Cold War divide.[5] After Yugoslavia broke with the Soviet Union in 1948, the U.S. government released frozen Yugoslav assets, including $47 million in gold, and later provided Tito's regime with loans, grants, and military aid in concert with aid from the World Bank and U.S. allies in Western Europe.[6] Between 1951 and 1960, "the United States extended to Yugoslavia $2.7 billion worth of military and economic assistance on a non-repayable basis," somewhat more than it provided to regimes in Spain, Portugal and Greece in the same period.[7] This influx of foreign aid promoted double-digit economic growth in Yugoslavia, as it did in Iberia and Greece, during the 1950s.[8] Economic recovery and growth was also assisted by the normalization of Yugoslavia's trade relations with the Soviet Union and Eastern Europe after Khrushchev repaired Soviet relations with Tito in 1955.[9]

But generous U.S. aid did not continue into the 1960s. As Cold War tensions eased in the late 1950s, the United States discontinued military aid to Yugoslavia and reduced financial aid, as it did in Spain and Greece, and balance of payment difficulties reemerged as a problem.[10] As it happened, regimes in Yugoslavia, Spain, Portugal and Greece all discovered they could solve their problems by exporting workers and importing tourists. Income from the remittances of Southern European workers employed in Western Europe, and money from Western European workers who spent their savings on holidays in Southern Europe, provided regimes with the hard currency they needed to promote economic growth and create domestic jobs. Earnings from emigré workers and vacationing tourists "made it possible [for these countries] to realize rapid growth without running up against the balance of payments problems that bring growth to a grinding halt in most poor countries."[11]

While the export of workers and import of tourists was called different things – 'Market Fascism' in Spain, and 'Market Socialism' in Yugoslavia – it was in fact a common economic strategy.[12] It was pursued with varying degrees of success by regimes in all four countries because they possessed a plentiful supply of workers, recently released from agriculture, and an abundant supply of sunny beaches, which attracted the tourist trade. By the early 1970s, Spain, Portugal, Yugoslavia and Greece each had nearly one million workers employed in Western Europe, and they returned nearly $1 billion in remittances annually.[13] This income provided hard currency for the government and income for worker households, which used the money to build houses and start businesses, thereby promoting local development.[14]

While Southern European states exported workers at about the same rate, which provided comparable remittance income, they were not equally successful at importing tourists. Spain far out-paced the others. The number of tourists visiting Spain increased from half a million in 1950 to 34 million in 1973 – more people visited Spain than lived in Spain – and tourist receipts contributed $3.3 billion to the economy in 1971.[15] In Yugoslavia, by contrast, the number of tourists grew from 41,000 in 1950 to six million in 1973, and they poured about $470 million into the economy.[16] Yugoslavia also relied more on Eastern European tourists, who spent less and paid in soft currencies.[17] In Yugoslavia, as in Spain, the regime assisted tourism by devaluing the dinar by half in 1965, making Yugoslavia a cheap vacation destination for tourists.[18]

This development strategy had important economic consequences in Yugoslavia. The government lifted the labor controls common in other communist states and allowed workers and businesses to keep some hard currency earnings, which they could use for savings, consumption, and investment.[19] By giving workers and businesses considerable control over economic decisions, and by promoting self-management in state controlled firms, the government provided incentives that helped increase productivity and promote development. As a result, Yugoslavia, like Spain, Portugal and Greece, recorded rapid rates of growth in the 1960s and early 1970s.[20]

In 1973–74, global economic developments brought an end to development based on the export and import of workers. Rising world food and energy prices triggered a recession in Western Europe, bringing an end to the decades-long economic boom. Faced with rising domestic unemployment, West Germany and other worker-importing states began returning workers to Southern Europe, and as many as one-third of Yugoslav workers returned in the next decade, reducing remittance income and increasing domestic unemployment.[21] And because recession, rising inflation and unemployment in Western Europe reduced incomes generally, fewer Western European workers vacationed in Southern Europe, thereby reducing tourist receipts in Yugoslavia and elsewhere.

This shared economic crisis was joined in Portugal, Greece, and Spain by political crises. In Portugal, military defeat in its African colonies triggered a military revolt; in Greece, military defeat in Cyprus led to mutiny and the dissolution of the junta; in Spain, the death of Franco (and assassination of his political heir Admiral Carrero Blanco) created a succession crisis. Simultaneous economic and political crises proved fatal for regimes in these countries, but not in Yugoslavia. There the regime survived because Tito remained in power and used his considerable personal authority to deflect potential political crises.[22]

It survived too because the regime, like dictatorships throughout Latin America and Eastern Europe, borrowed heavily to replace lost revenue and fund continued development.[23] Yugoslavia's debt doubled between 1971 and 1975, from \$2.7 billion to \$5.8 billion, and then rocketed to \$20.5 billion by 1981.[24] Indebted development in Yugoslavia deferred incipient crisis and promoted economic growth in the 1970s, as it did elsewhere. But economic crisis returned in 1979, when the United States raised interest rates. This increased borrower interest payments because loans had been tied to floating rates, and reduced borrower incomes because high interest rates also triggered a recession, which reduced demand for their goods.[25] The onset of renewed economic crisis in 1979–80 was joined by the political crisis that ensued after Tito's death in 1980.

During the 1980s, lenders forced dictators in Latin America, Eastern Europe and post-Tito Yugoslavia to make strenuous efforts to repay their debts. In Yugoslavia, as elsewhere, this meant devaluing the currency to increase exports, reducing imports and creating trade surpluses to provide the hard currency they needed to repay debt. And it meant curbing government spending – eliminating jobs and cutting food, transportation and energy subsidies – and raising taxes to create budget surpluses that could be used to repay debt.[26] In Yugoslavia, 'structural adjustment' resulted in widespread shortages of imported goods and energy, recession, rising unemployment, falling real wages and declining rates of growth.[27] Between 1985 and 1988, per capita GMP had declined from U.S. \$3,000 to \$2,400, and in 1991 the World Bank predicted that Yugoslavia would not regain 1989 income levels for more than a decade.[28]

In many countries, the debt crisis was a collective evil, joining different social groups in a common misery. And crisis typically joined them against a common political foe: the regimes whose heavy borrowing had made them vulnerable to global economic change. But in Yugoslavia, the debt crisis produced different results. Growing economic misery drove people in opposite political directions.

The response to the debt crisis in Yugoslavia was different because the regime's postwar development policies had divided Yugoslavia economically. At the beginning of the postwar period, economic development in Yugoslavia was unevenly distributed. But the regime's adoption of a worker-export, tourist-import development strategy in the 1960s exacerbated economic differences. Because most migrants were from Slovenia and Croatia, the money they remitted provided hard currency and economic benefits to households and businesses in these regions.[29] And because Slovenia, Croatia, and the Dalmatian coast were the primary tourist destinations, workers and businesses there were the principal beneficiaries of the tourist trade and the hard-currency earnings associated with it. There was also a synergy between the two: worker remittances helped finance private tourist-industry businesses for households in these regions.

Of course, the government recognized that its development strategy contributed to uneven development. And it used various mechanisms and insti- tutions–hard currency regulations, the development fund – to redistribute income, hard currency and wealth earned in the prosperous republics to the poorer republics.[30] But the economic gap between rich and poor republics widened nonetheless. And Tito recognized this: "To fight against social differences doesn't mean to fight for uniformity in poverty, but quite the reverse."[31] Other economic developments also contributed to economic differentiation. Currency devaluations – rapid ones in 1965 and 1981–83, and a slow one between 1975–81 – were generally good for the prosperous republics, but bad for the poor ones, and ongoing inflation, which was bad for everyone, was generally worse for people in the poor republics.[32]

As a result of the regime's domestic economic policies, the gap between prosperous and poor republics grew wider during the postwar period, particu- larly after 1965.[33] The prosperous republics had higher incomes and lower unemployment than the poor ones: "In 1947 the per capita social product in Slovenia was 2.4 times higher than in Kosovo, while in 1974 it was over 6 times higher."[34] Put another way, "the gross social product per capita of the Republic of Slovenia roughly approximates that of central Italy, while [that) of the autonomous province of Kosovo is comparable to that of Congo (Brazzaville), Ghana, or Liberia."[35]

In the 1980s, the political response to the debt crisis varied by region. People in the prosperous republics wanted to keep what they earned from remittances and tourism and use these resources to weather the crisis. They resisted redis- tributive policies that siphoned off their income to the poorer republics.[36] They saw economic advantage in political separatism.[37] As one Croatian hotelier explained: "The best thing is for us to live together with the Serbs, but with separate governments and separate accounting books. If the books could be kept separate, then I think we would get along."[38] Meanwhile, people in the poorer republics wanted to keep redistributive policies and institutions intact so they could use the resources they provided to weather the storm.[39] They saw economic advantage in political unionism. As a result, the prosperous republics began refusing to share their wealth; the poor began demanding that they redistribute it.[40] Because each saw economic advantage in different political solutions, economic crisis contributed to political crisis.

Like its postwar development policy, the regime's postwar political policies also contributed to conflict after Tito died.

REPUBLICAN POLITICS

During the postwar period, the regime adopted foreign and domestic principles and policies that facilitated a struggle for power between communist party factions based in the republics after Tito died. These policies grew out of Tito's attempt to address foreign and domestic problems.

Tito's communist regime came to power after the war largely as a result of its own efforts. In this regard, it was different from communist regimes in Eastern Europe, which were installed by the Soviets, but similar to communists who achieved postwar independence on their own initiative in China and in Vietnam. But the determination of communists in Yugoslavia and China to act independently brought them into conflict with the Soviet Union, which wanted to subordinate them to Soviet diplomacy.[41] Stalin's determination to crush Yugoslav insubordination prompted him to break with Tito in 1948, forcing Tito to reassert Yugoslav 'nationalism' as the regime's defense against domestic and foreign 'internationalists'.[42]

To prevent Yugoslavia from being isolated, a position that would make it vulnerable to domestic and foreign opponents, Tito purged the communist party in Yugoslavia of domestic opponents, requested U.S. economic and military aid, and then joined with leaders of post colonial states that were, like Yugoslavia, seeking to maintain an independent foreign policy as Cold War spheres hardened around them. Tito's decision to join with India and Egypt to promote 'non-alignment' was an important initiative because it provided a collective defense against superpower intervention and gave them, and later others, the opportunity to fashion independent foreign policies. Tito, Nehru, and Nasser made an unlikely trio, given the fact that they arrived at a common destination – non-alignment – from different directions: Yugoslavia had broken away from the Soviet sphere (de-aligned); Egypt had broken away from the U.S. sphere and joined the Soviet sphere (re-aligned); and India had refused to join either (non-aligned).[43] Although they came from different directions, they could agree on the need for two principles: 'self-determination' and 'non-alignment'. They supported self-determination, by which they meant the right of nations to secede from colonial empires, because they believed that a growing of independent states would help safeguard their own independence.[44] And they supported non-alignment to prevent the involuntary incorporation of independent post-colonial states into superpower spheres of influence.[45] "There is no justification at all for the view that small nations must jump into the mouth of this [Soviet] or that [U.S.] shark," one Yugoslav official explained.[46]

By developing foreign policies that promoted self-determination, Tito's regime would help legitimate domestic demands for self-determination. This was reinforced by domestic policies that legitimized secession and delegated considerable political power to the constituent republics.

The devolution of central political authority took place in stages, and for different reasons, during the postwar period. The adoption of constitutional provisions granting the right of self-determination and of secession by constituent republics was consistent with prewar party programs and modeled after Soviet constitutional provisions.[47] After Yugoslavia broke with the Soviet Union, Tito devolved considerable economic and political power to the republics in the 1950s and '60s. He evidently did so to distinguish the Yugoslav 'model' from centralist models, which enhanced the regime's international credibility, to prevent a resurgence of 'Serbian hegemony', which he saw as a defect of the royalist state before the war, to prevent rivals like Djilas and Rankoviç from using central institutions to mount a challenge to his authority, and to promote the regime's developmentalist strategy, which was regionally based.[48] The constitutional revisions of 1971 and 1974, which devolved even power to the constituent republics, were introduced in response to Croatian demands in 1970–71 for greater autonomy, and to create institutions that could survive Tito's death.[49] Because he had worked assiduously throughout the postwar period to prevent the emergence of political rivals, Tito could designate no heir. As an alternative, he divided his legacy between multiple estates.[50]

During the postwar period, Tito developed a foreign policy that cherished independence, promoted self-determination and promoted non-alignment. And he adopted domestic political policies that legitimized secession, promoted regional autarky, and provided for the creation of relatively autonomous political institutions in the constituent republics. While the economy was growing and Tito was still alive, these ideas and institutions functioned well, providing an alternative model of socialist economic and political development that contrasted sharply with Soviet and Chinese centralist models. But when debt crisis struck, and Tito died, the conditions that made this model function disappeared.

After Tito's death, a succession crisis ensued, a common problem in regimes where dictators died: China, South Korea, North Korea, Taiwan, and the Soviet Union. But the succession crisis in Yugoslavia most closely resembled the crisis that followed Brezhnev's death in 1982, mainly because in both Yugoslavia and the Soviet Union, potential successors representing factions of the communist party used or established political bases in the constituent republics to struggle for power.[51] In both cases, communists or former communists used the regime's political values (self-determination) and constitutional principles (the right of secession) to legitimize their demands for separate political authority, used political institutions in the republics to establish alternative political authority, used their opposition to central-government austerity programs to organize political support, and used ethno-national appeals to construct a 'popular front' or multi-class alliance that could contest for power on its own terms. In both cases, the struggle in the republics was initiated by factions of the communist party, though some non-communist dissident groups effectively participated in some republics (Slovenia and Croatia; Lithuania) and in both Yugoslavia and the Soviet Union, the roads to independence were paved by central government policies.

In Yugoslavia, the contest between Tito's successors within the communist party became a struggle between successors in constituent republics, as factions abandoned communist party labels and affiliations identified with the center and created 'nationalist' parties in the republics.[52] The regionalization and nationalization of communist party factions' began in Serbia, where Slobodan Miloseviç used nationalist themes to bid for power within the Serbian League of Communists and then demanded that Serbia obtain a greater share of power in the post-Tito state. But this antagonized factions based in other republics, and they began demanding greater political and economic autonomy.[53] As a result of their separate efforts, Yugoslavia democratized and also divided. Because the regime's ideas and institutions facilitated power struggles between communist party factions based in the republics, the struggle for secession ended in the division of power. Ironically, factions in the 1980s–90s inadvertently realized the communist party program of the late 1920s, when the party supported the right of self-determination for constituent nations and advocated the dissolution of the royalist state.[54] "Long live independent Croatia, Macedonia, Montenegro, Bosnia, Vojvodina and Serbia," the party proclaimed in the 1920s.[55] With minor changes, this program might easily have been adopted by the party's contemporary successors.

VIOLENCE AND WAR

In Yugoslavia, the economic and political crises of the 1980s led to democratization and division in the 1990s, as it did in the Soviet Union and Czechoslovakia. But why, in Yugoslavia, did it also lead to war? It resulted in war I think, because the regime's celebration of civil war in the 1940s legitimized violence, because military policy, which grew out of the regime's wartime experience in the 1940s, made it possible for individuals and republics to wage civil and uncivil war, and because the regime's popularity and the army's military strength made it possible for them to fight rather than surrender quietly, as they did throughout most of Eastern Europe.

It is often said that war contributes to state-building. This was certainly true in Yugoslavia. The communist party's successful partisan war against Axis invaders and its civil war against domestic fascist and royalist opponents during World War Two enabled it build a socialist state.[56] Its wartime efforts enabled the communist party to defeat domestic opponents, win both U.S. and Soviet support, establish an independent communist state, survive a break with the Soviet Union (the party's war record enabled it to garner widespread public support in its confrontation with the Soviet Union and also deter Soviet invasion after the break), and establish the regime as a credible independent force in international diplomacy.[57] Given its importance to the regime's postwar existence and success, it is not surprising that the regime celebrated partisan civil war, memorializing it as a heroic, purposeful activity, this despite the fact that it waged a ruthless partisan war and, at war's end, mercilessly slaughtered disarmed opponents.[58]

There are two problems with celebrating partisan civil war. First, it antagonizes co-residential groups who were defeated in war, cultivating bitterness as well as pride. Second, and perhaps more important, it not only sanctions state violence but legitimizes other kinds of violence. In their study on the relation between war and homicide, sociologists Dane Archer and Rosemary Gartner found that homicide rates increased in states that waged war, particularly in victorious states that sustained heavy combat losses.[59] They explained this phenomenon by arguing "that wars do tend to legitimate the general use of violence in domestic society" because they teach "the unmistakable moral lesson that homicide is an acceptable, even praiseworthy, means to certain ends . . . [a] lesson [that) will not be lost on at least some of the citizens in a warring nation."[60] They noted that murder rates were less likely to rise in countries defeated in war, arguing that because civilians in these countries viewed war as senseless, not purposeful, public defeat effectively discouraged private acts of violence.[61] By analogy, the Yugoslav regime's decades-long memorialization of brutal partisan war may have effectively legitimized civil war as a means of settling disputes in the 1990s.

Of course, other regimes also celebrated formative civil wars. In Spain, Franco long memorialized the Spanish Civil War. But civil war did not recur as Spain democratized after Franco's death, while it did in Yugoslavia after Tito's. It is important to note that Spanish elites were extremely worried about the possibility of renewed civil war – a limited civil war between Basque separatists and the state (fascist, royalist and then socialist) did erupt – and the threat of war became a kind of 'virtual reality' for participants in the democratization process.[62] The comparison is instructive because in both Spain and Yugoslavia civil wars joined fascists, royalist and republicans in brutal contests for state power.

But democratization was not accompanied by civil war in Spain, as it was in Yugoslavia, in part because Franco's portrayal of civil war as heroic and purposeful had, over time, been discredited by its association with fascism. This was not the case for the victors in Yugoslavia, who did not have their victory impugned by others. Instead, their success was acclaimed by others. Postwar attitudes helped de-legitimize civil war in Spain, but legitimize it in Yugoslavia. What's more, Tito's successors helped legitimate civil war. Tudjman's efforts to rehabilitate Croatian fascists effectively legitimated civil war for groups 'defeated' in war.[63] By trying to get rid of the 'Ustashe complex', Tudjman legitimized civil war as a purposeful activity for Croats, much as Tito's regime had already done for non-Croatian populations. Miloseviç, meanwhile, took steps to legitimize violence for Serbs.[64]

Tito's regime not only legitimized civil war, it adopted policies that prepared people to wage war. In the late 1950s, the regime adopted partisan war, what it called 'General People's Defense', as an important part of its military strategy, taking steps to conscript, train, and arm civilians for service not only in the regular army but in territorial defense forces based in the republics and in local militias, which were expected to wage guerrilla war against foreign invaders.[65] The regime first adopted and later extended these policies and practices because it had used partisan war as an effective weapon against foreign invasion during World War Two, because Soviet invasions of Hungary, Czechoslovakia and Afghanistan reinforced its fears of Soviet invasion, because initial Soviet successes in these campaigns persuaded it that the army alone could not withstand a massive and sudden assault, and because it thought that Swiss, Israeli and Vietnamese military strategies demonstrated the validity and continuing importance of an armed-nation approach to defense.[66] As a consequence of the regime's military doctrine, military weapons and communications were widely distributed among a trained citizenry.[67] As Adam Roberts noted in 1986, "the Yugoslav defense system rests on fragile social and political foundations. If those foundations fail, the idea of General People's Defense might be quickly forgotten; or, worse, it might be perversely misused for civil war."[68] As it happened, the regime's military policies prepared an armed populace to wage war on their own initiative.

Still, civil war might not have erupted had the army decided not to engage militias in the republics and forcibly contest the division of power. After all, the Soviet military decided, after several brief encounters and an attempted coup by elements in the army, not to contest the dissolution of the Soviet state or to initiate or participate in civil war (though the army in some successor states did, as the Russian army did in Chechnya). The Yugoslav army might have retired after its initial incursions were rebuffed in Slovenia and Croatia. Why did the residual regime, based in Belgrade under Milosevic's authority, and the army together wage war to contest the dissolution of the Yugoslav state? They did because the regime's residual popularity and the army's military strength made it possible for them to fight rather than surrender quietly, as they did in most of Eastern Europe.

Most communist parties in Eastern Europe had been installed by the Soviet Union. If they did not serve the Soviet Union, they were replaced, sometimes by force. This made them deeply unpopular with domestic populations. When crisis struck, regimes could rely on only a narrow base of domestic support, and on the Soviet army. But when Gorbachev withdrew Soviet military support, dictators faced the crisis alone, and the appearance of even small opposition movements were sufficiently strong to force them from power.

In Yugoslavia, by contrast, the regime had defeated foreign invaders and domestic opponents, coming to power without Soviet assistance, developed an independent foreign policy, and promoted domestic economic development that provided real, though unequal benefits to its citizenry. When crisis struck, Tito's successors in Belgrade could count on a considerable reservoir of domestic political support. In comparative terms, the regime in Yugoslavia closely resembled regimes in China, Vietnam and Cuba, where dictators had defeated domestic and imperial foes and developed foreign and domestic policies that won the support of important sections of the population, particularly the large rural peasantry. When economic crisis struck, these regimes survived, while neighboring dictatorships did not, largely because they could draw on substantial domestic support and, in the last resort, count on the army.[69]

In Yugoslavia, central authority remained fairly strong, largely because it enjoyed considerable popular support, and much of the army remained loyal, mainly because its leadership consisted of pro-central government Serbs.[70] What's more, the army itself enjoyed considerable prestige, having served with distinction as part of U.N. peace keeping forces around the world.[71] Unlike the Soviet army, which had been defeated in Afghanistan, the Yugoslav army's credibility and willingness to fight had not been undermined by defeat. Because the Belgrade-Milosevič regime and the army could claim to defend minority interests outside the boundaries of the residual Yugoslav state they could legitimize forcible efforts to contest the dissolution of the state.[72]

Political popularity and military strength, two legacies of Tito's postwar policies, stiffened the central government's determination to fight. As in China, the central government in Yugoslavia was strong enough to resist. But unlike China, it was not strong enough to win outright. They could not reassert central authority throughout Yugoslavia because support for the regime was more narrowly based in geographic and social terms than it was in China (it could count only on Serbia, Montenegro and minority populations in some other republics), because their opponents were better organized and armed, and because its opponents could count on financial and political support from émigré communities, particularly from Croatian émigrés in Germany and the United States, and from foreign states, first Germany and later the United States.[73] These developments legitimized violence and prepared people to wage uncivil war. When economic and political crises joined, they did just that.

CONCLUSION

The argument here is that the regime's economic policies resulted in uneven development and promoted opposing political responses to the economic crisis of the 1980s; that the devolution of central power to political institutions in the republics gave Tito's successors the means to struggle for power on their own terms; that the regime's celebration of partisan and its adoption as military doctrine legitimized and prepared factions to wage civil war; and that the regime's foreign and domestic policies gave it a strength that few other dictatorships possessed, giving Tito's successors in Belgrade the determination and ability to resist the division and dissolution of their authority.

NOTES

1. See Robert Schaeffer, *Power to the People: Democratization Around the World.* Boulder: Westview Press, 1997

2. Lenard J. Cohen, 'The Disintegration of Yugoslavia', *Current History*, November 1992, pp. 370, 373. If one argues, as many do, that ethnic antagonisms were the root cause of democratization, division and war, then one would hold Tito's regime as largely blameless for events that occurred. From this perspective, Tito tried for many years to contain simmering conflict, but failed in the end because ethnic identities were too strong and secular identities too weak to prevent conflict from erupting. But I think this treatment is unwarranted. First, it fails to treat the regime as a dictatorship, perhaps more moderate and popular than most, but a dictatorship nonetheless. Second, it minimizes the regime's role in the contemporary crisis. And third, because cause (ethnic conflict) and consequence (ethnic conflict) are the same, the explanation begins and ends in the same place, treating the intervening history as irrelevant.

3. Caglar Keyder, 'The American Recovery of Southern Europe: Aid and Hegemony', in Giovanni Arrighi, ed., *Semiperipheral Development: The Role of Southern Europe in the 20th Century.* Beverly Hills: Sage, 1985, pp. 141–42; David A. Dyker, *Yugoslavia: Socialism. Development and Debt.* London: Routledge, 1990, 94; William Zimmerman, *Open Borders. Non-alignment. and the Political Evolution of Yugoslavia.* Princeton: Princeton University Press, 1987, p. 115.

4. Ljubomir Madzar, 'The Economy of Yugoslavia: Structure, Growth Record and Institutional Framework', in John B. Allcock, John J Horton and Marko Milivojević, eds., *Yugoslavia in Transition: Choices and Constraints.* New York: Berg, 1992, p. 76.

5. Beatrice Heuser, *Western 'Containment' Policies in the Cold War: The Yugoslav Case. 1948–53.* London: Routledge, 1989.

6. Zimmerman, 1987, pp. 18, 22–23; Duncan Wilson, *Tito's Yugoslavia*. Cambridge: Cambridge University Press, 1979, pp. 68, 75; T. E. Vadney, *The World Since 1945*. London: Penguin, 1992, pp. 53–54.

7. Dijana Plestina, 'From "Democratic Centralism" to Decentralized Democracy? Trials and Tribulations of Yugoslavia's Development', in Allcock, Horton and Milivojeviç, 1992, pp. 132, 133; Dimitrije Djordjevic, 'The Yugoslav Phenomenon', in Joseph Held, ed., *The Columbia History of Eastern Europe in the Twentieth Century*. New York: Columbia University Press, pp 334–35; Constantine Tsoucalas, *The Greek Tragedy*. Harmondsworth, England: Penguin, 1969, p. 67; Sheelagh Elwood, *Franco*. London: Longman, 1994, pp. 151, 158, 163, 169; Kostis Papadantonakis, 'Incorporation is Peripheralization: Contradictions of Southern Europe's Economic Development', in Arrighi, 1985, p. 93.

8. Plestina in Allcock, Horton and Milivojevic, 1992, p. 132; Wilson, 1979, p. 125.

9. Wilson, 1979, p. 97.

10. Fred Singleton, *A Short History of the Yugoslav Peoples*. Cambridge: Cambridge University Press, 1985, p. 44. Poor harvests in 1960 and 1961 "necessitated food imports [and] placed additional strain on the balance of payments." Plestina in Allcock, Horton and Milivojevic, 1992, pp. 132, 133.

11. Raymond Carr and Juan Pablo Fusi Aizpurua, Spain: *Dictatorship to Democracy*. London: George Allen and Unwin, 1979, p. 57; Keyder in Arrighi, 1985, pp. 135, 145.

12 Giovanni Arrighi, 'Fascism to Democratic Socialism: Logic and Limits of a Transition', in Arrighi, 1985, p. 265, 268.

13. Spain and Yugoslavia each earned $1.4 billion from worker remittances in 1973; Greece and Portugal somewhat less. John Logan, 'Democracy from Above: Limits to Change in Southern Europe', in Arrighi, 1985, p. 164; Salustiano del Campo, 'Spain', in Ronald E. Krane, ed., *International Labor Migration in Europe*. New York: Praeger, 1979, p. 162. Maria Beatriz Rocha, Trinidade, 'Portugal', in Krane, 1979, p. 171; David D. Gregory and Cazorla J. Perez, 'Intra-European Migration and Regional Development: Spain and Portugal', in Rosemarie Rogers, ed., *Guests Come to Stay: The Effects of European Labor Migration on Sending and Receiving Countries*. Boulder: Westview, 1985, p. 237; George Yannopoulos, 'Workers and Peasants Under Military Dictatorship', in Richard Clogg and George Yannopoulos, eds., *Greece Under Military Rule*. London: Secker and Warburg, 1972, p. 121; Zimmerman, 1987, p. 114; Jasminka Udovicki, 'Nationalism, Ethnic Conflict and Self-Determination in the Former Yugoslavia', in Berge Berberoglu, ed., *The National question: Nationalism. Ethnic Conflict and Self-Determination in the 20th Century*. Philadelphia: Temple University Press, 1995, p. 292; Milan Mesić, 'External Migration in the Context of the Post-War Development of Yugoslavia', in Allcock, Horton and Milivojeviç, 1992, p. 180; Carl-Ulrik Schierup, Migration. Socialism and the International Division of Labor: The Yugoslav Experience. Aldershot, England: Avebury, 1990, pp. 18, 77.

14. Udovicki in Berberoglu, 1995, p. 292; Mesic in Allcock, Horton and Milivojevic, 1992, p. 180; Schierup, 1990, p. 18, 77.

15. Jaime Gama, 'Foreign Policy', in Kenneth Maxwell and Michael H. Haltzel, eds., *Portugal: Ancient Country. Young Democracy*. Washington, D.C.: Woodrow Wilson Center Press, 1990, p. 97; Gregory and Perez in Rogers, 1985, p. 236; Robert P. Clark, *The Basques: The Franco Years and Beyond*. Reno, Nevada: University of Nevada Press, 1979, p. 211; Carr and Aizpurua, 1979, p. 57.

16. Wilson, 1979, pp. 238–39.

17. Ibid., p. 232.

18.. Alvaro Soto Carmona, 'Long Cycle of Social Conflict in Spain' (1868–1986), Review, XVI, 2 (Spring 1993): 179; Ellwood, 1994, p. 180; Wilson, 1979, p. 156; Zimmerman, 1987, p. 78.

19. Dennison Rusinow, *The Yugoslav Experiment*. 1848–1974. London: C. Hurst, 1977, pp. 207–8; Zimmerman, 1987, p. 76; Mesic in Allcock, Horton and Milivojević, 1992, p. 178.

20. Laura D'Andrea Tyson, *The Yugoslav Economic System and Its Performance in the 1970s*. Berkeley: Institute of International Studies, 1980, p. 33.

21. Schierup, 1990, pp. 100–1; Tyson, 1980, p. 52; Zimmerman, 1970, p. 92; Plestina in Allcock, Horton and Milivojeviç, 1992, p. 146; Susan L. Woodward, *Balkan Tragedy: Chaos and Dissolution After the Cold War*. Washington, D.C.: Brookings Institution, 1995, p. 49.

22. See Schaeffer, 1997, Chapter 4.

23. Tyson, 1980, p. 95.

24. Christopher Bennett, *Yugoslavia's Bloody Collapse: Causes. Course and Consequences*. New York: New York University Press, 1995, p.69; Madzar in Alicock, Horton and Milivojević, 1992, p. 84; Udovicki in Berberoglu, 1995, p. 295.

25. Robert Schaeffer, *Understanding Globalization: The Social Consequences of Political. Economic and Environmental Change*. Boulder: Rowman and Littlefield, 1997, Chapter 5.

26. Schaeffer, *Globalization*, 1997, Chapter Five; Woodward, 1995, p. 51; John Walton and David Seddon, *Free Markets and Food Riots: The Politics of Global Adjustment*. Oxford: Blackwell, 1994, pp. 298–99.

27. David A. Dyker, *Yugoslavia: Socialism, Development and Debt*. London: Routledge, 1990, pp. 131–32; Woodward, 1995, pp. 52, 55, 82, 96; Walton and Seddon, 1994, p. 324.

28. Walton and Seddon, 1994, p. 321; Plestina in Allcock, Horton and Milivojevic, 1992, p. 152.

29. Mesic in Allcock, Horton and Milivojevic, 1992, p. 185; Chris Martin and Laura D'Andrea Tyson, 'Can Titoism Survive Tito? Economic Problems and Policy Choices Confronting Titoís Successors', in Pedro Ramet, ed. *Nationalism and Federalism in Yugoslavia*.
1963–1983. Bloomington, Ind.: Indiana University Press, 1984, p. 197; Plestina in Allcock, Horton and Milivojevic, 1992, p. 140.

30. Ibid., p. 140, 144; Pedro Ramet, 'Apocalypse Culture and Social Change in Yugoslavia,' in Ramet, 1984, p. 140.

31. John B. Allcock, 'Tourism and the Private Sector', in Allcock, Horton and Milivojevic, 1992, p. 404.

32. Martin and Tyson in Ramet, 1984, p. 189; Dyker, 1990, p. 101; Tyson, 1980, p. 76.

33. Singleton, 1985, p. 68.

34. Woodward, 1995, p. 293; Plestina in Allcock, Horton and Milivojevic, 1992, pp. 133–34; Schierup, 1990, pp. 166, 168.

35. Zimmerman, 1987, p. 4.

36. Woodward, 1995, p. 73; Susan Bridge, 'Some Causes of Political Change in Yugoslavia', in Milton Esman, E*thnic Conflict in the Western World*. Ithaca: Cornell University Press, 1975, p. 355.

37. Denitch, 1994, p. 71; Woodward, 1995, p. 64.

38. Blame Harden, 'A Body Blow to Croatia's Tourist Industry', *San Francisco Chronicle*, June 17, 1991.

39. Bridge in Esman, 1975, p. 355.

40. Woodward, 1995, pp. 69, 74, 115, 130.

41. Fernando Claudin, *The Communist Movement: From Comintern to Cominform*. Harmondsworth, England: Penguin, 1975, p. 487; Wilson, 1979, p. 50.

42. Zimmerman, 1987, p. 17.

43. Peter Willets, *The Non-Aligned Movement: The Origins of a Third World Alliance*. London: Frances Pinter, 1978, p. 6, 7; Ah E. Hillal Dessouki, 'Nasser and the Struggle for Independence', in William Roger Louis and Roger Owen, eds., *Suez 1956: The Crisis and Its Consequences*. Oxford: Clarendon, 1989, p. 33.

44. Robert Schaeffer, *Warpaths: The Politics of Partition*. New York: Hill and Wang, 1990, see Chapter 4.

45. Schaeffer, *Power to the People*, 1997. see Chapters 1–3.

46. Zimmerman, 1987, p. 21; Claudin, 1975, p. 488; Alvin Z. Rubinstein, *Yugoslavia and the Non-aligned World*. Princeton: Princeton University Press, 1970, pp. 24, 29, 77.

47. Aleksa Djilas, *The Contested Country: Yugoslav Unity and Communist Revolution*. 1919–1953. Cambridge: Harvard University Press, 1991, pp. 167, 168; Vojin Dimitrijeviç, 'The 1974 Constitution and Constitutional Process as A Factor in the Collapse of Yugoslavia', in Payam Akhavan and Robert Howse, eds., *Yugoslavia the Former and Future: Reflections by Scholars from the Region*. Washington, D.C.: Brookings Institution, 1995, p. 58.

48. Zimmerman, 1987, p. 28; Randy Hodson, Dusko Sekulic and Garth Massey, 'National Tolerance in the Former Yugoslavia', *American Journal of Sociology*, 99, 6 (May 1994): 1539; Joel S. Migdal, Strong Societies and Weak States: State Society Relations and State Capabilities in the Third World. Princeton: Princeton University Press, 1988, pp. 215–224.

49. Ibid., p. 1540; Ramet, 1984:125; Udovicki in Berberoglu, 1995, p. 291; Denitch, 1994, p. 105; Dimitrijeviç in Akhavan and Howse, 1995, pp. 71–2; George Schopflin, 'Political Decay in One-Party Systems in Eastern Europe: Yugoslav Patterns', in Ramet, 1984, p. 316; John Feffer, *Shock Waves: Eastern Europe After the Revolutions*. Boston: South End Press, 1992, pp. 258–59.

50. "Much has been written," he said, [to the effect) "that Yugoslavia will disintegrate when I go." He conceded that his death "could cause a very difficult crisis. . . because the question then would be pose who will take my place?" He concluded, therefore, that "we have to carry out this reorganization precisely so that our Yugoslav socialist community would not come to such a crisis." Ramet, 1984, p. 189; Zimmerman, 1987, p. 45.

51. Schaeffer, *Power to the People*, 1987, see Chapters 8 and 9.

52. Djordjević in Held, 1992, p. 338.

53. Bennett, 1995, p. 117.

54. Djilas, 1991, pp. 56, 68, 70, 76, 78, 84, 86, 97; Djordjeviç in Held, 1992, p. 321; Wilson, 1979, p. 15.

55. Djilas, 1991, p. 88.

56. Denitch, 1994, pp. 34–5.

57. Wilson, 1979, p. 39.

58. Robert Adams, *Nations in Arms: The Theory and Practice of Territorial Defense*. Houndmills, England: Macmillan, 1986, pp. 140, 142; Denitch, 1994, pp. 31–33; Woodward, 1995, p. 1; Djordjevic in Held, 1992, p. 324.

59. Dane Archer and Rosemary Gartner, *Violence and Crime in Cross-National Perspective*. New Haven: Yale University Press, 1984, pp. 79, 86.

60. Ibid., pp. 65, 66, 76, 92, 94–5.

61. Ibid., p. 86.

62. As Walter Lippmann observed long ago, "it is very clear that under certain conditions men respond as powerfully to fictions as they do to realities, and that in many cases they help to create the very fictions to which they respond. Whatever we believe to be a true picture [of the outside world], we treat as if it were the environment itself." Walter Lippman, *Public Opinion*. New York: The Free Press, 1922, p. 4. In this context, memorializing the civil war created a powerful contemporay political environment. It was a 'fiction' because it combined real memory with worried imagination about the present. And elites responded to this fictional environment, which they helped create, as if it were real.

63. Udovicki in Berberoglu 1995, pp. 299–300.

64. As Djilas observed, 'The Second World War in not over, not here anyway'. Michael Ignatieff, *Blood and Belonging: Journeys into the New Nationalism*. New York: Farrar, Straus and Giroux, 1993, p. 53.

65. Roberts, 1986, p. 137, 154–55, 172–73; James Gow, Legitimacy and the Military: The Yugoslav Crisis. London: Pinter, 1992, pp. 44–46; Woodward, 1995, p. 26.

66. Roberts, 1986, pp. 159, 161, 163–64; Zimmerman, 1987, p. 30.

67. Roberts, 1986, pp. 180–81, 215–16.

68. Ibid., p. 217.

69. Schaeffer, *Power to the People*, 1997, passim.

70. Udovicki in Berberoglu, 1995, p. 308.

71. Gow, 1992, pp. 56–7, 59, 72; Rubinstein, 1970, p. 143.

72. Robin Alison Remington, 'Political-Military Relations in Post-Tito Yugoslavia', in Ramet, 1984, p. 57; Roberts, 1986, p. 202.

73. Woodward, 1995, p. 137; Schaeffer, *Power to the People*.

3. YUGOSLAVIA: WHAT WENT WRONG? CONSTITUTIONAL ASPECTS OF THE YUGOSLAV CRISIS FROM THE PERSPECTIVE OF ETHNIC CONFLICT

Mitja Žagar

INTRODUCTION[1]

The former Yugoslavia became a synonym for a violent ethnic conflict, atrocities and suffering of people in the 1990s. Horrible pictures and tragic stories shocked the world, but slowly people got used to it and it sometimes seemed that the victims had learned to live with it.[2] The Dayton Accord and employment of International Forces (IFOR) in the beginning of 1996 brought a new hope for Bosnia-Herzegovina and the wider region.

This contribution analyzes the 'Yugoslav crisis' from the perspective of constitutional development in Yugoslavia and its impact on inter-ethnic relations. I pay special attention to ethnic, social, economic and political situation, conditions and processes in the 1980s and in the beginning of the 1990s. I find a few key factors and (especially political) developments that conditioned the 'Yugoslav crisis,' gave it ethnic dimensions, and triggered war. I analyze the role of the constitution in these developments. This analysis shows a controversial role that the constitution of 1974 played in the seventeen years of its

Research on Russia and Eastern Europe, Volume 3, pages 65–95.
Copyright © 2000 by Elsevier Science Inc.
All rights of reproduction in any form reserved.
ISBN: 0-7623-0280-1

existence. It decentralized the federation and increased the autonomy of federal units. It introduced certain forms of pluralism into the political system and opened social and political space, launching gradual social and political reforms. On the other hand, this constitution did not provide adequate mechanisms to prevent the escalation of social conflicts. The constitution of 1974, based on a specific Yugoslav Communist political ideology, failed to serve its main purpose: it did not resolve existing conflicts and assure peace and stability.

This chapter offers some proposals for the management of ethnic relations and conflicts. There is no universal constitutional model that could be used in all cases of ethnic conflicts. Nevertheless, we can indicate basic principles for the peaceful regulation of ethnic relations. The Yugoslav experience shows political mistakes that triggered the escalation of the crisis. Other multiethnic societies could use these findings to avoid similar mistakes.

THE FORMER YUGOSLAVIA: BASIC INFORMATION AND FRAMEWORK

States die hard. Ex-Yugoslavia was no exception. In the disintegration of this state, the official recognition of the independence of Slovenia and Croatia by the European Communities (European Economic Community) in January 1992 formally marked the end of the existence of the Yugoslav state created after World War I.[3] The new situation did not end the crisis but only gave it a new international dimension. In Croatia and Bosnia-Herzegovina the direct involvement of the Yugoslav army, now the army of a foreign country in newly independent states, transformed a civil war into an international aggression for a period in 1992 and the conflicts became a major problem for the international community. Explaining the Yugoslav crisis requires a perspective that needs to include (at least) some characteristics of the whole region and its historic development.

The Balkan Peninsula and its History[4]

The Balkan area is a mountainous peninsula in Southeast Europe. As a bridge between Asia and Europe on the way to Central and Western Europe, this region was a crossroads of peoples from prehistoric times on. Its turbulent historic development has been often disrupted by invaders moving in both directions. The mountainous landscape, sea coast, straits, rivers, valleys and mountain passes determined trade connections and the routes of movement of peoples. Yet some regions of the peninsula lived in relative isolation due to the difficulties of communication. These circumstances resulted in a specific ethnic diversity throughout history.[5]

The Balkan Peninsula was a cradle of democratic traditions and culture, considering the role of ancient Greece in their development. Politically, ancient Greece and its city-states were just a stage of historic development, preceded by a long period of mostly tribal prehistoric forms of social organization. After the Hellenic period this peninsula became an important part of the Roman Empire, and the center of the Eastern Empire after its division. The Byzantine Empire controlled this region for almost ten centuries. New states appeared in the Middle Ages. Ottoman Turks destroyed the crumbling Byzantine Empire that ceased to exist after the fall of Constantinople in 1453. For almost five centuries the Ottoman Empire ruled the Balkan Peninsula except for Istria, the Dalmatian coast, and islands that were controlled by Venice until the 18th century and later by the Habsburg Empire, Austria-Hungary. The decay of the Ottoman Empire, increasing influence of Austria-Hungary, the rise of modern Balkan nations[6] and creation of new independent states marked the history of the region in the 19th and 20th century. Frequent conflicts and the breakup of existing states were the main characteristics of this region in this period. These developments influenced the stability in Europe, and a new term 'balkaniza-tion' was invented just before World War I to indicate the breaking up of the region into mutually hostile political units. Throughout the history this region has also been a field of competition between great powers that wanted to assure their influence.

The Roman Empire had incorporated the whole territory of the former Yugoslavia and integrated it within the borders of one state. The division of the empire in the Fourth Century AD set the boundary-line in the territory of today Bosnia-Herzegovina and determined the future development of the region.

In the Fourth Century, Christianity spread and became the major religion. The fall of the Western Roman Empire in 476 further increased the role of Christianity and the pope. Invaders were Christianized and a distinctly Roman form of Christianity influenced their cultures. But the western and eastern forms of Christianity were becoming more different, leading to the formal schism in 1054. By then two cultural circles in the Christian World had been defined. Catholicism was a unifying factor, breaching the autarchy of feudal states and shaping a common ideology in the Christian world that adopted the Latin language that was its *lingua franca*.

The Byzantine Empire was heir to the Hellenistic civilization, and the Greek language was spoken and used in administration there. It was more culturally refined, commercialized, urbanized, and rich than feudal states in the West. The Christianity that developed in the Byzantine Empire was more mystical and liturgical than Roman Christianity. It was influenced by eastern mystical cults and religions. The Eastern Orthodox Church was less unified that the Roman

Catholic Church due to ethnic hostilities. This perhaps contributed to the success of Arab invasions in the seventh century. On the other hand, Orthodoxy's use of native languages in religious services contributed to the success of its missions among Slavic peoples in Eastern Europe, who converted to Christianity by the end of ninth century. The decline of the Byzantine Empire enabled the several new medieval Balkan states to form within the Eastern Orthodox cultural circle.[7]

The situation in the Balkan region changed with the invasion of the Ottoman Turks in the fourteenth and fifteenth century. Turks destroyed the Byzantine Empire and captured all Medieval Balkan states belonging to the Eastern cultural circle. They brought Islam, which dominated this region for five centuries, though without eliminating Orthodox Christianity in the region. Central and Western Europe faced Turkish intrusions for more than three centuries before the expansion was stopped.

Eastern and Western cultural circles were defined by the borders of Bosnia-Herzegovina to the West, and by the Sava and Danube Rivers to the North. This border continued to exist after the downfall of the Ottoman Empire had reduced the role of Islam in this region, and it has not disappeared in the time of ex-Yugoslavia.

The decline of the Ottoman Empire became obvious by the beginning of the nineteenth century when some parts of the empire were granted autonomy. Gradually this enabled the formation of new independent states: modern Greece (in 1830), Montenegro, Rumania and Serbia (in 1878), and Bulgaria (in 1878/1908).

In addition to the recognition of new independent Balkan states and autonomies within the Ottoman Empire, the Congress and the Treaty of Berlin of 1878 authorized the occupation of Bosnia-Herzegovina and the Sanjak of Novobazar by Austria-Hungary as well as Britain's occupation of Cyprus, though these provinces remained formally under Turkish suzerainty.

After Austria-Hungary annexed Bosnia-Herzegovina in 1908, two Balkan wars in 1912–1913 further changed the situation in the Balkan region. In the First Balkan war the members of the Balkan League (Bulgaria, Greece, Montenegro and Serbia) defeated the Ottoman Empire that lost most of its European territory under the treaty signed in London. The independence of Albania was also agreed on in principle following the uprising of Albanians in 1912. The Second Balkan war broke out in 1913 because of quarrels of Serbia, Greece, and Rumania with Bulgaria over the distribution of conquered Macedonia. The Second Balkan War enabled Turkey to recover a part of Thrace.

The assassination of Archduke Franz Ferdinand at Sarajevo by a Bosnian Serb Gavrilo Princip triggered World War I in 1914. This war again re-shaped the political 'architecture' of the region and changed the balance of power. The Austro-Hungarian Empire was dismantled, and a new independent state was established: the Kingdom of Serbs, Croats and Slovenes, later renamed Yugoslavia.

Ethnic and Cultural Reality

In the territory of ex-Yugoslavia (255,804 sq. km after World War II) people speak four main languages and more than twenty languages of traditional ethnic minorities. They use two different alphabets – Latin alphabet in the West and Cyrillic alphabet in the East. There are three main religions. Censuses show the following ethnic make-up of the population in this territory:

- *Six nations* of ex-Yugoslavia, all belonging to South-Slavic nations, are Slovenes, Croats, Muslims (Bosnian Muslims, Bosnians), Serbs, Montenegrins and Macedonians.
- The largest among more than twenty traditional *ethnic/national minorities* are Albanians, Hungarians, Roma (Gypsies), and Turks.[8]

Censuses included also a specific category, so-called 'Yugoslavs'. This category, introduced after World War II, enabled especially children from ethnically mixed marriages to opt for a new, broader 'ethnic identity', different from ethnic identity of their parents. The introduction of 'Yugoslavs by nationality' was conditioned by the Yugoslav communist ideology that foresaw a decreasing role of traditional ethnicity in the future. This political concept hoped to build a new common identity that would slowly abolish the existing ethnic diversity.

The ethnic situation in ex-Yugoslavia was very different from that of multi-ethnic states where the largest nation represents more than a half the total population. Serbs were the largest single ethnic group, but they represented only a bit more than one third of the total population in ex-Yugoslavia. They were in many ways the dominant nation, yet they were a minority in comparison with the total population.

Up to the war in the 1990s, economic migrations were the largest factor contributing to the changing ethnic situation and structure in ex-Yugoslavia. Migrations were influenced by the changing military, political, social, and economic conditions. The main reasons for migrations in different historic epochs were changing borders, frequent administrative changes, new administrations, wars, rebellions, fear of revenge after wars and/or rebellions ended,[9] famine and poor economic conditions, hope for a better life, and trade. The differences between two historic civilizational circles in the territory of ex-Yugoslavia,[10] though often hidden, have generated conflicts in certain historical periods.

Conflict and Cooperation: Traditions, Historical Myths and Ethnic Relations

Conflicts have positive and creative dimensions, releasing energy that can stimulate successful reforms and development, but inadequate procedures for the management of conflicts can endanger a plural society. The very existence of ethnic plurality in a certain environment inevitably generates ethnic conflicts, which may escalate.[11]

Ethnic relations in the territory of ex-Yugoslavia have been traditionally good despite the existing ethnic diversity. Small-scale conflicts have always existed, but only seldom did they escalate into violence. Formal and informal mechanisms for the management of ethnic conflicts usually prevented their escalation. These mechanisms were numerous and diverse. They included the traditional organization of life in multi-ethnic villages, generally accepted rules about social and ethnic relations in small local communities, and also the role of authorities, especially the police, of the ruling state. Traditional mechanisms were often undemocratic, sometimes even unfair or cruel, but they proved successful in specific communities because people accepted them. The actions of the state were less successful, because people perceived them as unjust repression.

Ex-Yugoslavia has undergone a tremendous transformation in this century. Modernization, urbanization, and democratization have changed social relations and destroyed traditional mechanisms for handling ethnic conflicts, without creating adequate replacements. No permanent mechanisms have been established that would democratically restrain ethnic conflicts. The absence of such mechanisms was a source of the 'Yugoslav crisis'.

Any successful mechanism would have had to take into account several factors that influenced ethnic relations in ex-Yugoslavia. Modern 'Yugoslav nations' emerged in the nineteenth and twentieth century, but their formation started much earlier. They trace their Slavic origin back to the great migration of peoples after the fall of the Western Roman Empire between the fifth and eighth centuries. They consider themselves heirs to their Medieval states, stressing their greatness and the historic continuities.[12] Though ethnic myths about these Medieval states often do not correspond to historic reality, they did help keep the culture alive during periods of foreign rule, build national identity, and enable modern nations to form within the existing empires. These myths also contributed to the struggle for national independence within the existing states.

THE YUGOSLAV IDEA AND THE CREATION OF THE YUGOSLAV STATE[13]

The Yugoslav idea was much older than the state itself. Influenced by Illyrianism and the Pan-Slavic idea and movement, the Yugoslav idea formed in the nineteenth century. It called for the cooperation among 'brotherly' South Slavic nations and for the creation of an autonomous state where they could live together. Several national politicians believed that a common state would foster the development of all South-Slavic nations together. The main objective of the Yugoslav idea was to end the Austrian/German, Italian, Hungarian and Ottoman hegemony and to establish these nations as free and equal European nations.

Initiatives to create a new South-Slavic state intensified just before and during World War I. Its advocates had differing ideas about the organization and nature of the future Yugoslav state. Their proposals ranged from an autonomous unit within the empire of Austria-Hungary or an independent state under the Habsburg Crown to an independent sovereign state; from a centralized unitary state to a confederacy; from an absolute or parliamentary monarchy to a democratic republic. Despite all these differences, the Yugoslav idea became the main unifying force that brought together political activists of all South Slavic nations, both at home and in emigration, on the platform of a common state. They presented the idea of a common Yugoslav state to the international public.

Already there were two conflicting concepts regarding the future organization of the Yugoslav state. On the one hand, there was a centralizing concept that advocated the creation of the unitary state, dominated by Serbs, with a strong center and a single Yugoslav nation. On the other hand, there was a decentralistic and federal concept that recognized the existence of ethnic pluralism and promoted broad ethnic autonomy and the equality of all South Slavic nations. Clashes between centralist/unitary and decentralist models continued through the formation of Yugoslavia, seven decades of its existence and the period of its dismantling.

Nikola Pašić, Serbian Prime Minister, and the government advocated the formation of a centralized unitary state, dominated by Serbia. The Declaration of Niš of 14 December 1914 stated the intentions of the Serbian government "to create out of Serbia a powerful southwestern Slavic state; all the Croats, and all the Slovenes would enter its composition."[14] Although this declaration reflected the Serbian expansionist politics of Regent Aleksandar and his government, its calls for the 'liberation' and 'unification' into one state of all Serbs, Croats, and Slovenes gained the support of many Croats and Slovenes who advocated the Yugoslav idea.

The main activity of the Yugoslav Committee was the opposition to the Treaty of London of 26 April 1915. This treaty, concluded by the Entente countries and Italy, promised Italy extensive territories, including Gorizia, a portion of Carniola, Trieste, Istria, northern Dalmatia and the Dalmatian Islands in exchange for an Italian declaration of war on Austria-Hungary. Members of the Yugoslav Committee, a group of anti-Habsburg Croat and Slovene politicians who had taken refuge in Allied or neutral countries, criticized the Treaty of London and tried to present actual interests of South Slavic nations. They advocated principles of national rights and self-determination that would justify these nations in claiming a state of their own.[15]

Dr. Anton Korošec, the president of the Yugoslav Club in the Austrian Parliament, presented the so-called May Declaration to the Vienna Parliament on 30 May 1917. The signers of this declaration, South Slavic members of the parliament, called for the unification of "all the lands in the Monarchy, inhabited by Slovenes, Croats and Serbs" into an independent state under the Habsburg crown. This was the first attempt to unify Slovenes, Croats and Serbs on the equal basis. Although this declaration did not foresee the unification with Serbia and Montenegro, it enabled the further development of the national movement.

The Corfu Declaration of 20 July 1917, signed by Serbian Prime Minister Nikola Pašić and President of the Yugoslav Committee Ante Trumbić, was a statement of basic principles and common intents regarding the creation of a new Yugoslav state. This declaration proclaimed the determination of all Serbs, Croats and Slovenes, brothers in blood, to form a united and independent state that would be a "constitutional, democratic, and parliamentary monarchy headed by the Karadjordjević dynasty." The new democratic constitution was to be prepared by the future Constituent Assembly, whose members had to accept it in its entirety, by a "numerically qualified majority."[16] The declaration recognized the equality of the three 'tribal' names and their right to national flags, religions, and two alphabets. It decided also on the new official name of a common state: The Kingdom of Serbs, Croats, and Slovenes. The declaration that called for the unification of all South Slavs regardless of the existing international borders, was called the 'Magna Carta' of the Yugoslav unification.[17]

The Corfu Declaration was a compromise that nobody liked much but everybody saw as an opportunity. It sought to postpone issues that were the sources of political and ethnic conflicts. The Serbian government was expected to officially notify the Allied governments of this declaration, but Pašić never did so. He later admitted that he had considered this declaration a political maneuver to buy a necessary time for the realization of his Great Serbian policy – the annexation of other South Slavic lands to Serbia.

At the end of World War I Austria-Hungary was collapsing and the independent State of Slovenes, Croats and Serbs was forming in its territory. This state united all South Slavic lands within the empire (i.e., Slovenia, Istria, Croatia with Dalmatia, and Bosnia-Herzegovina). The National Council of Slovenes, Croats and Serbs was established as a political representation in Zagreb on 6 October 1918. It called for the unification of all Slovenes, Croats and Serbs in their entire ethnic territory, regardless of existing international borders. The short-lived State of SCS was formally established on 29 October 1918 when the Croatian Parliament declared the independence of Croatia and its inclusion into the State of SCS. The National Council authorized the Yugoslav Committee to represent it abroad.

In the beginning of November 1918 representatives of the Serbian government, National Council, and Yugoslav Committee met in Geneva. Initially, Serbian Prime Minister Pašić claimed that Serbia alone represented all South Slavs. Finally he had to recognize the existence of the State of SCS and the role of its National Council "as the legitimate government of the Serbs, Croats and Slovenes, who live in the territory of the [former] Austro-Hungarian Monarchy."[18] A united state of all the Serbs, Croats and Slovenes was agreed upon and Montenegro was invited to join it. The Geneva Declaration established a common government to which the Serbian Government and the National Council of SCS would delegate six members each. In addition to the common government, the Serbian government and the National Council were to remain as equal partners until the first Constituent Assembly convened to adopt a new constitution determining the future political system by a two-thirds majority. The National Council insisted on such a solution to reduce the danger of domination by one side.

Serbia did not ratify this declaration, which defined the new state as a union of equal partners. The declaration conflicted with the views of Greater Serbian nationalists. Pašić and other Serbian political leaders saw Serbia as the 'liberator' of all the South Slavs. In this context, they wanted to create a new unitary state by annexing other territories to Serbia. They usually rejected any federalist solutions proposed by representatives of the National Council or Yugoslav Committee.

When the Grand Assembly of Montenegro adopted the decision to unite this kingdom with the Kingdom of Serbia on 26 November 1918, Serbia was the only internationally recognized partner and an important member of the victorious coalition. The Serbian government and Regent Aleksandar used this situation to assure their domination in the formation of a new state. They prevented the international recognition of the State of SCS as a member of the victorious coalition by the Allied governments. This eliminated the State of SCS (which was officially still considered a part of the former Austria-Hungary) from the negotiations that followed World War I. This Serbian policy probably damaged the position of the new Yugoslav state in its negotiations for its western borders. It resulted in the loss of an important part of Slovenian and Croat national territory to Italy and Austria, which was a tragic development for Slovenes and Croats. Nevertheless, this policy strengthened the internal position and power of Regent Aleksandar and Serbian political leaders.

The international situation, especially the fear that the Treaty of London would be realized in its entirety, unfavorable internal developments and the fear of socialist revolution (echoing contemporary developments in Hungary) forced the National Council to rush the process of unification. The National Council elected a delegation to inform Regent Aleksandar about its decision to realize the unification based on the Geneva Declaration.

The delegation went to Belgrade and presented the statement of the National Council to Regent Aleksandar, who then proclaimed the unification of Serbia with the State of SCS. This proclamation of 1 December 1918 marked officially the formation of the first Yugoslav state, the unitary Kingdom of Serbs, Croats and Slovenes. Contrary to the Geneva Declaration, which had authorized the Constituent Assembly to determine the organization and political system of a common state, the proclamation of Regent Aleksandar determined the common state as a unitary monarchy before the Constituent Assembly was even elected.

The creation of the first Yugoslav state was celebrated, but some were disappointed. They had expected a democratic federal state based on ethnic equality and cultural diversity, but they got a unitary centralized state that ignored ethnic and cultural diversity. The centralists and unitarians had political, economic and military power and, with Regent Aleksandar, dominated political life. Advocates of decentralization and federalism hardly had any political influence. Still, almost everybody agreed that the creation was a positive historic development and that it was in their interest to live together, at least for a time.

CONSTITUTIONAL DEVELOPMENT OF THE FORMER YUGOSLAVIA FROM THE PERSPECTIVE OF ETHNIC RELATIONS[19]

The constitutional development of ex-Yugoslavia could be divided into two main historic periods: (i) the period of the monarchy between the end of World War I and the beginning of World War II; and (ii) the period of the federal republic that started at the end of World War II and ended with the disintegration of the federation. Additionally, each of these periods could be divided into a number of stages.

The Yugoslav Monarchy

The proclamation of unification determined that the new state would be a unitary monarchy, but it took almost three years before the first constitution to be adopted.

According to the agreement between the National Council and Serbian government, the Interim National Legislature was to prepare legislation for the election of the Constituent Assembly and its agenda. This interim parliament, whose members were delegated, not elected by voters, functioned for 21 months, though its independence from the monarch was questionable. This interim period saw different limitations of political life and democracy, including the prohibition of the Communist Party (CPY), which did rather well in elections.

Following the elections in November 1920, the Constituent Assembly was convened. Boycotted by a few parties, the legitimacy and sovereignty of the Constituent Assembly were further reduced by the fact that its elected members had to take an oath to the monarch. Additionally, the standing orders of the Constituent Assembly introduced a simple majority vote for the passage of the constitution. This decision violated the Corfu Declaration that the constitution would be adopted in its entirety, by a "numerically qualified majority." Nevertheless, the centralist-dominated Constituent Assembly passed a centralist constitution by a simple majority.

The only constitutional proposal, that of the Pašić government, was inspired by Serbia's constitution of 1903. Its provisions on human and social rights reflected also newer trends, and to a certain extent followed the German (Weimar) Constitution of 1919. The Constitution of the Kingdom of Serbs, Croats and Slovenes was passed on 28 June 1921 with 223 votes (of 419 elected members of the Constituent Assembly) cast for it, far short of the qualified two-thirds majority requested by some autonomists. This constitution was usually called the Vidovdan Constitution because it was adopted on the feast day of Saint Vitus.

The Vidovdan Constitution defined the Kingdom of Serbs, Croats and Slovenes as a hereditary, parliamentary monarchy. It introduced a highly centralized unitary political system dominated by a monarch with strong prerogatives. The king appointed the government that had to win the vote of confidence in the parliament. Additionally, the weak parliament was controlled by the king, who was not politically accountable, and his government. The king had the right to dissolve the parliament and he exercised this right as he felt necessary. A strong Serbian dominated army, loyal to the king, was also an important political factor.[20] Democratic rights proclaimed by the constitution were often violated. This underdeveloped unitary monarchy could hardly be classified as a democratic state.

Unitarism was reflected in the constitutional concept of "one nation of three names" historically divided into three 'tribes' – Serbs, Croats, and Slovenes. This concept, based on Serbian expansionist tendencies, denied the existence of Macedonians, Montenegrins and Bosniaks as distinct ethnic groups. The constitution determined only one official 'Serbo-Croatian-Slovene language', which had never existed. The 'Serbian-Croatian-Slovenian' nationality of the individual was required for the exercise of certain political rights, such as the right to be elected to public offices, or to be employed as higher public servant. Other citizens had to comply with additional demands, such as ten-year permanent residence since the acquisition of the citizenship, to exercise these rights.

The Vidovdan Constitution did not protect the linguistic rights of numerous ethnic minorities. Nevertheless, Paragraph 13 of the Article 16 provided that minorities of other 'race and religion' had the right of elementary classes in their mother tongue under the conditions determined by law.

Territorial organization was also conditioned by centralism. There was no ethnic or regional autonomy, and the intention was to divide ethnic communities into several administrative units to decrease their internal ethnic coherence. According to Article 95 the largest administrative units were districts with up to 800,000 inhabitants. Other, smaller, administrative units were departments, counties, and communes. All these were defined also as units of local government, but their autonomy was very limited. For example, the head of each district was appointed by the king, and had the power to overturn the decision of the district council.

Although the Vidovdan Constitution had certain democratic potentials, especially regarding human and social rights, its provisions were mostly not realized in practice. Moreover, several constitutional provisions were ignored and the democratic institutions were often ineffective. The power of the king was growing.

On 6 January 1929, the king proclaimed his (personal) dictatorship and also formally abolished the Vidovdan Constitution, thereby consolidating the supreme royal rule. He blamed irresponsible and ineffective democracy for social, economic and political problems and the critical situation in the country. A package of laws and decrees proclaimed by the king further limited the right to association and other political rights, including the freedom of speech. All criticisms or initiatives to change the existing system were declared criminal activities, and were prosecuted. Centralization and unitarism were further strengthened.

The 'Law on the Name and Division of the Monarchy into Administrative Regions', passed in October 1929, introduced the new official name 'The Kingdom of Yugoslavia', thereby confirming the concept one 'Yugoslav nation'. 'Tribal names', ethnic life, religion, and political parties were forbidden. The law also determined a new territorial organization to strengthen the authority of the center and to promote the formation of a uniform nation. The new largest administrative-territorial units, replacing districts, became provinces, called 'banovine'.[21] Nine 'banovine' were established to increase authority of the center. Their borders were drawn to divide up historic ethnic communities between two or more different 'banovine'. Wherever possible, 'banovine' were designed to strengthen the share of the Serbian population in their total population. 'Banovine' were divided into smaller administrative units, namely, counties and communes, which were units of local government.

The Constitution of the Kingdom of Yugoslavia, granted by King Aleksandar in 1931, increased the power of the monarch again, and reduced the power of the parliament, the National Assembly. The constitution forbade every kind of political association on 'religious, tribal, or regional' basis, thereby restricting rights to association, gathering, freedom of speech, etc. The only trace of linguistic or ethnic pluralism in the Constitution was the definition of the official 'Serbian-Croat-Slovene' language, based on the recognition of three different languages.

Increasingly, the non-Serbian citizens became dissatisfied with the denial of the existence of ethnic pluralism, with Serbian domination and expansionism, with economic and social crises, with the restrictions of human rights and the curtailment of democracy, and with centralism and unitarism. The existing political system, determined by the prevailing unitary and centralistic ideology, was unable to solve the growing conflicts. The situation worsened after the assassination of King Aleksandar in 1934 and the threat called for political decentralization and federalization. The search for an acceptable compromise introduced cooperation and trade into the political process.

The endeavors resulted in the Cvetković-Maček Agreement, signed on 23 August 1939.[22] This Croatian-Serbian agreement recognized that ethnic differences would not disappear, and that national identities could not be transformed into a new Yugoslav national identity. The agreement anticipated the formation of the ethnically defined the 'Banovina of Croatia' with wide autonomy and elements of statehood. It emphasized the equality of Serbs, Croats and Slovenes in the common state as the foundation for resolving the national question in Yugoslavia.

The Banovina of Croatia was formed by a decree based upon the constitutional provisions on the state of emergency, and should have been confirmed later by the new People's Assembly. This never happened. Instead, the order on the state of emergency dissolved the Assembly, and the elections for the new Assembly did not take place until the beginning of World War II.

The formation of the Banovina of Croatia began the process of decentralization and federalization. In my view, claims for a similar autonomy, decentralization and federalization were to be expected also from all other nations, offering an opportunity for democratization and ethnic pluralism in Yugoslavia. On the other hand, this agreement could have been a political bargain between the two largest national elites to assure their domination.

The Yugoslav state and its army disintegrated within days following the attack on Yugoslavia in April 1941, thereby proving its fragility. Two years after the Cvetković-Maček agreement, the political system was still centralized and unitary, and democratization had not yet started. Such a political system, which failed to recognize ethnic plurality, regulate ethnic relations and equality, and develop democratic mechanisms for resolving ethnic conflicts, lacked the cohesion necessary to mobilize people for its defence. The territory of Yugoslavia was occupied and divided among the aggressors.

Four years of occupation saw different divisions of the territory and the creation and fall of Croat and Serbian quisling states. Nevertheless, not only did the National Liberation Movement liberate Yugoslavia, it was an important part of the anti-Hitler coalition. Although the Communist Party of Yugoslavia (CPY) led the uprising, the resistance was, above all, the National Liberation War – the struggle for ethnic survival and liberation of Yugoslav nations. The Liberation Movement united all patriots, regardless of their ethnic origin or political affiliation, in the common struggle against occupiers and their collaborators.[23] Additionally, this movement was in its nature multiethnic, built on an ideology of equality, equal cooperation, and the "brotherhood and unity of all Yugoslav nations." It respected ethnic diversity and was a coalition of national movements built on the federal model.

These principles laid the foundations for the new federation 'of equal and brotherly' nations, which was actually formed in 1943. The Democratic Federal Yugoslavia (DFY) represented cooperation between the partisan movement and the king's Yugoslav government in exile[24] and a consensual decision of Yugoslav nations for a common life in the future, with a right to self-determination. These principles contradicted the power of the CPY, with its hierarchical organization based on 'democratic centralism'. The CPY gained public support as the leading force in the national liberation war. This internal contradiction, which could also be described as the conflict between centralist and decentralist concepts within the ruling class, influenced post-war development. Despite all limitations and internal contradictions, the federation represented a step toward decentralization and democratization.

The Yugoslav Federal Republic

The Constituent Assembly,[25] adopted the Declaration on Proclamation of the Federal People's Republic of Yugoslavia (FPRY) on 29 November 1945 as a republican and federal form of government. The assembly continued to work on the constitution, and passed the Constitution of the FPRY in January 1946.

This constitution followed the rather centralized Soviet federal model establishing a 'people's democracy'. Nevertheless, following the national liberation war, it defined 'the Yugoslav federal republic' as a "community of equal nations, which, on the basis of their right of self-determination, including the right of secession, expressed their will to live together in a federative state." The FPRY was composed of six People's Republics (PRs): Slovenia, Croatia, Bosnia-Herzegovina, Montenegro, Serbia and Macedonia. The PR of Serbia included the autonomous province (AP) of Vojvodina and the autonomous region (AR) of Kosmet-Metohia. PRs were restricted in their sovereignty only with the rights transferred by the federal constitution to the federation. Each of PR adopted its republic constitution (in accordance with the federal constitution) that reflected its specific features. The People's Assembly, the federal parliament, could change the inter-republic boundaries only with the consent of the concerned PR. The federal constitutions guaranteed to all national minorities the right to cultural development and the free use of their language.

The People's Assembly, as the legislative branch of the federal government, had two chambers of equal rights and competencies. The Federal Chamber was elected by voters in general elections. The chamber of federal units, the 'Council of Nations' was to assure ethnic plurality and equality of the PRs and nations of Yugoslavia. Regardless of size each PR elected 30 representatives to this council, the AP elected 20 representatives, and the AR elected 15 representatives.

Following its pre-war orientation, the CPY insisted on ethnic equality and protection of minorities. Nevertheless, the actually existing monopoly of power of the CPY to a certain extent reduced the constitutionally provided autonomy of nations and PR. Considering the official ideology that the national liberation war and 'socialist revolution' resolved all ethnic conflicts, the constitution provided no formal mechanism for the management of ethnic relations and conflicts. If needed, these functions were performed by the CPY and its leadership in an informal way.

The constitutional law of 1953 introduced self-management and substantially changed the existing political system. It maintained that the introduction of self-management would slowly do away with social conflicts and enable full, direct social, economic and political participation of people, thereby replacing traditional political institutions. The constitutional law abolished the 'Council of Nations' as an independent chamber of the federal assembly, replacing it with the Chamber of Producers. The 'Council of Nations', as a kind of a 'half-chamber' with very restricted competencies, became a part of the Federal Chamber, a house of representatives. People's representatives, standing for the Council of Nations, were elected in such a way that each Republic's Assembly elected 10 representatives, the Assembly of the AP 6 representatives, and the Assembly of the AR elected 4 representatives.

A year later constitutional law again changed the system, and further decreased the role of the Council of Nations. The law provided for the calling of special sessions of the Council of Nations upon the request of its members, but such a special session has never been convened in the practice.

The formal introduction of self-management, no doubt, changed the system and transformed the official ruling ideology. To reflect its reformed nature the CPY renamed itself to the League of Communists of Yugoslavia (LCY) at its Seventh Congress in Ljubljana in 1958. The LCY paid special attention to ethnic relations and the official ethnic policy of the state. The Program of the LCY (1958) stressed the importance of the principle of self-determination of nations, which had been emphasized by the national liberation struggle as the basis for the creation of the Yugoslav federation. It noted that for the implementation of the actual equality of nations, economic equality was essential. It criticized "the remains of the bourgeois nationalism" as incompatible with self-management and democratic socialism. It proclaimed that developed self-management would resolve all class and social conflicts, including ethnic conflicts, thereby surpassing class society.[26]

The ideologically conditioned illusion that self-management would resolve social and inter-ethnic conflicts influenced also provisions of the Constitution of the Socialist Federative Republic of Yugoslavia (SFRY), passed in 1963. Nevertheless, ethnic conflicts accompanied the economic and social crisis in the late 1960s, proving that latent ethnic problems existed all the time.

The Constitution of the SFRY (1963) introduced 'socialist democracy' based on self-management, and changed the official name of the country. Although Yugoslavia was still defined as a multinational federation, the class component of the federation prevailed over the ethnic component in the Constitution. Six Socialist Republics (SR) are defined as "state socialist democratic communities, based upon the power of working people and self-management" with their own constitutions, which had to comply with the principles of the federal constitution. The constitution decided upon the equal status of two Autonomous Provinces (AP), Kosovo-Metohija and Vojvodina, defined as socio-political communities within the SR Serbia.

The Council of Nations, which was to reflect plural ethnic structure and to assure equality federal units and ethnic communities in the federal parliament, was still a 'sub-chamber' of the Federal Chamber in the five-chamber Federal Assembly. Its competencies were very limited.

Besides general provisions on the equality of languages, alphabets and nations, the constitution explicitly established the rights of members of each nation to education in their own language in the territory of another republic, in accordance with the laws of the respective republic (Article 42).[27] The federal Constitution also defined the rights of nationalities (national minorities) to education in their own language, while other rights of minorities were determined and regulated by the Constitutions and laws of the republics (Article 43). The term 'nationality' was introduced by the constitution as the synonym for the term 'national (ethnic) minority' to stress the official ethnic policy that treated national minorities as equal communities and ensured their equal status and rights.

In 1967, the Council of Nations was strengthened to assure the equality and influence of nations, nationalities, Socialist Republics and Autonomous Provinces. The Council of Nations was entitled to deal with all the matters important for the equality of republics, nations, and nationalities or referring to the constitutionally guaranteed rights of the republics. This chamber became thereby, within the framework of its competencies, equal to the Federal Chamber.

Following the aggravation of ethnic relations, in 1968 amendments defined the Chamber of Nations as the first chamber of the Federal Assembly and strengthened significantly its role and competencies. The Chamber of Nations also became the biggest chamber of the Federal Assembly.

The Constitutional Amendment XVIII (1968) noted the historic significance of the socialist autonomous provinces (SAP) of Vojvodina and Kosovo within the Socialist Republic of Serbia for the realization of national equality and for the development of self-management. The rights and duties of the AP and the competencies of its bodies were determined by its Constitutional Law in compliance with the federal and Serbian republic's constitution. Additionally, Amendment XIX determined the right to use minority languages in dealing with public institutions and in public activities, in accordance with the Constitutions and laws of the republics.

Constitutional amendments in the late 1960s strengthened decentralization and democratization by increasing the autonomy and independence of Socialist Republics and Socialist Autonomous Provinces and by stressing the importance of ethnic pluralism. However, the important role of the federal center was preserved throughout the period. Despite frequent changes, constitutional regulation did not introduce mechanisms for managing ethnic conflicts that would ensure the functioning of the system in times of aggravated social relations. This fact was proved by the subsequent course of events.

The Constitution of the SFRY (1974) followed the trends started by Constitutional Amendments in the 1960s. The constitution emphasized ethnic pluralism and to some degree allowed it to be manifested in the political system. It further decentralized the federal system and increased the autonomy of federal units, SRs and SAPs. Although the constitution did not open the political space much, it started the process of democratization.

The federal constitution defined SRs as nation "states based on the sovereignty of the people and self-management by the working class and working people." As such, they were "socialist, self-managing communities of working people and citizens and of nations and nationalities having equal rights." Both SAPs within the SR of Serbia were not defined as states, but only as "autonomous, socialist, self-managing democratic socio-political communities" which were to provide for the ethnic equality and preserve the ethnic plurality of these communities.

Both chambers of the Assembly of the SFRY, The Federal Chamber and The Chamber of Republics and Provinces, reflected the federal structure on the basis of parity. Each republic, irrespective of its size elected 30 delegates into the Federal Chamber, and each SAP elected 20 delegates. The Chamber of the Republics and Provinces was composed of delegations of Assemblies of SRs and SAPs. Each republic assembly delegated 12 delegates, and each assembly of the SAP 8 delegates. The delegates elected to the Chamber of the Republics and Provinces retained their tenure in the respective republic of province assembly. These solutions were introduced to ensure the greatest possible measure of equality of the republics, as well as all the nations and nationalities in both Chambers of the Assembly of the SFRY. The constitution defined which matters were to be decided in the Chamber of Republics and Provinces on the basis of the consensus of the Assemblies of all SRs and SAPs. If a consensus was not reached, with the exception of urgent measures for the period of one year, the decisions could not be adopted. A form of the minority veto given to the SRs and SAPs was to ensure their equality.

In order to ensure equality of federal units, nations, and nationalities, the parity structure was introduced also in the Presidency of the SFRY. The principle was adopted that the federal and ethnic structure should be considered also in the formation of the Federal Executive Council, which was the federal government.

The fact that the constitution (1974) defined SRs as nation-states based on the sovereignty of the people and nations with nationalities, was especially important for the future development. The republics were given practically all the attributes of statehood: from entirely symbolic ones (the national anthem, the coat of arms, the flag, etc.) to their own constitutions as well as the entire structure of state authority bodies, that are sovereign within their own competencies. The rights, duties and competencies of republics were determined by the republic constitutions, that were only not to be in contradiction with the federal constitution.

The uniform territory of the Yugoslav federation was composed of the territories of the SRs. The borders and territory of a certain SR could only be changed with the consent of that republic, a fact that gave each one a special status similar to that of international borders.

Although formally decentralized, even with the introduction of certain confederal solutions, ex-Yugoslavia was more centralized than anticipated by the constitutional system. President Tito and the LCY were the main integrative factors. Their functioning was mostly informal, since the constitution did not determine their role, nor did it create any other formal institutions that could have assumed their informal role in the management of conflicts and assuring the necessary cohesion. This deficit became obvious in the 1980s and the beginning of the 1990s, following Tito's death and the dismantling of the LCY.

The deepening crisis in ex-Yugoslavia was becoming more and more obvious during the 1980s. The ruling regime, for a long time, did not acknowledge or deal with the crisis. It was not until the end of the 1980s, that a decision was made to amend the federal constitution. Although the changes were urgent, there was no consensus on how to do so, for the constitution could only be amended by the consensus of all federal units. Growing differences and conflicts between two options, the centralized and decentralized concept of the federation, made a comprehensive reform impossible. Finally, a compromise solution was adopted that did not solve the problems. The Amendments of 1988 began a wider democratization and the necessary economic and political reforms proposed by the federal government of Prime Minister Marković, but they soon proved insufficient. The proposed reforms failed because of the lack of political consensus.

These amendments introduced certain elements of centralization at the level of the federation, such as special stable financing of the Yugoslav People's Army, and slightly changed competencies of federal bodies. The level of centralization was much smaller than advocated by centralists, but these amendments did not include proposals by decentralists.

Subsequent attempts in 1989–1991 to amend the federal constitution and reform the political system all failed for lack of consensus. Conflicts between the two different political options were constantly growing, until finally they paralyzed the existing political system.

Differences among SRs and SAPs became especially evident in the different level of democratization in every federal unit. Individual federal units started to reflect the political views of their respective political leaders, creating a different political system in every federal unit, especially regarding the level of political pluralism and the introduction of a multi-party system. The disintegration of Yugoslavia had already started at the end of the 1980s, while unsuccessful attempts to reform the federation were still going on.

The constitutional reforms that were to harmonize the republic constitutions with the federal constitution, as amended in 1988, led to changes exceeding the initial aim. The adoption of the new republic constitutions of Serbia and Croatia, and almost a hundred constitutional amendments to the constitution of Slovenia changed relations in the federation substantially in the beginning of the 1990s.

Serbia adopted its new constitution in September 1990, before the first multi-party elections took place. This constitution introduced the presidential system and diminished the autonomy of autonomous provinces by abolishing their constitutions and replacing them with provincial statutes. Provincial assemblies could adopt provincial statutes only with previous consent of the Serbian Republic Assembly. Although the Serbian constitution meant a significant encroachment upon the Constitution of the SFRY (1974), Serbia maintained that the federal bodies could not be changed and that autonomous provinces, now totally controlled by Serbia, should remain represented in these bodies, thereby ensuring the Serbian control over these bodies.

The Constitution of the Republic of Croatia was adopted in December 1990, when Croatia was still a constitutive part of the Yugoslav federation. This constitution introduced the multi-party political system and a specific variant of the presidential system. However, it did not define Croatia as a multinational state, which aggravated the relations and conflicts with the Serbian minority (that wanted to be a constituent nation and not only a national minority), finally resulting in the civil war (1991–1992) and division of the country. During that war the Croatian Parliament passed the special constitutional law on human rights and freedoms and on special rights of ethnic minorities. In addition to certain traditional minority rights, minorities were given the right to special cultural autonomy and their proportional participation in the representative bodies was assured, together with the possibility of the founding of local communities and regions with special autonomous status.

Slovenia adopted almost a hundred amendments to its republic constitution in 1989–1991. These introduced political pluralism, furthered democratization, and strengthened the autonomy and independence of Slovenia. This process enabled a peaceful transition and social stability. The constitution of Slovenia was adopted in December 1991, just before the official recognition of Slovenia's independence in January 1992.

THE CRISIS: WHAT WENT WRONG?[28]

The Constitution of the SFRY of 1974, as amended 1981 and 1988, was obvi-ously unable to handle the conflicts that tore down the Yugoslav federation in the beginning of the 1990s. It did not provide adequate mechanisms for the democratic management of ethnic relations. Nevertheless, the constitution was not solely to blame. A number of factors contributed to the escalation of the tragic historic development. I shall analyze and present only a few circumstances that were crucial for the generation of this crisis, focusing only on the ethnic issues.

(1) In ex- Yugoslavia, despite occasional conflicts, ethnic communities have usually lived together in tolerance. Peaceful coexistence was possible because of traditional mechanisms for the management of ethnic conflicts. There were traditional networks and channels of communication, formal and informal authorities, recognized individuals and/or institutions, that intervened in a conflict. Often these traditional mechanisms were undemocratic by today standards. Nevertheless, they usually managed to regulate ethnic relations and to provide social stability.

In the past, states did not intervene in ethnic relations much. As time passed, states intervened more obviously and directly, introducing new processes that sometimes conflicted with traditional ones. Several traditional mechanisms for the management of ethnic relations and were not adequately replaced. Some were destroyed by states and were even prohibited when they conflicted with the existing law.

Anyway, states are traditionally not very successful in dealing with ethnic relations and conflicts. Repression can be successful in the short-run while aggravating conflicts in the long run.

(2) Politicians tend to underestimate the importance of a state's ethnic policy. Instead of using the creative potential of multiculturalism, they fear ethnic diversity as a possible source of conflict that could destroy the state. In this context, ethnic diversity, by itself, becomes a problem. Policy is usually based on competition, with each side trying to subdue the other. In a multiethnic environment conflicts of interests often have an ethnic dimension.

(3) In the constitutional development of ex-Yugoslavia from its establishment to its disintegration, ethnic relations played an important, though indirect, role. During the monarchy, the existence of ethnic pluralism was denied. Nevertheless, diversity often conditioned political bargaining, as it was the case with the Cvetković-Maček Agreement, which started democratization, albeit too late.

Ethnic diversity was officially recognized during the federal republic. Nevertheless, in the absence of political pluralism, there were severe limitations also to ethnic pluralism. Based on the ideology of a non-conflict society, the constitutions did not allow for the expression and coordination of different interests. This strengthened centrifugal forces of disintegration.

(4) The actual situation of ex-Yugoslavia always differed substantially from the existing constitutional model. Some provisions in all constitutions were never realized or applied. This gap between a normative system on the one hand and the objective reality on the other sides becomes a serious problem when the existing normative system can no longer regulate social processes.

(5) In ex-Yugoslavia informal channels traditionally played a crucial role. The role of King Aleksandar and a few political leaders during the monarchy, and the role of the President Tito and the CPY/LCY later illustrate this point. The important decisions were made actually by President Tito and the CPY/LCY, then formally confirmed.

Because the existing constitutional system did not assure the necessary cohesion, informal centers of power had to act as centripetal forces to provide the necessary cohesion. President Tito was especially successful in managing ethnic relations. He seldom used repression to that end. Mostly, he would use his personal charisma, his ability to attract and mobilize people by presenting them certain common interests. He promised to find solutions to the immediate problems, and usually he would deliver. After his death in 1980 and the dismantling of the LCY in 1990 there was no mechanism that could have successfully replaced their cohesive and uniting role in ex-Yugoslavia.

(6) One of the permanent characteristics of ex-Yugoslavia was a permanent conflict between advocates of centralism and advocates of decentralization. Most of the time centralism was stronger than decentralization. Centralism usually combined with domination by a certain ethnic or political elite, continuously generating resistance from those who demanded decentralization and autonomy.

Decentralization was introduced into the Yugoslav federation especially in the 1970s. It opened social and political space, and enabled a degree of democratization. In the 1980s the tension between the centralistic and decentralized options increased again. Disputes regarding democratization, the introduction of political pluralism and market economy added to this conflict, just while the struggle for Tito's legacy was going on.

Two conflicting concepts proved to be incompatible. On the one hand, there was the concept, advocated mostly by Serbia, Montenegro and the federal army, of the strong, centralized and monolithic federation, without political pluralism.[29] On the other hand, the second concept, advocated initially by Slovenia and Croatia, called for the decentralization, increased autonomy of federal units, introduction of political pluralism and market economy, and the democratic reform of political system.[30] The advocates of the first concept refused to even talk about decentralization. While advocates of decentralization did not accept centralistic proposals. The absence of the will to negotiate in a democratic process paralyzed the system.

The fact that the proponents of two conflicting concepts came from different republics and were of different ethnic origin lent ethnic dimensions to this conflict. Proponents of both concepts were members of the communist ruling elite, although they belong to very different factions of the LCY, which was organized as a kind of federation of republic organizations.

(7) Political pluralism in different parts of the former Yugoslavia was a result of a process of democratization that started in the 1980s, enabling the first multi-party elections in all Yugoslav republics in 1990. Although we could describe this as a positive development, we need to analyze certain of its problems.

It usually takes generations to develop the necessary level of political culture and socialization for democracy. The idea that it could be established by the simple formation of competing political parties does not recognize the real situation in such environments where political monism existed for a long time. There are no democratic political traditions, and most politicians were politically socialized in a totalitarian system, within a former ruling party. Traditional political ideologies are unknown to most people. In such conditions political leaders and parties desperately search a way for politically mobilizing people successfully. Ethnic identity and especially nationalism proved to be the most effective in this context.

The introduction of political pluralism in an ethnically plural environment without democratic traditions and socialization could lead to division of such a society along ethnic lines, as politicians and parties use ethnic identification of people for their political situation. This was what happened in Croatia and, most tragically, also in Bosnia-Herzegovina.

To prevent such a development in other similar cases, there would be a need to introduce certain mechanisms in a transition period that would prevent direct political confrontation along ethnic lines. A system of government should provide for sharing of powers to assure ethnic equality and protection of minorities.

(8) In retrospect, we can detect several warnings of the coming escalation of conflicts. No mechanisms existed to collect such indicators, analyze them and present them to the political leaders, together with proposals for measures for the prevention and resolution of conflicts. The earlier a conflict is detected, the easier is it prevented or resolved.

The main indicators of worsening ethnic conflicts in ex-Yugoslavia were: the lack of information on the situation and developments in other parts of Yugoslavia, or existence of only one-sided information; growing intolerance; political mobilization along ethnic lines; upsurging nationalism in different nations and presentation of nationalist programs arguing for exclusion or domination; lack of communication and cooperation; absence of common interests; and calls for increasing autonomy and independence.

Most of these conflicts had already existed before they became ethnic conflicts. It is especially important to determine how different social conflicts are translated into ethnic conflicts.

CONCLUSION: LESSONS TO BE LEARNED

The case of the former Yugoslavia shows how a relatively successful multi-ethnic state can turn into a disaster in a short time. It shows the importance of the adequate regulation of ethnic relations in multiethnic environments. Constitutions and legal systems should be adapted to the changing situation to assure good ethnic relations in a plural environment.

The constitution and legal system could not succeed in this task alone. A network of institutions, organizations and a movement that connects a state and civil society should be established. An important, if not central, part of this network is a system of early warning that could detect indicators of a possible conflict and activate appropriate mechanisms for handling conflict.

I would especially like to stress the important role of education and mass media in this context. Not only should they provide the necessary information on ethnic diversity, knowledge about 'others,' opportunities for meetings and cooperation, and education for the management of ethnic conflicts, but they should especially work on the promotion of the ideology of cooperation. This should complement or even replace the prevailing ideology of competition.

NOTES

1. Bibliographical note: I am listing used books, documents and articles in the list of references at the end of my contribution. References are included in footnotes only in case of direct citations or when they are directly referred to in the footnote.

2. The confusion of politicians contributed to such a situation. Over time, a few atrocities reported by media attracted attention and shocked the public in the West. These events increased the pressure for international intervention to stop the war. We should have the greatest respect for all activists and non-governmental organizations who searched for a peaceful solution.

3. Five independent successor states were established in the territory of the former Yugoslav federation: the Republic of Slovenia, the Republic of Croatia, the Republic of Bosnia-Herzegovina, the Federal Republic of Yugoslavia consisting of Serbia and Montenegro, and the (Former Yugoslav) Republic of Macedonia.

4. For more information on Balkans, its historic development and Balkan states see e.g.: Christopher Cviic (1991), Remaking the Balkans, Council on Foreign Relations Press, New York; Misha Glenny (1990), The Rebirth of History: Eastern Europe in the Age of Democracy, Penguin Books; London.; Barbara Jelavich (1983), History of the Balkans, Cambridge University Press, Cambridge, London, New York, New Rochelle, Melbourne, Sydney: Volume 1: Eighteenth and Nineteenth Centuries & Volume 2: Twentieth Century; Charles Jelavich, Barbara Jelavich (1977), The Establishment of the Balkan National States, 1804–1920, Vol. 8 of A History of East Central Europe, University of Washington Press, Seattle.

5. Among the oldest historical peoples known in this territory in the Antiquity there were Illyrians, Vlahs (also called Vlachs, Wallach, Wallachians), ancient Greeks (Hellenic peoples, including ancient Macedonians), and Dacians. The Slavs began to penetrate the Balkan Peninsula at the end of the 5th century AD. Avars made their incursions southward in the 6th century. Proto-Bulgarians, Turkic people originating from the area between the Ural and the Volga River who had come via the steps north of the Caspian Sea, followed in the 7th century AD. Proto-Bulgarians coalesced with the Slavs, and by the end of the 9th century they all spoke a Slav-based language. Roma (Gypsies, Romanies), originating from the Northern India, began to move into the region from the 10th century. A massive invasion of Ottoman Turks followed in the 14th and 15th century. Many other ethnic groups have passed this territory or moved into the region.

6. A modern nation can be defined as "a stable, historically developed community of people with a territory, (specific) economic life, distinctive culture, and language in common." (*Webster's New Universal Unabridged Dictionary* (1983), Deluxe Second Edition, Dorset & Baber, USA, 1983, p. 1196.) Such a community should have its distinct national/ethnic identity, defined as the identification of its members and the community as a whole with national myths, common history, heroes, values, language and culture, their leadership, etc. (See: Philip Schlesinger (1987), 'On National Identity: Some Conceptions and Misconceptions Criticized', *Social Science Information/ Information sur les sciences sociales* (London, Paris), Vol. 26, No. 2, 1987, pp. 250–254.) Some authors do not distinguish national identity form nationalism. (See: Anthony Giddens: *A Contemporary Critique of Historical Materialism* (1981), Macmillan, London & Basingstocke, pp. 13, 45–46; & The Nation-State and Violence: A Contemporary Critique of Historical Materialism (1985), Vol. 2, Polity Press, Cambridge, pp. 116, 215–219.) The emergence and development of modern nations are conditioned by the existence of modern nation states. Nations are products of the historic epoch of capitalism. (See: Ernest Gellner (1983), *Nations and Nationalism*, Cornell University Press; Ithaca & London, pp. 6–7, 53–58; & Eric J. Hobsbawm (1990), *Nations and Nationalism since 1789: Programme, Myth, Reality*, Cambridge University Press, Cambridge, p. 9.)

7. Medieval states in the Balkan region often existed as independent political entities only for a short time; their borders and political status were changing almost constantly. Often they were under the suzerainty of great powers, especially the Byzantine Empire.

8. The last reliable census in ex-Yugoslavia in 1981 showed the following ethnic situation: The total population was 22.4 million, and among them there were 8.1 million Serbs (36.3% of the total population of Yugoslavia in 1981), 4.4 million Croats (19.75%), two millions Muslims (8.92%), 1.75 million Slovenes (7.82%), 1.3 million Macedonians (5.97%) and 0.58 million Montenegrins (2.58%). The largest four of more than twenty 'traditional ethnic minorities' were Albanians (1.7 millions - 7.72%), Hungarians (0.4 million - 1.9%), Roma/Romanies/Gypsies (0.15 million - 0.75%) and Turks (0.1 million - 0.45%). The largest Albanian ethnic minority was more numerous than two nations of ex-Yugoslavia in 1981; considering ethnic projections, it became the fourth largest ethnic entity in this territory by 1991. (Data from: *Statistički bilten* (Statistical Bulletin - Federal Statistical Office, 1982), Br./No. 1295, Savezni zavod za statistiku, Beograd 1982.) ,

The last census for the whole territory of the former Yugoslavia was prepared in 1991 when the country had already started to disintegrate. I consider this census irregular. Namely, Albanians in Kosovo boycotted the census; there was a war in Croatia; people were frightened by the growing nationalist pressure in different parts of ex-Yugoslavia, etc.

9. The most known example, was the migration of Serbs in the 17th century following the defeat of their rebellion against Turks that was supported by Austria. Serbs feared the Turkish revenge. Several Serbs escaped to today's Vojvodina; others, mostly from the Kosovo region, retreated with the Austrian army and came to the territory of nowadays Croatia, especially to the Krajina region. They were settled in this region as a military shield against the possible Turkish invasion. After the independence of Croatia was declared in 1991, the Krajina region became the center of Serbian rebellion.

The withdrawal of Serbs from Kosovo, on the other hand, enabled Albanians converted to Islam to settle in this territory. This changed the ethnic structure in Kosovo. The escalation of conflicts in Kosovo in the 1980s and in Krajina in the 1990s was an important generator of the Yugoslav crisis.

10. In the Ottoman Empire, cultural diversity was facilitated there better than in the less tolerant Western Circle dominated by the Roman Catholic Church. After the defeat of Medieval Balkan States, Turks respected local traditions. The Ottoman Empire was above all an Islamic state. Non-Muslim peoples were divided into five religious communities (Millets): Orthodox, Gregorian Armenian, Roman Catholic, Jewish, and Protestant. Any individual who converted to Islam could have joined the ruling group, and many of them made successful careers. The Ottoman Empire was tolerant to non-Muslims, preserving the plurality of the region. Certain social groups often occupied certain professions, contributing to changes in the ethnic structure. Towns and cities were ethnically and religiously plural, but rural communities reflected a local peasant culture and were small and ethnically homogeneous.

11. Ethnic conflicts are conflicts of interests between distinct ethnic communities and/or their members. (See, e.g., Donald L. Horowitz (1985), *Ethnic Groups in Conflict*, University of California Press, Berkeley, Los Angeles, London, pp. 4–16, 95–96, 139–140, 216–228; participating authors in Joseph V. Monteville, ed. (1990), *Conflict and Peacemaking in Multiethnic Societies*, Lexington Books, Lexington, Toronto.)

12. These Medieval states are the main origin of their identity even if they were short-lived. Slovenes refer to the Medieval state of Carantania/Karantanija in the 8th century (in the territory of nowadays Austria) and Great Panonia in the 9th century (nowadays Hungary). Croats stress the importance of the Medieval Kingdom of Croatia. Bosnian Muslims refer to the Medieval Bosnian state. Serbs consider themselves heirs to the Medieval Serbian Empire that was independent from the 11th to the 14th century. Montenegrins are especially proud of the actual independence of the Montenegrin state that was established in the 18th century. Macedonians stress the importance of the independence of a short-lived Samuel's state at the end of the 10th century that was recaptured by the Byzantine empire in 1018.

13. Writing on ex-Yugoslavia, I used especially: Ivo Banac (1993, c 1984), *The National Question in Yugoslavia: Origins, History, Politics*. Cornell University Press, Ithaca, London; Aleksa Djilas (1991); Dimitrije Djordjević, ed. (1980), *The Creation of Yugoslavia, 1914–1918*, Clio Books, Santa Barbara & London; J. B. Hoptner (1962), *Yugoslavia in Crisis, 1934–1941*, Columbia University Press, New York & London; Barbara Jelavich (1983); Svein Mønnesland (1992), *Før Jugoslavia og etter: Nye stater - gamle nasjoner*; Sypress Forlag, Oslo; Janko Pleterski (1986), *Narodi, Jugoslavija, revolucija (Nations, Yugoslavia, Revolution)*, ČZDO Komunist-TOZD Komunist, Ljubljana, Državna založba Slovenije, Ljubljana; etc.

14. The English translation cited in Banac (1993, c 1984), p. 116.

15. The main leaders of the Yugoslav Committee, Franjo Supilo and Ante Trumbić, advocated the formation of a unitary state and considered Serbs, Slovenes and Croats one people. Nevertheless, they were aware of existing differences and did not want to accept Serbian expansionism but believed that the principles of self-determination entitled the South-Slavs to establish the state of their own. They opposed the unification as a completely one-sided act that would establish a total domination of Serbia; they believed that people in the western part of the territory would feel such a unification as a conquest by Serbia, rather than liberation.

16. A 'numerically qualified majority' for the passage of the constitution was a mechanism to prevent the domination of the largest single group. A majority bigger than a simple majority of members of the Constituent Assembly, that is 50% of (all) votes plus one, had to vote in favor for the passage of the constitution. The declaration did not decide the exact size of this qualified majority normally defined as a share/percentage of all or voting members, e.g., two-thirds majority, three-fifths majority, etc.

17. Prime Minister Pašić favored centralism in a unitary state, though he did not completely reject some ideas regarding decentralization and local autonomies. Most participants in the Corfu Conference, including delegates of the Yugoslav Committee, agreed on the unitary concept and took an antifederalist stand. The declaration itself did not include any provisions on the internal form of government to avoid a possible conflict. The Yugoslav Committee accepted this compromise because of their great fear of Italian expansion following the conclusion of the Treaty of Paris. The recognition of the crown of Karadjordjević decided the future political system in a new state to a much larger extent than initially supposed.

18. The English translation cited in Banac (1993, c 1984), p. 134.

19. On constitutional and political development of ex-Yugoslavia see, e.g.: Jovan Djordjević (1980), *Ustavno pravo (Constitutional Law),* Savremena administracija, Beograd; Stevan V. Djordjević (1967), *O kontinuitetu država: S posebnim osvrtom na medjunarodnopravni kontinuitet Kraljevine Jugoslavije i FNRJ (Continuity of States: A case of international legal continuity of the Kingdom of Yugoslavia and the Federative People's Republic of Yugoslavia)*, Naučna knjiga, Beograd; Aleksandar Fira (1979), *Ustavno pravo (Constitutional Law)*, Beograd; Vladimir Goati (1991), *Jugoslavija na prekretnici - od monizma do gradjanskog rata (Yugoslavia at the Turning-Point: From Monism to Civil War)*, Jugoslovenski institut za novinarstvo, Beograd 1991; Denison Rusionow, ed. (1988), *Yugoslavia: A Fractured Federalism*, The Wilson Center Press, Washington, D.C.; Majda Strobl, Ivan Kristan, Ciril Ribičič (1986), *Ustavno pravo SFR Jugoslavije (Constitutional Law of the SFR Yugoslavia),* časopisni zavod Uradni list SR Slovenije, Ljubljana.

20. This army played an important role in the process of formation of the new state. Regent Aleksandar and Serbian political leaders used it as a powerful political weapon in the negotiations. After the unification, the former Serbian army included some 3,500 officers of the former Austro-Hungarian army and a few hundred officers of the former Montenegro army. The Serbian generals and higher officers were especially devoted to the Karadjordjević throne.

21. 'Ban' was a title of historic Croat rulers, governors of Croatia, Slavonia and Dalmatia. 'Banovina' (pl. 'banovine') was his territory, the ban's province.

22. Prime Minister Dragisa Cvetković was the leading Serbian politician at the time, and Vlatko Maček was the president of the Croatian Peasants Party and the leader of

'The Peasant Democratic Opposition'. It was an agreement on cooperation between the ruling elites of the two biggest nations.

23. During WW II, Croat Ustashe, Serbian Chetniks, Domobranci (militia), etc. actively collaborated with occupiers. Not only did they fight against partisans, they also collabo-rated in administering the territory with quisling states. On the other hand, there was a civil war along ethnic lines, especially in today's Bosnia-Hercegovina. Croat Ustashe and Serbian Chetniks usually did not fight each other directly. Nevertheless, they terrorized the local population and both terrorized political opponents, Gypsies, etc.)

24. The Second Conference of the Antifascist Council of National Liberation of Yugoslavia on 29 and 30 November 1943 was the conference of national liberation movements. It established the Democratic Federal Yugoslavia (DFY). All the national minorities in Yugoslavia were ensured all national rights by a special decree. A deci-sion was made that, the future form of government would be established after the war based on voting. (For these documents in Serbo-Croat language, see: Petranović, Zečević (1987), pp. 791–801. Citations translated by M. Ž.)

25. The Constituent Assembly was composed of the Federal Chamber, a house of representatives where one representative was elected per 40.000 voters, and the Chamber of Nations representing the ethnic plurality of the Yugoslav community. Each of the six federal units, irrespective of its size, elected 25 representatives, whereas citizens of Vojvodina elected 15, and citizens of Kosovo and Metohija 10 representatives to the Chamber of Nations. Such a regulation was to ensure equality of nations and national-ities or federal units, respectively, of Yugoslavia in the drafting and adopting of the new constitution. (See: Zakon o ustavotvorni skupščini - The Law on the Constituent Assembly, Službeni list DFJ/ Uradni list DFJ - Official Gazette of the DFY, No. 63/1945)

26 See: *Program Zveze komunistov Jugoslavije, sprejet na 7. kongresu ZKJ (The Programme of the LCY: Adopted at the 7th Congress of the LCY)*, Komunist, Ljubljana, 1978, pp. 157–166. (Citations translated by MŽ.)

27. Article 42 introduced an exemption to the principle of equality of languages and scripts of the nations of Yugoslavia: In the Yugoslav People's Army only the Serbo-Croat language is used in the commanding, military education and administration.

28. More on the Yugoslav crisis, see, e.g.: Bogdan Denitch, *Limits and Possibilities: The Crisis of Yugoslav Socialism and State Socialist Systems* (1990), University of Minnesota Press, Minneapolis, MN., and *Ethnic Nationalism: The Tragic Death of Yugoslavia*, University of Minnesota Press, Minneapolis, 1992; Branka Magaš, The *Destruction of Yugoslavia: Tracing the Break Up, 1980–1992*, London: Verso, 1992.

29. The Presidency of the SFRY prepared a proposal for the reform based mostly on the Serbian concept of centralized federation. This proposal supported by federal insti-tutions, including the federal army, was advocated also by Serbia. (See: 'A Concept for the Constitutional System of Yugoslavia on a Federal Basis', *Review of Interntional Affairs, Vol. XLI, No. 974*, Belgrade, 5 November 1990, pp. 15–18.)

30. Advocates of decentralization first proposed the transformation of the existing Yugoslav federation into an asymmetrical federation. This model was designed to enable coexistence of different political models and systems, which already existed in different republics, within one federation. When this proposal was refused, a proposal for A Confederate Model among the South Slavic States was made by Slovenia and Croatia, but it was also immediately refused by the advocates of centralistic solutions. (See: 'A Confederate Model Among the South Slavic States', *Review of Interntional Affairs, Vol. XLI, No. 973*, Belgrade, pp. 11–22.)

REFERENCES

Banac, I.. (1993). *The National Question in Yugoslavia: Origins, History, Politics.* Copyrighted 1984. Ithaca, N.Y.: Cornell University Press.

Bilandzić, D. (1985). *Zgodovina Socialistične federativne republike Jugoslavije: Glavni procesi (The History of the Socialist Federal Republic of Yugoslavia: Main Processes).* (3rd ed.). Zagreb: Školska knjiga.

Bobbio, N. (1989). *Democracy and dictatorship: The nature and limits of state power.* Oxford: Polity.

Cviic, C. (1991). *Remaking the Balkans.* New York: Council on Foreign Relations Press.

Clissold, S. (Ed.) (1960.) *A Short History of Yugoslavia from Early Times to 1966,* Cambridge: Cambridge University Press.

Culjic, S. (1989). *Narodnosna struktura Jugoslavije i tokovi promjena (The Ethnic Structure of Yugoslavia and the Trends of Changes),* No.108. Zagreb: Ekonomski institut.

Dahl, R. A. (1989). *Democracy and its Critics,* New Haven: Yale University Press.

Denitch, B. (1990). *Limits and Possibilities: The Crisis of Yugoslav Socialism and State Socialist Systems.* Minneapolis: University of Minnesota Press.

Denitch, B. (1994). *Ethnic Nationalism: The Tragic Death of Yugoslavia.* Minneapolis: University of Minnesota Press.

Djilas, A. (1991). *The Contested Country: Yugoslav Unity and Communist Revolution, 1919–1953.* Cambridge, Mass.: Harvard University Press.

Djordjević, D. (Ed.) (1980). *The Creation of Yugoslavia, 1914–1918.* Santa Barbara, Calif.: Clio.

Djordjević, J. (1980). *Ustavno pravo (Constitutional Law).* Beograd: Savremena administracija.

Djordjević, S. V. (1967). *O kontinuitetu država: S posebnim osvrtom na medjunarodnopravni konti-nuitet Kraljevine Jugoslavije i FNRJ (Continuity of States: A case of international legal continuity of the Kingdom of Yugoslavia and the Federative People's Republic of Yugoslavia).* Beograd: Naučna knjiga.

Durakovic, N. (1984). Istorijski korijeni identiteta Muslimana (Historical Roots of the Muslim Identity). *Razprave in gradivo/Treatises and documents, No.17,* 39–45. Ljubljana: Institute for Ethnic Studies.

Fira, A. (1979). *Ustavno pravo (Constitutional Law).* Beograd: Savremena administracija.

Fisher, D. H. (1989). *Albion's Seed: Four British Folkways in America.* New York: Oxford University Press

Gellner, E. (1983). *Nations and Nationalism.* Ithaca, N.Y.: Cornell University Press.

Giddens, A. (1981). *A Contemporary Critique of Historical Materialism.* London: Macmillan.

Giddens, A. (1985). *The Nation-State and Violence: A Contemporary Critique of Historical Materialism.* Vol. 2. Cambridge: Polity Press.

Glenny, M. (1990). *The Rebirth of History: Eastern Europe in the Age of Democracy.* London: Penguin.

Goati, V. (1991). *Jugoslavija na prekretnici - od monizma do gradjanskog rata (Yugoslavia at the Turning-Point: From Monism to Civil War).* Beograd: Jugoslovenski institut za novinarstvo.

Hague, R., Harrop, M., &, Breslin, S.. (1992). *Comparative Government and Politics: An Introduction.* (3rd ed.). London: MacMillan.

Hobsbawm, E. J. (1990). *Nations and Nationalism since 1789: Programme, Myth, Reality,* Cambridge University Press, Cambridge.

Hoptner, J. B. (1962). *Yugoslavia in Crisis, 1934–1941.* New York: Columbia University Press.

Horowitz, D. L. (1985). *Ethnic Groups in Conflict.* Berkeley: University of California Press.

Jelavich, B. (1983). *History of the Balkans.* Volume 1: Eighteenth and Nineteenth Centuries & Volume 2: *Twentieth Century.* Cambridge: Cambridge University Press.

Jelavich, B., & Jelavich, C. (1977). *The Establishment of the Balkan National States, 1804–1920*. Vol. 8 of *A History of East Central Europe*. Seattle: University of Washington Press.

Jovičić, M. (1977). *O ustavu (About the Constitution)*. Belgrade: Savremena administracija.

Kržišnik-Bukić, V. (1996). *Bosanska identiteta: Med preteklostjo in prihodnostjo (Bosnian Identity between Past and Future)*, Ljubljana: Inštitut za narodnostna vprašanja. (Especially English Summary, pp. 115–121.)

Lerotić, Z. (1989). *Jugoslavenska politička klasa i federalizam (Yugoslav Political Class and Federalism)*. Zagreb: Globus.

Lucarelli, S. (1995). *The International Community and the Yugoslav Crisis: A Chronology of Events*. EUI Working Paper RSC No. 95(8). Florence: European University Institute.

Magaš, B. (1992). *The Destruction of Yugoslavia: Tracing the Break Up, 1980–1992*. London: Verso.

Mønnesland, S. (1992). *Før Jugoslavia og etter: Nye stater - gamle nasjoner*; Oslo: Sypress Forlag.

Montevile, J. V. Ed. (1990). *Conflict and Peacemaking in Multiethnic Societies*. Lexington and Toronto: Lexington Books.

Murray, W. S. (1967). The Making of the Balkan States. *Studies in History, Economics and Public Law 102*, edited by the Faculty of Political Science of Columbia University. New York: AMS.

Pavlowitch, S. K. (1971). *Yugoslavia*. Nations of the Modern World series. New York and Washington: Praeger.

Petranović, B., & Zečević, M. (1987). *Jugoslovanski federalizam: Ideje i stvarnost, Tematska zbirka dokumenata, Prvi tom 1914–1943 (The Yugoslav Federalism: Ideas and Reality, Thematic Collection of Documents.Vol. I, 1914–1943)*. Beograd: Prosveta.

Pirjevec, J. (1995). *Jugoslavija, 1918–1992: Nastanek, razvoj ter razpad Karadjordjevićeve in Titove Jugoslavije (Yugoslavia, 1918–1992: The Formation, Development and Downfall of Karadjordjevic's and Tito's Yugoslavia)*. Koper, Slovenia: Založba Lipa.

Pleterski, J. (1986). *Narodi, Jugoslavija, revolucija (Nations, Yugoslavia, Revolution)*. Ljubljana: Državna založba Slovenije.

Poulton, H. (1991). *Balkans: Minorities and States in Conflict*. Foreword by Milovan Djilas. London: Minority Rights Group.

Ramet, P. (1984). *Nationalism and Federalism in Yugoslavia, 1963–1983*. Bloomington: Indiana University Press.

Ramet, P. (1985). *Yugoslavia in the 1980s. Westview Special Studies on the Soviet Union and Eastern Europe*. Boulder, Colo.: Westview Press.

Ristelhuber, R. (1971). *A History of the Balkan Peoples*. Edited and translated by Sherman David Spector. New York: Twayne Publishers.

Rusinov, D. I. (1977). *The Yugoslav Experiment, 1948–1974*. Berkeley: University of California Press.

Rusinov, D. I. Ed. (1988). *Yugoslavia: A Fractured Federalism*. Washington: The Wilson Center Press,.

Sartori, G. (1987). *The Theory of Democracy Revisited*. Catham, N.J.: Catham House Publishers.

Schlesinger, P. (1987). On National Identity: Some Conceptions and Misconceptions Criticized. *Social Science Information/ Information sur les sciences sociales, .26*(2), 219–264.

Schlesinger, P. (1988). *Statistični godišnjak Jugoslavije (The Statistical Yearbook of Yugoslavia - 1988)*. Year XXXV. Beograd: Savezni zavod za statistiku.

Strobl, M, Kristan, I., & Ribičič, C. (1986). *Ustavno pravo SFR Jugoslavije (Constitutional Law of the SFR Yugoslavia)*. Ljubljana: Časopisni zavod Uradni list SR Slovenije.

Sugar, P. F., & Lederer, I. J. Eds. (1969). *Nationalism in Eastern Europe*. Seattle: University of Washington Press.

Vucinich, W. S. (1965). *The Ottoman Empire: Its Record and Legacy.* Princeton, N.J.: Van Nostrand.
Vucinich, W. S. (1989). In World Directory of Minorities, edited by Minority Rights Group. London: Longman International Reference.
Zametica, J. (1992). The Yugoslav Conflict: An analysis of the causes of the Yugoslav war, the policies of the republics and the regional and international implications of the conflict. *Adelphi Paper 270.* London: International Institute for Strategic Studies and Brassey's.
Žagar, M. (1994). Nation-States, Their Constitutions and Multi-Ethnic Reality: Do Constitutions of Nation-States correspond to ethnic reality? *The Journal of-Ethno-Developmentt, 3*(3), 1–19.
Žagar, M. (1995a). Constitutions in Multi-Ethnic Reality, *Gradiva in razprave / Treatises and Documents. No. 29–30,* 143–164. Ljubljana: Inštitut za narodnostna vprašanja / Institute for Ethnic Studies.
Žagar, M. (1995b). Nekaj hipotez o kvadraturi kroga: Ustava SFRJ in proces osamosvajanja Republike Slovenije - Etnična dimenzija osamosvajanja Slovenije (The Constitution of the SFRY and the Struggle for the Independence of the Republic of Slovenia: Ethnic dimensions of the independence). *Gradiva in razprave / Treatises and Documents. No. 29–30,* 231–260. Ljubljana: Inštitut za narodnostna vprašanja / Institute for Ethnic Studies.

4. TERRITORIES WITH MIXED POPULATIONS: ARE THEY GOOD OR BAD FOR PEACE?

ENCLAVES AND ETHNIC TOLERANCE

Dusko Sekulic, Randy Hodson and Garth Massey

The horrible events in Yugoslavia have been explained in many different ways. Economic differences and 'internal colonialism' theories;[1] 'cultural explanations' referring to the incompatibility of cultural groups; and versions of Huntington's 'clash of civilizations' have all been invoked at different times. Another popular account at the time of the breakup of Yugoslavia was the 'ethnic hatred' theory. Its basic assumption was that it was solely the communist dictatorship that had kept the groups off each other's throats; when the pressure of dictatorship eased, latent intolerance broke out and embroiled everyone in mortal struggle against everyone else. Many authors questioned this explanation on the grounds of historical and political analysis.[2] The main indicators creating doubts about this explanation were: the non-existence of 'communal' conflicts, and of residential segregation, and the increase of 'mixed marriages' and of self-declared 'Yugoslavs' in the censuses.[3] All this did not present a picture of latent conflict just waiting to explode.

We know that during World War II the territory of Yugoslavia experienced a bloody civil war and that many accounts remained unsettled from that time. The basic political divisions in Yugoslavia were traditionally along national lines. This was last evident in 1971. In that year there culminated a conflict over the direction of economic reform and the degree of political liberalization

Research on Russia and Eastern Europe, Volume 3, pages 97–110
© 2000 by Elsevier Science Inc.
ISBN: 0-7623-0280-1

among the republic's Communist Party leaderships. In the process, the leader-ship's started to make alliances with the more traditionalist nationalist political forces which was most evident in Croatia. Intervention by the center, at that time led by Tito, led to the dismissal of political leaderships in Croatia, Serbia, Slovenia and Macedonia. When in 1987 Milosevic created the same type of alliance the powerful centre with capability of intervention disappeared and Yugoslavia started its road to national fragmentation and disintegration.[4] From the sixties onwards, greater emphasis was placed upon the federal elements in Yugoslavia, where the basic justification for the 'federalization' was to accommodate the different nationalities of Yugoslavia. But did these political divisions and cleavages really reflect deep existing animosities? Was intolerance and ethnic hatred something that characterized the population of the former Yugoslavia?

We are in a unique position to check how widespread ethnic tolerance was on the eve of the breakup of Yugoslavia. In the winter of 1989–90 the Consortium of Social Research Institutes of Yugoslavia conducted interviews in workplaces and households in all the republics of Yugoslavia, utilizing a multi-stage random cluster sampling design. The completed survey included 13,422 adults ages 18 and older distributed across republics and autonomous provinces (Kosovo and Vojvodina) in accord with the size of the population of each. Ethnic tolerance was measured on a five point-Likert scale with five propositions: (1) Nationality should be a central factor in choosing a marriage partner; (2) Nationally-mixed marriages are more unstable than other marriages; (3) Every nation should have its own state; (4) People can feel completely safe only when the majority belong to their nation; (5) Among nations it is possible to create cooperation, but not full trust. All items had item-total correlations above 0.4 (We started also with the sixth item – "Without leaders every nation is like a man without a head" – which was dropped from further analysis because 61.5% of the respondents were grouped in the 'strongly agree' category)

HOW TO EXPLAIN THE DIFFERENCES?

The first thing that we can look for is the absolute level of tolerance. If the 'ethnic hatred' theory is correct we can expect relatively high levels of intolerance. Because this theory presupposes a causal connection between attitudes and actions we can expect that the intensity of tolerance is negatively connected with the occurrence of conflict. If the unit of analysis is the republic then we can expect that the republics which later experienced civil war (Bosnia and Herzegovina, Kosovo, Croatia) would have a higher level of intolerance than the republics spared the civil war (Slovenia – having a very short clash with the Yugoslav Army – and Macedonia and Montenegro – spared at least for now).

Another explanation of the differences could be a version of a 'modernization' theory. We could expect that the tolerance would be the highest in the most economically developed and most urbanized republics and lowest in the least developed republics. Consequently we will expect the highest tolerance in Slovenia and the lowest in Kosovo.

Finally we can look for the explanation in a variant of the 'contact' theory.[5] The main claim of the 'contact' hypothesis is that contact among members of the different group, under certain conditions reduces prejudices and increases tolerance. (The conditions much be as follows: the contacting people must be of equal status; contact must be in the pursuit of an active goal-oriented effort and common goal; attainment of the common goal must be an inter-dependent effort based on cooperation without group competition; and it must have support of authorities, law or customs.)[6] As Forbes[7] has shown, the contact hypothesis describes processes on an individual level. What is valid on the individual level does not necessarily hold on the aggregate, group level. An increase of tolerance of individuals in contact with the out-group can increase resentment of the rest of the group and increase intolerance in general.

Our study can be classified as 'ecological' or a study of proportions, to use Forbes's terminology. In this type of study, the ethnic/national heterogeneity/ homogeneity of the city, neighborhood, or the republic/province in our case is correlated with the attitudes of the people. Does heterogeneity increase tolerance as would be implied by the contact hypothesis expanded to the aggregate level? Or does contact increase intergroup intolerance, as hypothesized by Forbes? The results of previous research is mixed. On the one hand Kalin & Berry,[8] White & Curtis[9] and Curtis & White[10] have found that tolerance of French Canadians toward English speaking Canadians increases with their increased presence and vice versa. But the opposite relations holds for the tolerance toward immigrants; as their number increases intolerance increases.[11] The votes for the National Front in France were positively correlated with the share of 'colored immigration' to the constituency.[12]

What holds for immigrants holds also for inter-racial relationships. American studies show increased prejudices toward Blacks as their proportion increases.[13] A similar finding is reported for Australia by Mitchell,[14] indicating that the 'white' people in areas with most Aborigines voted most negatively in the constitutional referendum for giving to Aborigines the right to vote. Forbes[15] concludes from these and other studies that contact on the group level does not reduce prejudice and intolerance.

On the other hand we can conclude that the contact among culturally similar people (French and English speaking Canadians) does decrease prejudice and intolerance. Contact in a situation of the relatively speedy change of the ethnic composition (immigration) creates a sense of threat which may be better explained by the competition theory. Also, contact among culturally very different and/or physically different people, can also increase prejudices and intolerance.[16] If that is so we can predict that increased heterogeneity in the former Yugoslavia will increase tolerance and decrease prejudices because the cultural difference among the groups was small and settlement patterns relatively stable.

THE BASIC RESULTS

Comparison of the average tolerance levels in different republics. (Tolerance is spread on the scale from 1 to 5 where 1 means absolute intolerance and 5 maximum tolerance).

Table 1.

Republic	Average Tolerance
Bosnia and Herzegovina	3.88
Vojvodina	3.83
Croatia	3.63
Montenegro	3.45
Serbia	3.28
Slovenia	2.67
Macedonia	2.53
Kosovo	1.71
Average	3.28
N	13.422

What can we conclude from these data? First, that the level of economic development and level of tolerance are not correlated. The most developed republic, Slovenia, is among the republics which are intolerant, although the second most developed part of the former Yugoslavia, Croatia, is among the tolerant republics. On the other side of the scale the least developed republics Kosovo and Macedonia are extremely intolerant – as modernization theory would predict – but the other two economically underdeveloped parts of Yugoslavia do not fit the predictions of modernization theory. Underdeveloped Bosnia and Herzegovina is the most tolerant republic and Montenegro is on the upper part of the tolerance scale. Putting all this together we cannot find a meaningful explanation for tolerance level in the level of economic development.

Second, as we have showed in another paper,[17] the closest explanatory variable is national heterogeneity. The republics that have more nationally mixed populations also have a higher level of tolerance. We know that Bosnia and Herzegovina and Vojvodina are the most 'mixed' parts of the former Yugoslavia and they are on the top of the tolerance scale. The low position of Slovenia on the tolerance level can be explained by its homogeneity. It is the most homogenous of the former Yugoslav republics, with Slovenians comprising 90.5% of Slovenia's population in 1981 and 87.6% in 1991.

If we look at the top of the tolerance table Bosnia, Herzegovina and Vojvodina are the most tolerant and at the same time the most heterogeneous parts of the former Yugoslavia. At the same time we can add that they differ significantly in levels of economic development, with Bosnia being underdeveloped and Vojvodina developed. (In 1989 GDP per capita in dinars (1972 prices) was 20.063 for Vojvodina and 11.424 for Bosnia and Herzegovina.[18]

On the other hand, at the bottom of the tolerance level we have the two least economically developed parts of Yugoslavia, Kosovo (4.317 din. GDP per capita) and Macedonia (10.891 din GDP per capita) and the most developed Slovenia (33.103 din. GDP per capita). By analyzing Slovenia, Kosovo and Macedonia, the least tolerant parts (if we use a two-tailed t-test by contrasting the mean for each republic and autonomous province with the mean for all republics and provinces, these three units are clearly less tolerant than the average) we can reach two interesting conclusions. First, the two least tolerant units, Kosovo and Macedonia, involve Albanians – in Macedonia as a substantial minority and as majority in Kosovo. (In 1991 it is estimated that Albanians constituted around 90% of the Kosovo population. This is an estimate because Albanians mainly refused to participate in the census. In Macedonia they were 21% of the population). It must also be noted that within these two units the Albanians are significantly less tolerant than corresponding groups. In Kosovo the average rating for Albanians is 1.79 and for Serbs 1.99. For Macedonia the average tolerance rating for Albanians is 1.84 and for Macedonians 2.61. In both cases this difference is statistically significant.

One explanation for this phenomenon may be the cultural difference between the Albanians and the other national groups – in this case Serbs and Macedonians. The Albanians are the only non-Slavic group, speaking a non-Slavic language, and being predominantly of Muslim religion. Whereas a difference in religion also exists between the Bosnian Muslims, Serbs, and Croats, they are identical in language and the religiosity of the Bosnian Muslims was not very intensive (The religiosity of Bosnian Muslims is probably much higher now after the permanent religious mobilization performed by the SDA – Party of Democratic Action – and experiences of war). The Kosovo Albanians are consequently 'different' on two accounts, religion and language, from the rest of the population. The fact that they are more intolerant than their 'neighbors' probably arises from the status that they occupied in the Yugoslav states and an even longer history of persecution dating from the expansion of Serbian nationalism in the nineteenth century.[19] Even without a historical explanation, the fact of difference is known to contribute to identity and conflict. Sociologists working in the tradition of Robert E. Park, focusing on cultural similarity together with manner and time of entrance of the ethnic group, the concentration-dispersion of the group, and its visibility,[20] recognized that cultural difference is one of the obstacles to assimilation.[21] Also, as Gurr formulated the situation, "the greater the cultural and social differences between a communal group and others with which it interacts, the greater will be the strength of group identity."[22] Obstacles to assimilation and the strength of group identity are prerequisites for group conflict. On the basis of this knowledge we can argue that the reason why intolerance in Kosovo and Macedonia is even greater than we expected on the basis of its heterogeneity (because Kosovo and Macedonia are more heterogeneous than Serbia and Slovenia but are less tolerant) is because the intergroup differences between Albanians and Slavs (Serbs and Macedonians) are greater than among different Slavic groups in other parts of Yugoslavia.

As for Slovenia and its level of intolerance, it is significant that it is the most economically developed republic of Yugoslavia but also the most homogeneous. There is no direct 'threat' nor a history of domestically rooted ethnic conflict. But Slovenia is an 'immigrant country' thanks to its level of economic development. The 'ethnic homogeneity' is in constant decline as the result of the immigration from the poorer south of the country. Slovenes in 1961 constituted 95.6% of the population of Slovenia and, in 1991, 87.6%. Let us remember for the moment previously mentioned research results from Canada. The contact between French speaking and English speaking Canadians reduces prejudices but contact between Canadians with immigrants increases the prejudice. We could argue that there is a difference in the mechanisms that produce intolerance in Kosovo and Macedonia on the one hand and in Slovenia on the other. Slovenians' intolerance is more of the 'western type' – it is directed against immigrants in the same way as was mentioned in research in Canada or as it is partially confirmed for 12 European countries by Quillian.[23] (living near the immigrants increases intolerance – working with them increases tolerance). On the other hand the reasons for intolerance among other groups are of a different 'type' because the relationship among these groups is based on the long-standing life side by side. This is then transformed into disputes over territory. Whether

Kosovo is Serbian or Albanian territory is the same question posed around the world wherever competing groups claim the same territory. One of the most famous examples of this is the dispute among Palestinians and Israelis, which is the most similar dispute to the existing one in Kosovo.

The main conclusion from the analysis where the units were republics and autonomous provinces of the former Yugoslavia is that greater ethnic heterogeneity leads to greater tolerance.

TOLERANCE AND CONFLICT

With the benefit of hindsight, one of the main conclusions could be that the level of tolerance measured through attitudes does not predict behavior (conflict). The bloodiest and most protracted conflict happened in Bosnia and Herzegovina, yet we see that Bosnia and Herzegovina was the most tolerant of the former Yugoslav republics. Before the war in Croatia, again if we look at Table 1, the level of tolerance in Croatia was above the Yugoslav average. On the other hand the explosion in Kosovo was predictable on the basis of the highest intolerance among the republics and provinces of the former Yugoslavia. The basic conclusion is obviously that the conflict was not correlated with the diagnosed intolerance levels.

This conclusion should probably not surprise us. Attitudes and behavior are not the same thing. As we remember from the classical analysis of Robert Merton[24] prejudice (attitude) and discrimination (behavior) are not in perfect correlation. Depending on the social context we can have prejudiced non-discriminators and non-prejudiced discriminators. Social pressure and tendencies toward conformity are interposed between attitudes and behavior. As we showed in another study,[25] the authoritarianism which was/is the central feature of the political culture of the former Yugoslavia can be taken as the key explanatory variable. As we mentioned, the diagnosis of authoritarianism is based on the fact that the large majority of respondents in all republics and provinces of the former Yugoslavia accepted the statement that "without leaders every nation is like a man without a head." If in such an authoritarian situation the leaders instigate conflict, the followers will follow, regardless of their individual tolerance. Conformity to the leader's call (without whom the nation is like the man without head) can produce conflict overriding individual tolerance.

Even without authoritarianism as the intervening variable the situation can force people to act aggressively, regardless of their individual attitudes. If your village is attacked by a paramilitary formation of the other ethnic group, you don't have much choice but to fight. As one Croat member of the Serbian unit engaged in systematically killing Muslims after the fall of Srebrenica, testified in The Hague, "it was kill or be killed." As we know from history, only a few heroic German soldiers refused to shoot Jews, women, and children in Ukraine, Russia and the whole of Eastern Europe. Were they Hitler's 'willing executioners' or simply people put in a situation where they had to choose to kill rather than to be killed?

ENCLAVES AND TOLERANCE

It would be to simple to dismiss our results by stating that attitudes do not predict behavior. Another element must also enter the picture and this is the phenomenon of enclaves. As we know from the recent ethnic conflicts around the world, the start of the conflict and the most intense fighting happened around the 'enclaves' – concentrations of minorities surrounded by the majority. Nagorno-Karabach, Krajina, and Northern Ossetia took their places on the front pages of the newspapers beside the 'old' inhabitants of it, such as Northern Ireland or the Punjab. The existence of such 'enclaves' contradicts the basic assumptions of the 'contact' hypothesis. Enclaves offer more opportunities for contact between the minority and majority, unless we refer to the kind of enclave that does not contain a mixture but an ethnic composition opposite to that of the surrounding territory. If contact automatically reduces prejudices, and if the chance of contact among different groups in enclaves is intensified because they are more heterogeneous, then we should expect that the prejudices and intolerance should be lower in enclaves. We know that the conflict is more intense in enclaves, but what about prejudice?

In order to answer this question we took as the measurement units, opcina (the smaller territorial units – 'boroughs') and not republics as in our first analysis.[26] In most opcinas the republic's majority is in the majority (for example Croats in Croatia). But in some opcinas the republic's minority is in the majority (for example in 'Krajina' in Croatia, Serbs were in the majority and Croats in the minority). We operationally defined such opcinas as 'enclaves.' Operationally therefore, the enclave is any opcina where the republic's minority is a local majority (for example, for Croatia all the opcinas in which Serbs are in the majority).

From the regression model of tolerance on the set of demographic, social status, participation, religiosity, and national composition variables (for more details see Massey, Hodson & Sekulic, 1999, Table 2) we give here the coefficients indicating the importance of the arrangement of minority/majority relationships. We use a set of dummy variables measuring the specific nature of each group's status as a majority/minority group. The reference categories

Table 2. Unstandardized Regression Coefficients of Tolerance on National Composition

Independent variable	Unstandardized Regression coefficients
Dominant majority	−0.260***
(ex. Croats in Croatia, Serbs in Serbia etc.)	
Local Minority	0.134 **
(ex. Serbs in Croatia, Croats in Serbia etc)	
Enclaved majority	−0.390***
(e.g. Croats in Croatia living in Serb dominated areas)	
Eclaved minority	−0.127**
(e.g. Serbs in Serbian dominated parts of Croatia)	
Majority in mixed area	−0.162*
(Croats in Croatia living in opcina where around half of population are Serbs)	
Minority in mixed area	0.129 **
(e.g. Serbs in Croatia living in area where the other half of population are Croats)	
Secondary minorities	0.000
(baseline)	

***p ≤ 0.001   **p ≤ 0.01   *p ≤ 0.05 (2-tailed t-test)

used are 'secondary minorities' which are those having no republic in which they are dominant. The categorical measure of majority/minority status allows us to examine patterns of tolerance across different structural settings, including enclaves and nationality-mixed locales (opcinas in which neither of the ethnic groups is in majority).

If we analyze the data on that level, one of the conclusions that could be drawn is that *people in enclaves are less tolerant than outside the enclaves.*[27] The most intolerant are the 'enclaved' majority members when they live as a minority in enclaves. The difference between the enclaved majority living as a minority and majority living as a majority is, for example, most visible in Montenegro. The Montenegrins living as a majority have an average tolerance level of 3.35, but when they live as 'enclaved' – in the case of Montenegro that means in *opcinas* dominated by Muslims or Albanians – their tolerance level is significantly lower – 2.76. For the whole of Yugoslavia, we can see that the greatest coefficient pointing in a 'negative' direction (more intolerant than the 'baseline') is for the enclaved majority (−0.390). The minority that is the majority in the enclave is also less tolerant compared with their co-ethnics who live dispersed among the majority (regression coefficients −0.127 compared with 0.134)

This result puts us closer to the possibility that attitudes are predictors of behavior. Ethnic conflicts started in enclaves and as we can see from the data, ethnic tolerance in enclaves are lower than outside the enclaves. This contradicts the 'contact' hypothesis because the possibility of contact in enclaves among different ethnic groups is higher than outside the enclaves but the actual tolerance is lower. We can argue that the enclave – the territory in which the minority (example Serbs in Croatia) is local majority (Serbs in Krajina where they are local majority) – creates a situation where the prediction of the contact hypothesis does not hold. Contact does not decrease intolerance, but rather, increases it. Our finding is corroborated by the findings of McIntosh et al., for Rumania.[28] Their finding indicates at first glance, a contradictory result – that increased heterogeneity increases tolerance in Rumania but decreases it in Bulgaria. The explanation given by the authors is that the 'mixed' areas in Bulgaria are dominated by a minority; Turks are in the majority in such areas and Bulgarians are in the minority which corresponds to our definition of 'enclaves.' On the other hand, 'mixed' areas in Rumania are still dominated by the majority; Romanians are still dominant and Hungarians constitute less than half of the population.

If we combine the findings of these two studies we can reformulate the contact hypothesis in a non-linear fashion. Increased chances for contact (measured as the minority's proportion of the total population) increases tolerance – but only to a certain level. When 'minority' proportions approach or become the majority, the contact-tolerance relationships reverses and further increase in the proportions of the minority decreases tolerance.

This reformulation also allows us to reinterpret some earlier research. For example in his classical study Williams[29] found in three cities that contact decreases prejudice but that it did not hold for the fourth city, where there was no correlation between contact and prejudice. The fourth city was in the South which means that the number of blacks was much greater. In that sense the whole 'South' could be interpreted as an enclave.

It must be emphasized that this increased intolerance holds not only for the majority but also for the minority. For both groups – minority and majority – in enclaves tolerance decreases.

If we leave the enclaves there is another important conclusion that can be drawn from our data. *Majorities are less tolerant than minorities.* In every one of our three demographic situations ('normal situation' where the majority is majority and minority is minority, in 'mixed' situation where the majority and minority are in equal proportions, and in 'enclaved' situation where the majority and minority reverse their positions) the majority is less tolerant than the minority. Even in enclaves, regardless of the fact that the minority (which is now the 'local' majority) is less tolerant than when dispersed (regression coefficients –0.127

compared with 0.134) they are still more tolerant than the majority outside of enclaves (regression coefficient –0.260). And in enclaves the majority which is now a local minority (for example Croats in Krajina) are less tolerant than the local majority (e.g. Serbs in Krajina) For example absolute average levels of tolerance in Croatia for Croats living as dominant majority is 3.53. For Serbs living in Croatia as a minority it is 3.67 (which is statistically significantly higher). But for Serbs as local majorities in enclaves it falls to 3.36. In Serbia Serbs living as the majority have the average level of tolerance of 3.10. When they live in enclaves it falls to 2.96. When the minorities are living dispersed among Serbs in Serbia their average level of tolerance is 3.32 but it falls to 3.10 when they are enclaved and live as 'local majority.'

CONCLUSIONS

The main conclusion from our research is that in enclaves (defined as local territories dominated by a group that is a minority within the larger state – e.g. Quebec, Northern Ireland, what was called Krajina in Croatia, and which after 1995 is mostly 'empty land') the average level of individual ethnic tolerance is lower than outside the enclaves. The analogy to these findings could be found in the literature on nationalism and in discussions of 'peripheral' nationalism. Hitler was an Austrian, Milosevic is Montenegrin, etc. When applied to the relations between local minorities and majorities of the larger state, the periphery is a similar concept to our enclave.

We argue that the 'contact' hypothesis holds outside enclaves. An increased proportion of the minority group increases tolerance, but in the enclaves that does not hold. In this way we can argue that we are adding another 'condition' under which the contact hypothesis is valid.

A practical application of this finding could be that the creation of enclaves should be avoided. Here we should keep in mind the possibility that the enclave as we defined it here may not be the same as the ghetto. The ghetto may be a territory where the group that is a minority within the larger society constitutes the local majority with no presence whatever of the group that is the larger majority living 'around' but not 'within' the ghetto. Whether intolerance in such a ghetto follows the same pattern as in the enclave remains to be seen.

NOTES

1. M. Hechter. *Internal Colonialism: The Celtic Fringe in British National Development 1536–1966.* (Berkeley: University of California Press, 1975.)

2. Dusko Sekulic, 'Nationalism versus Democracy: Legacies of Marxism'. *International Journal of Politics, Culture and Society*, Vol. 6, No. 1, 1992.; Bogdan Denitch, *Ethnic Nationalism: The Tragic Death of Yugoslavia'*. (University of Minnesota Press, 1994); L. Cohen, *Broken Bonds*.(2nd ed.). Boulder, Colorado: Westview, 1995); Susan Woodward, *Balkan Tragedy: Chaos and Dissolution After the Cold War* (Washington, D.C. Brookings, 1995); L. Silber and A. Little, *Yugoslavia: Death of a Nation* (TV Books, 1996); Craig Calhoun, *Nationalism* (University of Minnesota Press, 1997).

3. D. Sekulic, G. Massey and R. Hodson, 'Who Were the Yugoslavs? Failed Sources of a Common Identity in the Former Yugoslavia'. *American Sociological Review*, 59, 1994.

4. Dusko Sekulic, 'Nationalism versus Democracy. Legacies of Marxism'. *International Journal of Politics, Culture and Society*. Vol.6. No.1, 1992.

5. Gordon Allport, *The Nature of Prejudice*, (New York: Doubleday, 1954).

6. Thomas Pettigrew, *'Intergroup Contact Theory'* Paper presented at the Flinders University Seminar, 1998.

7. D. H. Forbes, *Ethnic Conflict: Commerce, Culture and the Contact Hypothesis.* (New Haven and London: Yale University Press, 1997).

8. R. Kalin and J. W. Berry, 'The Social Ecology of Ethnic Attitudes in Canada,' *Canadian Journal of Behavioral Science*, 1982, v. 14.

9. P. G. White & J. E. Curtis, 'Language Regions and Feelings Toward Outgroups. Analysis for 1968 and 1984,' *Canadian Journal of Sociology*, 1990, Vol. 15.

10. J. E. Curtis and P. G. White, 'Proximity or Regional Cultures: A Re-examination of Patterns of Francophone-Anglophone Liking of Each Other', *Canadian Journal of Sociology*, 1993, vol. 18.

11. D. T. Studler, 'Social Context and Attitudes Toward Coloured Immigrants,' *British Journal of Sociology*, 1977, Vol. 28.

12. S. Taylor, 'The Incidence of Coloured Populations and Support for the National Front', *British Journal of Political Science*, 1979, vol. 9.

13. R. M. Williams, *Strangers Next Door: Ethnic Relations in American Communities.* (Englewood Cliffs, N.J.: Prentice Hall, 1964); M. A. Fossett and K. J. Kiecolt, 'The Relative Size of Minority Population and White Racial Attitudes', *Social Science Quarterly*, 1989, Vol. 70.

14. I. S. Mitchell I.S. 'Epilogue to a Referendum', *Australian Journal of Social Issues*, 1968, vol. 3.

15. Forbes, 1997.

16. Robert E. Park, *Race and Culture.* (New York: Free Press, 1950).

17. R. Hodson, D. Sekulic and G. Massey, 'National Tolerance in the Former Yugoslavia', *American Journal of Sociology*, 1994: Vol. 99, No. 6.

18. *Statistical yearbook of Yugoslavia*, 1991. (Beograd, 1991) pp. 442, 475.

19. N. Malcolm, *Kosovo: A Short History* (New York University Press, 1998).

20. N. M. Marger, *Race and Ethnic Relations: American and Global Perspectives* (Wadsworth, 1994), pp. 125–27.

21. Pierre Van den Berghe, *The Ethnic Phenomenon* (New York: Elsevier, 1981).

22. T. R. Gurr, 'Why Minorities Rebel: A Global Analysis of Communal Mobilization and Conflict Since 1945', *International Political Science Review*, 1993, Vol. 14) p. 174.

23. L. Quillian, 'Prejudice as a Response to Perceived Group Threat: Population Composition and Anti-Immigrant and Racial Prejudice in Europe', *American Sociological Review*, 1995, Vol. 60.

24. Robert K. Merton, 'Discrimination and the American Creed', in *Discrimination and the National Welfare.* ed by Robert MacIver, (New York: Harper and Row, 1949).

25. J. Zupanov, D. Sekulic and Z. Sporer, 'A Breakdown of the Civil Order: The Balkan Bloodbath', *International Journal of Politics, Culture and Society*, 1996, Vol. 9, No. 3.

26. Hodson, Sekulic, Massey, 1994.

27. See G. R. Massey, R., Hodson, and D. Sekulic, 'Ethnic Tolerance and Enclaves,' *Social Forces*, forthcoming.

28. M. E. McIntosh, M. A. McIver, D. G. Abele, D. B. Nolle, 'Minority Rights and Majority Rule: Ethnic Tolerance in Romania and Bulgaria', *Social Forces*, 1995, Vol. 73.

29. R. Williams, 1964.

REFERENCES

Allport, G. (1954). *The Nature of Prejudice*. New York: Doubleday.

Calhoun, C. (1997). *Nationalism*. University of Minnesota Press.

Cohen, L. (1995). *Broken Bonds*. (2nd ed.). Westview.

Courtis, J. E., & White, P. G. (1993). Proximity or Regional Cultures ? A Re-examination of Patterns of Francophone-Anglophone Liking of Each Other. *Canadian Journal of Sociology*, 18.

Denitch, B. (1994). *Ethnic Nationalism. The Tragic Death of Yugoslavia*. University of Minnesota Press.

Forbes, D. H. (1997). *Ethnic Conflict. Commerce, Culture and the Contact Hypothesis*. New Haven and London: Yale University Press.

Fossett, M. A., & Kiecolt, K. J. (1989). The Relative Size of Minority Populaton and White Racial Attitudes. *Social Science Quarterly*, 70.

Gurr, T. R. (1993). Why Minorities Rebel. A Global Analysis of Communal Mobilization and Conflict since 1945. *International Political Science Review*, 14.

Hechter, M. (1975). *Internal Colonialism: The Celtic Fringe in British National Development 1536–1966*. Berkeley: University of California Press.

Hodson, R., Sekulic, D., & Massey, G. (1994). National Tolerance in the Former Yugoslavia. *American Journal of Sociology, 99*(6).

Kalin, R., & Berry, J. W. (1982). The Social Ecology of Ethnic Attitudes in Canada. *Canadian Journal of Behavioral Science*, 14.

Malcolm, N. (1998). *Kosovo. A Short History*. New York: University Press.

Marger, N. M. (1994). *Race and Ethnic Relations. American and Global Perspectives*. Wadsworth Publishing Company.

Massey, G., Hodson, R., & Sekulic, D. (1999). Ethnic Tolerance and Enclaves. To be published in *Social Forces*.

McIntosh, M. E., NcIver, M. A., Abele, D. G., & Nolle, D. B. (1995). Minority Rights and Majority Rule. Ethnic Tolerance in Romania and Bulgaria. *Social Forces* 73.

Merton, R. (1949). Discrimination and the American Creed. In R. MacIver (Ed.), *Discrimination and the National Welfare*. Harper & Row.

Mitchell, I. S. (1968). Epilogue to a Referendum. *Australian Journal of Social Issues,* 3

Park, R. E. (1950). *Race and Culture*. New York: Free Press.

Pettigrew, T. F. (1998). *Intergroup Contact Theory*. Paper presented at the Flinders University Seminar

Quillian, L.(1995): Prejudice as a Response to Perceived Group Threat: Population Composition and Anti-Immigrant and Racial Prejudice in Europe. *American Sociological Review*, 60

Sekulic, D. (1992). Nationalism versus Democracy. Legacies of Marxism. *International Journal of Politics, Culture and Society*, 6.(1)

Sekulic, D., Massey, G & Hodson, R. (1994). Who Were the Yugoslavs? Failed Sources of a Common Identity in the Former Yugoslavia. *American Sociological Review* 59.

Silber, L., & Little, A. (1996). *Yugoslavia. Death of a Nation*. TV Books.

Statistical yearbook of Yugoslavia 1991. Beograd.

Studler, D. T. (1977). Social Context and Attitudes Toward Coloured Immigrants. *British Journal of Sociology*, 28.

Taylor, S. (1979). The Incidence of Coloured Populations and Support for the National Front. *British Journal of Political Science*, 9.

Van den Berghe (1981) *The Ethnic Phenomenon*. New York: Elsevier.

White, P. G., & Curtis, J. E. (1990). Language Regions and Feelings Toward Outgroups. Analysis for 1968 and 1984. *Canadian Journal of Sociology*, 15.

Williams, R. M. (1964). *Strangers Next Door. Ethnic Relations in American Communities*. Englewood Cliffs. N.J.: Prentice Hall.

Woodward, J. (1995). *Balkan Tragedy: Chaos and Dissolution After the Cold War*. Brookings.

Zupanov, J., Sekulic, D., & Sporer, Z. (1996). A Breakdown of the Civil Order. The Balkan Bloodbath. *International Journal of Politics, Culture and Society*, 9(3).

5. CIVIL SOCIETY, DEMOCRACY, AND THE YUGOSLAV WARS

Sonja Licht

The Federal Republic of Yugoslavia broke apart violently mainly because the preconditions of transition to democracy had not taken root there, and because the political elites of every single federal unit were more eager to destroy the federal state than to give up power by facing the first really free and fair democratic elections. Federations work only if they are democratic, and democracy works only in countries where there is a genuine civil society. To explain the crisis, first, and then to suggest ways of moving past it, I shall offer some historical observations, primarily from the perspective of an activist in the pro-democracy movement from the mid-1960s, and then discuss the retarded development of Yugoslav civil society, and its effectiveness or lack thereof.

SOURCES OF THE YUGOSLAV CRISIS

By omitting many links in the causal chain, we can, without much distortion, also attribute the breakup of Yugoslavia to unfinished business left over from World War II. The former Yugoslavia is probably the only place on earth where World War II is still going on; it is being fought by people who continue to call each other Communists, Chetniks, and Ustashas.

Of course, Yugoslavia's World War II was unique in that more of the fighting consisted of a civil war than of a war against the Axis powers. Many more people were killed in the ethnic war than were killed by the Germans and their allies. It is a resurgence of that civil war that has marked the conflict in the Balkans during the 1990s. But why was World War II not finished in that

Research on Russia and Eastern Europe, Volume 3, pages 111–124.
ISBN: 0-7623-0280-1

country? I shall argue that it was a lack of democratic structure that best explains this situation.

The Communists – and especially the Partisan movement – played a great part in ending the Civil War by crushing their enemies. Tito saw that after such a terrible internecine struggle the only way to establish orderly rule over the country was to overcome nationalism as much as possible. He had two intelligent ideas about how to do this. The first was to insist, as a firm practice, upon a commitment to 'Brotherhood and Unity' throughout the country. The second was to stress internationalism as a doctrine, though initially he approached it in terms of the prevailing Soviet ideological framework. Whatever the Communists did, they did for the sake of maintaining their own control – not at all to establish the control of the majority of their countrymen as a democratic right.

Perhaps the 'Brotherhood and Unity' idea would have worked if it had followed a long, deep process of reconciliation. This would have been possible only after a debate about what had actually happened in Bosnia, Croatia, Serbia, and Montenegro in 1945, which could and should contribute to a genuine search for peace and reconciliation. However, Tito's obsession with control meant that the peoples of Yugoslavia were not allowed to discuss these topics in a normal way. Shutting down the topic created a fermentation of all kinds of fears and frustrations among the people who still remembered. They were not given a chance to face the facts. Nothing was done to bring them together or help them overcome the wounds they had suffered. So, though the Brotherhood and Unity project was a good idea, it was done less than half way.

The Partisans were fully committed to communism, but this did not keep them out of a conflict with the Kremlin. In fact, Tito was not opposed to Stalin and his ideology, but he had the backing of a genuine revolution and naturally he wanted more control and Independence than Stalin was prepared to permit. Their fight was simply a conflict over power. In fact, the first Congress of the Yugoslav Communist Party held after the split with Stalin was oriented toward proving to Big Brother that Yugoslavs were better communists than the Soviets themselves. They started collectivizing the land after 1948 (i.e. the break with the Soviet Union and the Communist International) just to prove this point.

After the break with Soviet Union, the leadership recognized their need for some ideological doctrine that would appeal to their own people. Here they encountered a dilemma. They could emphasize the idea that we are a people on our own, that we want our own nation. But this was only four years after the conclusion of an extremely bloody civil war and any reference to nationalistic ideas would be dangerous. Instead, they adopted a Proudhonian principle – the notion of economic self-management, of self-government, as the distinguishing characteristic of the newly emerging Yugoslav political system. One of the best historians of that period, Andrija Kresic, argues that the decision in favor of self-management was taken precisely to avoid the danger of the most logical alternative: any form of nationalism. Self-management did have a basis in reality at the level of the individual enterprise, or the local 'commune', but there was no genuine self-governance for larger, non-local units. Self-management of any kind contradicted the regime's main objective – to keep control – but it was less dangerous to them than nationalism and certainly less dangerous than real democracy because they were not empowering individuals, but rather small units, which are easier to control.

Even so, self-management was kept in low-profile to minimize the danger, and whenever there was any sign of an intention to broaden it, the regime would immediately act to nip democracy in the bud. For example, there were occasions in the sixties when two candidates were nominated for the same post. Invariably, both candidates would be ousted from the party, which meant an end to their political life. The regime purposely maintained a deep gap between state and society – 'state' meaning themselves, and 'society' meaning all the others who had to be controlled. In such a system, civil society cannot develop because society is treated as an immature entity. Moreover, there are also contradictions between this, state controlled and regulated system and economic reforms. Genuine, effective economic reforms require a degree of political pluralism and the opening up of a political space, but these normal developments were ruled out during Tito's long years in power.

NATIONALISM AND SELF-DETERMINATION

By the 1990s it became clear that political elites throughout the whole federation had abandoned Tito's approach and had actually started playing on nationalism as a way of saving their own power. However, this change did not actually begin in the 1990s, but as long ago as 1968, when the student movement made it apparent that the older, original communist political approach, based on the absolute control of the party state, was losing ground.

In 1971 there was an important political movement with democratic overtones widely depicted as the 'Croatian spring'. This was actually an invented legend, for the Croatian spring did not arise spontaneously from the grass roots. Instead, it was the result of processes that took place within the Communist Party. Tito himself started playing with the fact that by empowering one republic against another, he would be able to maintain his rule as before.

In 1967 the Slovenian leadership had decided to divide some state funds so as to use a portion of the money for road construction. Toward this end, they argued in favor of a statist decentralization of power. Tito could see that by empowering the national leaderships in their own fiefdoms one by one, he could allow them to play the role of champion among their own people and thereby gain more support. And at the same time, the control from above became easier. This is how the whole process developed; decentralization was used to create states within states.

Two things were going on at the time: the quest for democracy and the quest for local control. Tito's regime concluded (correctly) that the quest for democracy was more dangerous to their own power, so they chose to go the other way – toward decentralization. It is a mistake to recall warmly the good old days of Josef Broz Tito, because he was one of the main sources of the crisis that has been going on during the 1990s. It was he who began to preempt the claims of authentic nationalists – the 'bourgeois' nationalists. In fact, the Serbians who had been nationalists earlier were horrified by this resurgent nationalism, which was exactly the kind of nationalism created by communists throughout Yugoslavia, with the exception of Alija Izetbegovic, the new leader of Bosnia and Herzegovina personally (though most of the people around him were loyal Party members).

During the mid-eighties the first steps toward political pluralism were beginning to shape up, and the established politicians always argued this: Political pluralism is very dangerous because it will divide the country into ethnic parties. They were right. That is exactly what happened. But the predecessors of the ethnic parties were not varied political parties but the communist party itself – or rather, the communist parties, because at one point we had eight of them. Every single Communist Party throughout the federation, i.e. the republics and the provinces, was a party of its own. We had a pluralism of Communist Parties under one hat, which became an ethnic Communist Party. And every one of these parties created a politics of its own, according to the nationalism prevailing in its republic. This is how the country started falling apart. The Slovenian party left first, but I know from a number of insiders who have not written this down yet, that if they had not left, the Serbian Communist Party was ready to do so. A decision had clearly been reached that the Party Congress in 1990 should fall apart, though it was obvious that whenever the Party fell apart, the country would also fall apart.

From 1945 onward, Yugoslavia was held together by the Communist Party, which did not allow Yugoslavia to develop as a democratic federation. The leaders were preparing for the country to fall apart – not intentionally, perhaps, but they were already dividing the infrastructure up. For example, every republic had to have its own steel mill and sugar factory, whether or not it owned the raw materials for such plants. Most of these factories never worked. We needed one single plant for the whole country, but the leaders were preparing for the breakup.

It was the Serbian politician Slobodan Milosevic who rode the wave of nationalism in the most outstanding way in the late 1980s though he had never been a nationalist in his earlier years. He was obsessed only with power, and he understood at one point that nationalism was the only card left for him to play. It proved, unfortunately, to be the right card and others were ready to play it too. Milosevic's special situation was that he was politically active in Serbia, which had a built-in source of division – the two autonomous provinces Vojvodina and Kosovo. He simply took away the autonomy of those two provinces, and he had a strong argument for doing so: Why should there be autonomous provinces only in Serbia? Why not within Croatia? In fact, this was logical; there were many Serbs in Croatia and if they had had autonomy, we might have avoided a war. The largest diaspora consisted of Serbs living outside their home republic. Serbia itself was divided into three, of which only Serbia proper was under the control of Serbian politicians.

In the federal parliament and in the federal presidency, each republic had two votes. Serbia had three, but of these three, one was for Kosovo, another was for Vojvodina, and one was Serbia proper – so in fact they had one against two. The Vojvodina vote might support Hungarian concerns and the Kosovo vote would represent Albanians. After 1974, these autonomous provinces almost had the status of republics. This structure gave arguments to everyone. The Croats could say that Serbia has three votes, while Serb nationalists actually felt under-represented.

The Serbian side liked to present itself as wanting to keep Yugoslavia together. This is basically untrue. They complained about the Serbs being divided among at least three republics that were becoming more and more independent as states. According to their version, the Serbs were paying the price that started even before World War II when Vojvodina was created. In the beginning of the eighties, as Serb nationalism became visible, they wanted Serbs everywhere to get together in their own separate state; they tend to deny this now. This is because traditional Serbian nationalism of the period was not militant against Yugoslavia, so the communists and nationalists were picking up bits and pieces of the problem that they wouldn't touch. By about 1980 there was a segment of Serbian nationalism that was becoming resistant toward Yugoslavia. This arose, I think, from the decentralization that followed the defeat of the Serbian liberalizing party leadership in the beginning of the 1970s. In 1971 Tito decided to destroy the nationalist movement in Croatia. A few months later he decided to destroy the reformist/liberalist movement in Serbia too. There were reform communists in Serbia and nationalist communists in Croatia, and he got rid of all of them. He was very careful never to hit one without hitting the other. This was his system of 'checks and balances'. (Today when we talk about mirror images within Croatia and Serbia, it was created in those times by this policy of balancing.) But the Croatians, in particular, hadn't expected it and felt it as a shock. There was a real mass movement there. The removing of the Croatians in 1971 was a bigger shock than the removal of the liberals from power in Serbia.

In Croatia, Serbia, and Slovenia there was an intensive dissident intellectual activity going on. Unfortunately, a great many intellectuals played a major part in stirring up enemy images and chauvinistic hatred. There were tense meetings, for example of the associations of writers.

Still, the truly blameworthy people were not the intellectuals but the old leaders of the republics – especially Milosevic. They were ready to destroy the country in order to preserve their own power. On the other hand, it is true that in Croatia, a grass roots quest for independence was growing and that was an important factor. What Milosevic wanted was to destroy the federal government of Yugoslavia. There cannot be a real federation without democracy, but he and the Serbian federalists wanted to recreate the state in an undemocratic way. The separatists were convinced that no possible form of Yugoslavia could be acceptable. The Slovenes were in between these two positions, but the Croatian leaders managed to manipulate them. Whatever Croatia decided to do, it had to happen one day after Slovenia took the same step. They wanted to hide behind Slovenia because there is such a bad mark in the history books against an independent Croatia, which had been a puppet state of Nazi Germany. They did not want to be seen as the first republic leaving Yugoslavia.

The nationalists used each other's problems. For example, they expressed sympathetic concern in Croatia and Slovenia for the fate of the Albanians – until the point when they separated. It was instrumental; they were using Albanians. In 1989 I went to Ljubljana on a fact-finding mission for Helsinki Watch. I met a number of Albanians who couldn't take the repression of the Serbian regime anymore and had come to Slovenia expecting to find a friendly environment. Already they were kept there in a ghetto.

If we look at the hierarchy in 1999 we see that the top people are those who had been in power before, with Izetbegovic as the only exception – and he is surrounded by former Communists. All the presidents who were running around the country in 1990–91, debating how to tear the country apart, are in power in 1999.[1] They broke Yugoslavia down simply to be able to stay in power. On the other hand, they would not have been able to do so if there had not been structural conditions that tended to destroy the country's unity, and if they had not received serious support from abroad, including some major foreign politicians.

STRUCTURAL FACTORS IN NATIONALISM

The most serious of these structural factors was the economic one. The leaders of Croatia and Slovenia genuinely believed that their economic level was so high that they could easily do without Yugoslavia, which they believed was holding back their development.

Economic contradictions had existed since the very beginning of Yugoslavia – especially between north and south – and became even worse over time. At the end of World War II, Slovenia was four times as developed as Kosovo in its GNP and other indicators. By about 1990, it was 13 times more developed than Kosovo. Such a north-south gap would be a terrible problem in any country, but this was shocking. In Switzerland, when one goes from the Italian to the German part, one can see a developmental ratio of perhaps one to two – but not one to thirteen. Naturally, this exacerbated the tensions in Yugoslavia. In every one of these nationalisms – Croat, Serbian or whatever – people believed that the others were stealing something from them, such as taking their currency or some such thing. If Yugoslavia had been a market economy throughout the post-war period, inequality would not have been worse. But this was a communist government, so why did they not bridge the gap? People rarely discuss that question when they are explaining the destruction of the country.

Kosovo played an extremely significant role in stimulating Serbian nationalism The Albanians there had been slow in developing their own quest for nationhood, for statehood, probably mostly because of the lack of a real intelligentsia. After World War II, there were no more than a dozen educated people and as late as 1968, according to my Albanian friends, only a handful of people there were dissatisfied. As they became emancipated, there was a tendency to start thinking of themselves as a people. They are not Slavs and there was serious police repression against them.

Tito had wanted Albania to join Yugoslavia. In 1945 he believed that this might actually happen, so he and his colleagues made a number of moves to please Enver Hoxha (1908–85), Albania's party boss, who was a real hard line, violent communist dictator. Albania is situated in a good position on the Mediterranean, and Tito sought to persuade Hoxha to join in a confederation that might also include Bulgaria. Many historians believe that this explains what happened in 1948 in the conflict with the Soviet Union: Tito could not control his appetite for his neighboring countries. Stalin became concerned that he was building an independent communist power in the Balkans, which the Soviets might not be able to control well enough.

In the years following the mass demonstration in Kosovo in 1968, the desire of Albanians for independence and unity increased. Many Kosovar Albanians began dreaming of Albanians coming together. As the repression of the Serbian and Yugoslav police became more overwhelming, especially after Milosevic took over Serbia it became virtually the only political agenda. The real question was, how to go about accomplishing this agenda?

In the belief that violence would be too costly, most Kosovars accepted Ibrahim Rugova's policies, which were based entirely on peaceful methods.

Only after the Bosnian war ended (with no recognition extended to Rugova at all) did public opinion became impatient and start to shift toward favoring a guerrilla movement.

Finally, another institution that was important in the Yugoslav social structure was religion. The different churches, especially the Orthodox and Catholic churches, were long marginalized by the communist government, but in the 1990s they were deeply implicated in stirring up nationalism. Many religious leaders were involved in creating enemy images and the perception that each nation was being threatened.

INTERNATIONAL FACTORS

Foreign states and foreign social movements have had an impact on Yugoslavia's fate since the end of World War II. When Stalin talked with Churchill at Yalta, they drew a map on a napkin and Stalin signed it, saying that he accepted the proposal that Yugoslavia should be half within the Western sphere of influence and half within the Eastern sphere. Stalin kept his pledge, I think because he knew that to invade Yugoslavia would provoke a war between the Soviet Union and the West. He would not hesitate to invade Czechoslovakia and Hungary, but he did not challenge the fifty-fifty arrangement concerning Yugoslavia. It was because of this that Yugoslavia was able to retain a position of non-alignment throughout the Cold War.

When Tito was dying, journalists and others from the West kept asking whether Yugoslavia would stay together or fall apart. We were naive enough to consider that a stupid question; the country had a lot of reasons to stay together and we believed that it would do so. But the West could see the problem from a different perspective and understood the real prospect of disaster. The international community (the European Union and the United States) made it clear, once the possibility of breakup began to loom, that some groups would have a chance for self-determination – the Croats, the Slovenes, and the Bosniacs – whereas others, the Serbs and the Albanians, would not be permitted self-determination. The combination of these positions continues to keep this region unstable.

What about the people who were in the front line of political life in the international community? Many people from the left, even members of the peace and democracy movements – people who had been extremely effective in supporting dissidents – became confused about these problems. In many cases, they regard self-determination as a basic human right and as the answer for everything. But I used to ask them: Are you for self-determination of federal units or self-determination of peoples? Usually they could not understand the question, though it is the main conundrum behind our crisis. Without under-standing what was entailed, they generally favored self-determination of Slovenia, Croatia, Bosnia, and Kosovo. I asked them: What about the Serbs in Croatia and Bosnia? They did not favor self-determination for the Serbs because they saw self-determination as a right belonging to federal units, not peoples, and could not accept any change of borders.

The peace movement had been a single-issue movement. It was easy to oppose nuclear weapons because we had the support of almost the whole population. But when this Yugoslav business emerged, there were all sorts of interests involved, and peace activists could not analyze the situation. In the United States, for example, the majority of our leftist friends were on the same side on this issue as Jesse Helms and Robert Dole. They could not explain it. Moreover, during the Gulf War they also lost their orientation and some of them favored military intervention instead of a long term support to the demo-cratic opposition, which was far more able to get rid of Saddam's regime. Instead of lobbying in support of disarming the whole Balkan region, they started lobbying for arming and for lifting of arms embargoes. I had been deeply involved in the international peace movement, but while Yugoslavia was falling apart, the peace movement was dying. It does not have enough strength anymore to come out clearly on any issue.

BUILDING AN EFFECTIVE CIVIL SOCIETY

In my opinion, the key to creating a stable democracy is to enable a civil society to flourish. Much of my work has been devoted to fostering this objective, which I want to review historically throughout the remainder of this chapter. I have already mentioned the repression that took place in the early 1970s against nationalists in Croatia and reform communists in Serbia. These ill-fated move-ments were preceded by the emergence of certain opposition groups, as early as the beginning of the 1960s, but we can date civil society's beginning at about 1968. A certain ferment had already been taking place within the party, and even within the party in other communist countries – renegade commu-nists. The best known of these renegades was Milovan Djilas, who criticized the party in the beginning of 1950s from within for becoming a 'new class'. He was sent to jail as punishment. His dissidence was matched elsewhere in the country, notably by Mihailo Mihailov, who had started a small group in about 1963 and, a little after Djilas, also went to jail, though as a non-communist.

Another Marxist group, the Praxis group, started in 1964, again as renegade reformers who still believed in Marxism, but in a non-dogmatic way. They consisted of academics, mainly philosophers and sociologists, who criticized the dictatorial, non-democratic nature of the party-state and adopted a libertarian attitude. In Belgrade and Zagreb they had a serious impact on students and broader intellectual circles as well. Indeed, this was the intellectual background of the student movement of 1968, which criticized the party leadership and the economic reforms implemented without democratic changes. I was involved in the student movement of 1968 when we occupied all the university buildings in Belgrade. The slogan was: Down with Bureaucracy from the Party Top. Actually, in 1966 there had already been the first large demonstration in Belgrade against the war in Vietnam. The police were brutal, which stirred up student resistance.

In 1968, movements of this sort were happening all around Europe and the United States. The first night when we occupied the Belgrade University in June 1968 we intended to send a delegation to Zagreb, where there were a few protests, but not very strong ones. In Zagreb the politicians tried to portray our protest as nationalist, but soon our colleagues found out the truth. Tito was smarter than the leaders of other countries because he gave his support to the students. He succeeded in breaking up the student movement in that way. He said in a televised speech that the sudents were right in their demands and criticism and that only five to 10% were infected by 'hostile ideology'. This meant us at the Belgrade Faculty of Philosophy who, in fact, were at the core of the movement together with our professors – mainly members of the Praxis group.

Soon some of the student leaders were arrested, including my husband. There were several political trials in the first half of 1970s and most of them were sent to jail. At that time people started questioning the very nature of the regime and insisting on changing it toward pluralism and toward a genuine respect for human rights – the principles in the Human Rights basket of the Helsinki Accords. This was really going beyond the party framework and beyond the Praxis movement. We were now debating the party as such, questioning the core functioning of the state and trying to bridge the gap between state and society.

The petition movement started in Belgrade in 1976-77. People were signing up on political issues, such as the death penalty or asking about the fact that the police had the discretion to deny citizens their passports and their right to travel abroad. This was 1977, and the participants were the same people as in the 1967 and 1968 movement, including some professors from the Praxis group who were involved in the Free University. Each petition had 50 to 300 signatures. The petition movement opened up a new stage, which was taken over in debates by the Slovenians on such issues as reproductive rights, gay and lesbian rights, and nuclear energy.

One of the first serious anti-nuclear energy movements was initiated in Belgrade in 1986 and '87. There were several hundred thousand signatures, largely from high schools and universities, and they overturned the decision to build a nuclear power plant.

Then the first real feminist groups and draft resisters and the first serious criticism of the army started in Slovenia. We were puzzled because they started to talk about a 'national civil society'. At that time we thought it was a marginal matter. Who cares? We were impressed by the power of the groups who criticized the federal army. That took real courage. The army was living a life of its own and eating up money. We in Belgrade didn't have their courage. (Unfortunately, since Slovenia got its independence it has been building up its own army and now spends four times as much on the army as before, but the people who were active during that period have become silent about Slovenia's new army.)

The first peace groups started in Slovenia, along with the first peace institute, the first debate, and books published on peace movement. But the activists were not interested in disseminating their projects across the borders of republics. They held meetings and discussions, but the concept of a national civil society was limiting them to local activism.

In 1984–1985 foreign peace activists started coming to Belgrade and we formed a small group for peace and democracy. A number of people whom we approached and asked to join it refused; they did not want the group to be called 'Belgrade Group for Peace and Democracy'. Democracy was 'too political'. The fear of politics played a devastating role in the upcoming events.

We could not yet understand what was going on, especially in Slovenia. I was cooperating closely with the peace groups in Ljubljana, which had a petition movement as well. It was fascinating how the issues of peace and democracy were not on people's minds in other parts of the country, including Serbia. They were still impressed by the Yugoslav policy of non-alignment, and believed that the issue of war and peace was simply unimportant for Yugoslavia as a whole.

Because we could not find a real response in Belgrade, we discussed with some friends from Croatia the necessity of creating something more political. The nationalistic debate was stirring up and we had to face the fact that there were two lines of activity – one was civic and the other was nationalism. By 1988 a number of people formed an organization for a Yugoslav democratic initiative. We set up branches in Kosovo and all over the country. This was during the time of transition in Poland, Hungary, and Czechoslovakia. In Croatia the authorities would not register the group because they said it was political. In Belgrade as late as at the end of 1988 an influential communist political figure promised our parliament that "pluralism will never find a way here in this country." (He still holds one of the most important places in Milosevic's Socialist Party.)

Our new organization was not a political party but something short of that. The first political parties were created in 1989, most of them limited within the framework of each federal unit. There were just a few, the first one the Social Democratic Union, which tried to be really Yugoslav by nature. The communist party was already divided on national lines. However, this trend was not yet fully taking over the civil society.

In 1991 an independent association of women from all over Yugoslavia was created in Zagreb. It never took off. The Croat members were insisting that it should be in Zagreb, but after the HDZ won the elections there, it became impossible for it to function in Zagreb and they moved it to Sarajevo. It was obvious that they were moving the headquarters from one place to another in an attempt to prevent the looming conflict.

The first human rights groups emerged from the petition movement and the political trials in Belgrade and Zagreb. The most vivid human rights group was the Yugoslav Helsinki Committee for Human Rights, which had three chairpersons. The Writers Union had a human rights committee, which defended Adem Demaci, the Kosovar nationalist who spent time in prison.

In April 1984 the police came to the Free University, where Djilas was giving a lecture. They arrested 28 and released them after two or three days, except for one worker who was killed by the police. This arrest was followed by a trial known as the trial of 'Belgrade Six'. The trial was very different from the ones in 1970–73 because they had more media coverage than could have been dreamed about earlier. This one created a huge protest outside the country. More than 2,000 articles were published about the trial. There was a coalition behind it built by the federal police and the Belgrade Party organization, which was headed by Slobodan Milosevic.

Civil society did grow substantially for a time, but the concern we all faced was this question: Was there any hope for civil society during a civil war? The law professor and human rights activist Zdravko Grebo was one of the people who never gave up hope, even while war was raging throughout Bosnia and Herzegovina. When the first barricades were built in Sarajevo, at the beginning of March 1992, he used a small independent radio station to call the citizens of Sarajevo to rise at the barricades – and they did. For a moment many of us started believing in miracles. But it proved to be wishful thinking. There was much more power concentrated on the opposite side – on the side of conflicting nationalistic political elites who were willing to destroy the country in order to prevent genuine change in the political system.

The representatives of civil society organized inter-ethnic dialogues, advocated nonviolence, and raised their voices against intolerance and the politics of hatred. When the war broke out they organized all kinds of anti-war organizations and actions, supported draft resisters, and faced more difficult times than the dissidents in the communist era, since they often had to oppose not merely the regime but almost their whole nation as well.

What did they accomplish? On the most important issue they achieved almost nothing; they could not prevent the war and the victimization of civilians. They put much effort into mobilizing international public opinion, but the international media and policy-makers ignored them and their proposals, including the suggestion to establish a U.N. protectorate over Bosnia and Herzegovina and to oppose the recognition of nation states based on purely ethnic principles. They even failed to convince parts of the international peace movement that the principles of democracy should not be subordinated to the principle of self-determination. The latter principle was often used as the ideological justification for ethnic cleansing and the creation of ethnic states.

Civil society did, however, manage to preserve 'islands of civility and resistance' in the face of war and, in the case of Serbia, in the face of severe international sanctions, which added an external isolation to the isolation already imposed by our own government.

Radio 'Zid' in Sarajevo, Radio B92 in Belgrade, 'Feral Tribune' in Split, the Croatian Helsinki Committee, the Civic Forum in Tuzla, the Women in Black and the Center for Cultural Decontamination of Belgrade, to name just a few of those 'islands' that resisted the chauvinist-nationalist-fundamentalist madness, were repeatedly accused of being traitors to their 'national' community, even on those rare occasions when they were recognized by it. At the same time they were for years ignored by the international community. This was the same international community which claimed to be supporting the values that media and civic groups were fighting for. There was not a single serious conference organized by high level political and media people to debate how these organizations should be supported or how to demilitarize the region. But there were hundreds of debates devoted to the questions of whom should they arm and how to undertake an efficient military intervention.

During the war and the years of sanctions, numerous women's groups were formed in patriarchal regions, opposing both war and patriarchy. These were the most vigorous part of both the anti-war movement and civil society in general.

With the street protests of 1997 in Serbia, a genuine transition toward democracy was obviously starting and winning the support of a majority of citizens. Without the survival of those "islands of civility and resistance," this majority would have had no group to turn to. It is important for the international community to recognize the importance of participatory democracy and economic development. If future ethnic conflicts are to be prevented, if the values of peace and democracy are to be taken seriously, then the international community must finally recognize that democracy is not only an internal issue, but also a major transnational one. And it must, finally, develop a genuine political will to act in preventive ways (including major support to development of civil societies), since it has been proven too often that comprehensive sanctions (i.e. isolation of particular countries, and military interventions) neither are productive in ousting dictatorial regimes nor contribute to international peace and security.

NOTES

1. Franjo Tudjman died in December 1999, after the writing of this chapter.

Sonja Licht is the President of the Board of the Fund for an Open Society in Belgrade, a part of the Soros foundations network, and a long time member of democratic opposition in Serbia. She was one of the founders and the co-chair of the Helsinki Citizens Assembly. She is also chairing the Task Force on the Future of FRY, organized by the East West Institute, New York/Prague, in July 1999. The Task Force is the driving force behind a complex set of activities that became known as the Bratislava Process.

6. THE LEGACY OF HUMPTY DUMPTY: THE ROLE OF RECOGNITION IN THE DISSOLUTION OF YUGOSLAVIA

Corey Levine

INTRODUCTION

Humpty Dumpty sat on a wall (between East and West).
(Once the wall fell and Humpty Dumpty came crashing down)
All the King's Horses and All the King's Men,
Couldn't put Humpty Dumpty together again.

The history of the territory known for the better part of last century as Yugoslavia, is a history that has been given over to 'balkanization' – the process of fragmentation. This history also seems a mirror in which the major events of the twentieth century have been reflected. Thus, Yugoslavia (born 1918, died 1991) played a significant role in the political evolution of the twentieth century, particularly in the post-Cold War transition. In its violent and protracted death lay the crumpled dreams of a 'new international order' based on integration and co-operation, for Europe in particular, but also for the world.

The Yugoslav War of Dissolution (1991–1995) was ultimately about statehood, sovereignty, self-determination and the meaning of the concept of "nation."[1] It was the fallout from these issues, which was at the heart of the Yugoslav crisis, that provided the greatest repercussions for the 'new international order' and defined international involvement in its dissolution.

Research on Russia and Eastern Europe, Volume 3, pages 125–143.
2000 by Elsevier Science Inc.
ISBN: 0-7623-0280-1

> In its bitter dissolution, Yugoslavia . . . became the subject of international experimentation.
> It was the litmus test of a 'new world order' in which the old world would be made to
> work better through international co-operation and in which new elements would be forged
> in the Yugoslav crucible.[2]

The litmus test for this 'new world order' began with the question of diplomatic
recognition for the various republics that constituted the state of Yugoslavia.

While often taken to be a fundamentally legal concept in international relations, it is arguable that recognition is ultimately a political concept since often
the decision whether or not to recognise a state is dependent more on political
considerations than exclusively on legal factors.[3] As one legal scholar has pointed
out, "recognition is basically a political judgement, although it has been clothed
in legal terminology."[4]

Thus, what role did the 'political judgement' of recognition have to play on
the centrifugal forces that ultimately fractured Yugoslavia beyond repair?

While the impact of recognition on the dissolution of Yugoslavia remains
contentious in the various analytical discourses of the conflict, most analysts
generally agree that diplomatic recognition had some role to play in the country's
disintegration. This is due to the problematic fact that the international community recognised the former republics of Yugoslavia as sovereign states before
such contradictory issues as the inviolability of borders, the right to self-determination and minority rights had been effectively resolved.[5] Although whether,
as one analyst puts it, "diplomatic recognition probably had the effect of
prolonging the crisis at a lower level of engagement and discouraging a more
rapid conclusion at a higher level,"[6] remains a highly contested issue.

Yet, why and how did the international community come to the conclusion that
granting recognition to the republics was the best course of action to follow with
Yugoslavia? What impact did this decision have on the events that were to follow? And most important, what lessons can be culled from the disaster that was
the dissolution of Yugoslavia and the international community's role in it?

THE CONCEPT OF RECOGNITION

In current legal circles there are essentially two competing, though overlapping,
theories as to how states achieve recognition.

> The constitutive theory [of recognition] maintains that it is [only through] the actual act of
> recognition by other states that creates a new state and endows it with an international
> personality and not the process by which it actually obtained independence. The declara-
> tory theory maintains that recognition is merely an acceptance by states of an already existing
> situation. A new state will acquire [international] capacity not by the consent of others [but]
> by virtue of a particular factual situation.'[7]

However, there are serious difficulties with both forms of recognition. With the constitutive theory, the status of a state recognised by one actor but not by another, and therefore "apparently both an 'international person' and not an 'international person' at the same time," is not only a 'curiosity' in legal terms.[8] It can also set up potentially volatile dynamics in the sphere of international relations, such as China pre-1971 or Taiwan today.

Although the declaratory theory is generally regarded by legal analysts as the more authoritative approach to achieving recognition,[9] the political implications of an entity declaring its independence and as such, having its 'declaration' of sovereignty considered legally binding and politically valid is extremely potent. The 'Turkish Republic of Northern Cyprus' is one example of the potential difficulties the declaratory road to recognition can lead to.

Recognition can also be granted explicitly (*de jure*) or occur by implication (*de facto*).[10] With *de facto* recognition, although an entity has not achieved formal diplomatic recognition, it nonetheless can have important consequences. Given that there is no legal framework in which to interpret the signals implied by *de facto* recognition, vastly different conclusions can be drawn by the various actors directly involved in the secession or creation of a state. However, it is the political intention of the actors bestowing *de facto* recognition that lies behind its importance.[11]

As we shall see, *de facto* recognition, as well as both the constitutive and declaratory forms of recognition all played a significant role in the disintegration of Yugoslavia and the various permutations created by its meltdown.

SELF-DETERMINATION, THE 'INVIOLABILITY OF BORDERS' AND THE BREAK-UP OF YUGOSLAVIA

When the crisis in Yugoslavia first broke out, it was the international community's 'intention' to preserve the unity of the country.[19] "The alternative, it was felt, could have potentially nightmarish consequences for the 'new international order': an overflow of refugees across international borders; ethnic conflict spill-over to neighbouring countries; or the potential for setting a dangerous example for other multiethnic countries in the newly liberated Eastern Bloc, particularly the Soviet Union."[12] The international community reasoned that there was plenty to fear from the break-up of Yugoslavia, in any of these scenarios. Thus, in the early days of the crisis, the international community actively asserted their support for the "territorial integrity and preservation of the unity of Yugoslavia."[13]

In committing itself to the principle of a unitary Yugoslavia, the international community also committed itself to one of the primary principles of the sovereign state, the 'inviolability of borders'.[14] The legal framework guiding this principle stems from the Helsinki Final Act of 1975: no changes to international borders except by the consent of all parties concerned[15] (known in legal terminology as *uti possedetis juris*).

At the beginning of the conflict, the international community chose to apply the principle of *uti possedetis* to Yugoslavia over unilateral efforts to secede from an established state (the principle of the right to self-determination).[5] Although the international community did later tackle the crisis through the prism of the right to self-determination, they retained their commitment to the 'inviolability of borders'; however, applying it to the pre-existing internal borders of Yugoslavia.[5]

This created an interesting situation because the principle of *uti possedetis* had never been previously applied, either from the vantage point of borders created for internal political and administrative purposes, or outside the colonial context.[16] The application of these two contradictory principles, (both of which are enshrined in the United Nations Charter), was to have major repercussions both on the nature and evolution of the conflict. It also created one of the central tensions in the international community's involvement in the Yugoslav crisis.

Many analysts feel that the application of *uti possedetis* by the international community gave Serbia, (acting under the colour of the federal government which it dominated), at least at the beginning of the conflict, an implicit green light to crush the unilateral declarations of independence of the seceding republics, on the grounds that they represented an attempt to alter the country's borders without full consent of all parties;[17] thus setting in motion the train of events that led to the protracted brutality of the war.

However, once the international community gave up on the idea of a unified Yugoslavia, (or rather not so much gave up on the idea, as abdicated the role they had undertaken in maintaining the status quo of a unitary state), the progression of events leading to formal recognition (and beyond) became a march of 'discordant policy' and ineffective action on the part of the international community with much of this 'march' being guided by the European Community (later the European Union).[19]

It was the prevalent consensus that the Europeans should take the dominant role in the Yugoslav crisis. This is attributable to a number of factors. First, the UN viewed the problem, especially in the early stages of the conflict, through the 'self-denying optic' of non-interference in internal affairs.[18]

As well, European leaders, particularly the German and French, spurred by their inability to make a united response to the Gulf War, were especially eager in the run-up to the Maastricht summit on European economic and political union, to demonstrate that they had come of age with regard to foreign policy.[19] They wanted to test the emerging machinery for foreign policy co-operation in the EC, the envisaged 'Common Foreign and Security Policy' (CFSP), espe- cially on an issue so close to home. Finally, the limited interest of the superpowers, particularly the United States, tended to push the Europeans to the foreground.[20] Thus by virtue of proxy and proximity, the Europeans became the stand-in for the international community in the Yugoslav conflict.

THE EUROPEAN COMMUNITY, THE BADINTER COMMISSION AND THE BREAK-UP OF YUGOSLAVIA

The fact that the Europeans initially took the lead role in the Yugoslav crisis shaped future international involvement, in part because of the framework for recognition they created for the fractious republics. However, this framework for recognition was "neither clearly understood, nor clearly implemented [or] supported."[21] This was ultimately to have repercussions in the way the conflict played out and whose reverberations can be found in the Dayton Peace Agreement.

The ongoing constitutional crisis in Yugoslavia and the unwillingness of the various factions to respond to the limited and confused incentives offered by the EC negotiators forced the international community to reassess their polit- ical objectives.[22] Although the EC had been reluctant to accept the right to self-determination for the Yugoslav republics, it was faced with a federal govern- ment that was prepared to use force to maintain the internal status quo. Thus the EC had three choices: allow Serbia to assert control in the name of non- interference in internal matters; intervene to protect the rights of non-Serbs in Yugoslavia, or recognise the right to self-determination.[23]

The EC negotiators charged with dealing with the Yugoslav crisis proposed a constitutional plan that called for a loose association among the Yugoslav republics. All of the republics except Serbia and Montenegro accepted the proposal. On November 8, 1991, the EC voted to impose economic sanctions against Yugoslavia, but to exempt any republic that agreed to the EC proposal. On December 2, 1991, the EC restored aid to Bosnia, Croatia, Slovenia and Macedonia, leaving only Serbia and Montenegro subject to the sanctions.[24]

By separating the parts from the whole in applying sanctions, the EC, in effect, granted *de facto* recognition to the republics as independent states,[25] although they continued to officially only recognise the federal state of Yugoslavia. This effectively gave Slovenia and Croatia (the only 2 republics which had thus far formally 'declared' their sovereignty) the green light to continue to assert their independence. At the same time, it provided cover to

Serbia to continue the war against secession since officially the international community continued to apply the principle of *uti possedetis* over the principle of the right to self-determination in the conflict.[22]

As negotiations dragged on, a few EC members, particularly Germany, began to push for formal recognition of the republics basing their arguments on the principle of the right to self-determination although they also claimed that it would halt the fighting.[26] However, the pressure for 'speedy recognition' was initially resisted and not only within the EC itself.[27]

The chief EC negotiator, Lord Carrington, the United States and the United Nations all opposed recognition fearing it "would lead to an increase in violence and might frustrate the peace talks."[28] The UN Secretary General at the time, Javier Perez de Cuellar, expressed his concern that "any early selective recognition could widen the present conflict and fuel an explosive situation."[29]

These fears were also expressed inside Yugoslavia as well. Bosnian President, Alija Izetbegovic, "requested both the German Foreign Minister and [Lord Carrington], not to precipitate him into a situation where the Bosnian authorities, unable to accept a Serbianised rump-Yugoslavia, [if Slovenia and Croatia were formally to secede through recognition], would be forced to request recognition."[30]

However, Germany pressed ahead and threatened to recognise Slovenia and Croatia unilaterally if the EC as a whole did not agree. Due to the political pressures of the Maastricht Treaty negotiations which were taking place at that time, and anxious to hold together their foreign policy mechanisms, the EC agreed to establish a process for deciding on whether (and which) republics would receive recognition.[31] This procedure, known as the Badinter Commission, after the French judge chairing it, played a central role in the political and legal response of the EC to the dissolution of Yugoslavia.

Germany had a number of reasons for pushing ahead with recognition and not all of them related to the "cultural and historical ties" it had with both Slovenia and Croatia. Domestic politics also played an important role.[32] Close on the heels of their emotional reunification after forty years the German government was vulnerable to the charge that just having restored the right to self-determination to East Germans, they "had no grounds on which to deny this to the peoples of Yugoslavia."[32]

The chairman of the Christian Democratic Party, as early as July 1991, was quoted as saying:

> We won our unity through the right to self-determination. If we Germans follow a status quo policy and do not recognise the right to self-determination in Slovenia and Croatia, then we have no moral or political credibility. We should start a movement in the EC to lead to such recognition.[33]

It was a powerful selling tool, not only for the German public, which was putting strong pressure on the German government to recognise Slovenia and Croatia; but also as a tool Germany used in which to convince its EC counterparts to come aboard the recognition bandwagon.

In order to hold up their faltering 'common' foreign policy and preserve the delicate Maastricht negotiations, both of which needed Germany's compliance, the EC issued a set of guidelines on recognition for the republics of Yugoslavia.[34] The conditions these guidelines laid out, included "respect for the inviolability of all frontiers which can only be changed by peaceful means and by common consent (*uti possedetis*); guarantees for the rights of ethnic and national groups and minorities; commitment to settle by agreement . . . all questions concerning state succession."[35]

Along with the guidelines, the EC confirmed that it "would not recognise entities 'that are the result of aggression'."[36] They then invited the republics "to state by December 23, 1991 whether: (1) they desired to be recognised as independent states; (2) they agreed to the commitments in the guidelines above; (3) they accepted the provisions of the [EC] draft convention, especially those on human rights and national or ethnic groups; and (4) they approved the involvement of the [UN] and the continuation of the EC [in] Yugoslavia."[37]

This extensive catalogue of criteria, far in excess of traditional standards for recognition of statehood, confirms that the EC recognition requirements "were not meant to be a criteria of statehood as such, but rather introduced political criteria concerning the possibility of establishing diplomatic relations."[38]

The applications of those republics "replying positively to the EC statement" were to be submitted to the Badinter Commission for a decision regarding eligibility for recognition "before January 15, 1992, the date of the proposed EC decision on recognition."[39] However, on December 23, 1991, Germany unilaterally moved to formally recognise Croatia and Slovenia.

Thus, in the space of six months (from the time Slovenia and Croatia declared independence on June 25, 1991), the EC had moved from "a unified position on maintaining the political and territorial integrity of the Yugoslav state through a growing dispute on how to handle the crisis to a discordant policy on recognition of seceding republics."[40]

The EC's inability to effectively back their policy options with either a stick or a carrot led to a miscalculated proposition, particularly on the part of Germany, that the only way to handle the crisis then, was through diplomatic recognition. Although some analysts claim that "the German analysis was probably right (in that) recognition was inevitable,"[41] other analysts, drawing upon the same information, came to opposite conclusions.

THE DIFFERENT OPTIONS

Was recognition inevitable? Whether alternative courses of action might have headed off the conflict is still debatable. However, there were at least two other options available. The international community could have accepted the dissolution of Yugoslavia from the beginning and worked to ensure that the split was more peaceful.[42]

Alternatively, "the international community could have retained it's policy preference for a single state, but made it clear that use of force to maintain the federal state would be met with a swift international response – politically, economically, or even militarily."[43]

Instead, the EC, as the stand-in for the international community, had created a paper trail of commitments to certain principles and guidelines that sent confused and competing signals to the various parties. With Germany's recognition of Slovenia and Croatia, the conditions by which the carrot of recognition was to be granted were rendered virtually meaningless. Thus, the principles of the inviolability of borders and the protection of minority rights became nothing more than a hollow shell of political rhetoric since Germany's premature action effectively removed the mechanisms by which the international community could enforce their commitment to these principles.

Part of the problem was the inability on the part of the international community to agree on the nature of the crisis. There was disagreement as to whether the break-up of Yugoslavia was due to the "inevitable dissolution of a state that never engendered a single nation; the inevitable clash of culture and religion, particularly between the Christian and Muslim worlds; a civil war of sequential secession to avoid minority status, or a calculated land grab."[44]

While there are elements of each of these perspectives that played a role in the break-up, the fact that the international community could not agree on the nature of the conflict, severely limited their ability to develop an effective policy of intervention in the crisis. Although the international community retained their commitment to *uti possedetis* throughout – from their initial commitment to a unitary state to their subsequent commitment to self-determination; they were not prepared to actively enforce this commitment under any comprehensive or cohesive political, economic or military framework.

Yet, the question remains: would Slovenia and Croatia (and thus later Bosnia-Hercegovina and Macedonia) have continued to claim independence if the international community had either continued their insistence on a unitary state, or withheld recognition until conditions for a peaceful dissolution had been met (backed-up with effective mechanisms for maintaining either policy)? While this is a highly debatable and contentious issue, several factors remain clear.

It is well documented that Slovenia began the process of 'sequential secession' and in the beginning was the republic most anxious to cede, particularly because it was the most committed to the democratisation process, which Serbia consistently blocked. If Slovenia had backed down in its demand for independence in the face of united opposition, it is unlikely the other secessionist republics would have followed through with their claims, even Croatia.

As well, Slovenia was anxious to be embraced by the West and integrated into its structures and institutions. Thus, whether they would have continued to claim sovereignty in the face of a strong, united and consistent Western opposition to its independence is therefore contestable.

Croatia's attitude to the West was more ambivalent and unclear, although post-Dayton there are indications they are willing to make concessions to Western demands in exchange for much needed infusion of Western economic aid. However, with premature recognition by Germany, Croatia did not have to worry about fulfilling the EC criteria for statehood.

There had previously been concerns in the international community about Croatia's failure to provide adequate protection to minorities in its new Constitution. This was articulated by the Badinter Commission, which recommended withholding recognition until Croatia could provide guarantees for minority rights. Before this could be addressed, Germany accorded recognition, thus thwarting any attempt to provide an effective response to the findings of the Badinter Commission.

While protection for minorities is at heart a legal question and therefore, not necessarily reflective of political realities; it remains that the opportunity to address the political issues that lay behind this legal question before recognition would be considered was lost when recognition was granted outside the EC's formal review process.

While the act of constitutive recognition cannot be held accountable for the carnage in Yugoslavia, as we have seen above, there are indications that the tortured and protracted process of dissolution was exacerbated by the political tool of recognition and nowhere was this more true in the former republics than in Bosnia-Hercegovina and the events that subsequently unfolded there.

THE ACT OF RECOGNITION AND REPUBLIKA SRPSKA

As previously noted, the Bosnian authorities had already indicated that to remain in a 'Serbianised rump-Yugoslavia' was an untenable position for them. Once it was (more or less) a foregone conclusion that Slovenia and Croatia were to be granted formal recognition, Bosnia-Hercegovina, which had not yet declared its independence, also applied to the Badinter Commission for recognition.

The Commission however, noted that the "popular will for an independent state had not been 'clearly established' [and suggested that] this conclusion could be changed if an internationally supervised referendum, open to all citizens without discrimination, were held."[45]

On March 1, 1992, Bosnia held a referendum in which 63% of the electorate participated, although the vast majority of Bosnian Serbs boycotted the referendum. 99% of those voting cast their ballot for independence and on March 3, 1992, Bosnia-Hercegovina declared itself a sovereign state which was formally recognised a month later by the international community.[46]

In response to the Bosnian declaration of independence, the Bosnian Serbs 'declared' the independent Serb Republic of Bosnia-Hercegovina – Republika Srpska, although it was never accorded *formal* (constitutive) recognition as a sovereign state.

At this point, it is worth noting that the Badinter Commission was the first official body to state authoritatively that "the effects of recognition by other States are purely declaratory."[47] This statement was the first to give legal credence to the declaratory theory of recognition (that a state becomes recognised as an international personality by 'declaring' itself so, i.e. – Republika Srpska) over that of constitutive recognition.

Yet, the Commission also determined that, on the basis of *uti possedetis*, "the Serbian population in Bosnia-Hercegovina [was] entitled to all the rights accorded to minorities under international law,"[48] but this did not include the right to secede. Thus, what did Republika Srpska's 'declaratory' act of recognition actually mean, given that it was not accorded recognition by any other state?

There are two cases that addressed the legality of the declaratory recognition of Republika Srpska – one directly (Kadic vs. Karadzic, 1995), the other one indirectly (Prosecutor vs. Tadic, 1995).

In Kadic vs. Karadzic, a Bosnian woman, "on behalf of herself and all others similarly situated",[49] began legal proceedings against Radovan Karadzic, (while he was in New York as an invitee of the United Nations), as 'President' of Republika Srpska, for acts of rape, torture, etc. committed by military forces under his command.

The judge, in his ruling on the case, determined that Republika Srpska fit "the definition of a state in international law,"[50] and concluded that,

Although the definition of statehood requires the capacity to engage in formal relations with other states, it does not require recognition by other states . . . Srpska is alleged to control defined territory, control populations within its power, and to have a president, a legislature and its own currency. These circumstances readily appear to satisfy the criteria for a state in all aspects of international law.[51]

Whether Republika Srpska could have fit the legal criteria for statehood is disputable (although space limitations deter us from exploring this question). Although upholding Republika Srpska's declaration of independence, the judge points to the contradictory fact that the defendant, Karadzic, "advances the [conflicting] position that he is not a state actor, even as he asserts that he is the President of the self-proclaimed Republic of Srpska."[52] By implying he cannot be held accountable as a state actor, although asserting he is a state agent, Karadzic effectively repudiated Republika Srpska's ability to function as an international legal personality, (which recognition bestows).

The same issues, although in a slightly different context, arise in the second case. In Prosecutor vs. Tadic, the defendant, Dusko Tadic, was tried for 'genocide', 'war crimes' and 'crimes against humanity' at the International War Crimes Tribunal of the former Yugoslavia (ICTY). In the case, he argued that the ICTY had no legal jurisdiction in which to prosecute him, claiming that that only Bosnia-Hercegovina, as an independent state, had "the competence to establish jurisdiction to try crimes that have been committed on its territory."[53]

This argument is interesting because not only does an admitted state actor of Republika Srpska (as a commander in the Bosnian Serb Army) recognise the domestic sovereignty of Bosnia-Hercegovina (over that of Republika Srpska) in trying claim Bosnian domestic jurisdiction over his alleged crimes; but its territorial sovereignty as well. Since the crimes the appellant was being tried for took place in the vicinity of Prijedor, a town on the territory claimed by Republika Srpska (both pre- and post-Dayton); this argument also implicitly repudiated the existence of Republika Srpska as an international personality.

Although two state agents of Republika Srpska in judicial proceedings in effect claimed that it did not constitute a state; nonetheless, the Dayton Peace Agreement recognised Republika Srspka as a legitimate entity, albeit within the Constitutional framework of Bosnia-Hercegovina. This act of recognition effectively legitimised the advantages wrought by four years of war and "reflected the surrender of important principles for the international community: to resist ethnic purification and to oppose the acquisition [of territory] through [the use of] force."[54] As well, by politically and legally legitimating the existence of Republika Srpska, the Dayton Agreement, effectively partitioned Bosnia-Hercegovina into two officially autonomous entities, thus continuing the process of 'balkanisation' even further.

THE DILEMMA OF KOSOVO

The further balkanisation of Yugoslavia is currently being played out in Kosovo. Acting under the authority of Chapter VII of the United Nations Charter which states that the UN has a right to interfere in the internal matters of a sovereign state if those matters are deemed to constitute a "threat to international peace and security,"[55] UN Security Council Resolution 1244 (hereafter 1244) of 10 June 1999, establishes "an interim administration for Kosovo as a part of the international civil presence under which the people of Kosovo can enjoy substantial autonomy . . ."[56]

While ultimately recognising the sovereignty of the Federal Republic of Yugoslavia over the territory of Kosovo,[57] 1244 effectively gives power for Kosovo over to the international community, and its representative in Kosovo, the Special Representative of the UN Secretary-General (SRSG).[58] "The framework through which this is accomplished is the regulation issued on 25 July 1999 by the SRSG, vesting all legislative and executive authority in UNMIK (the United Nations Interim Administrative Mission in Kosovo), of which the SRSG is the head."[59]

In the foreseeable future, it is unlikely that the international community will confer official recognition on Kosovo as an international legal personality (think Canada and Quebec or Spain and the Basque region and the implications for such countries with secessionist-oriented territories if Kosovo were to gain diplomatic recognition). However, since the passing of 1244, the international community's actions in the territory has been on a steady, incremental process towards *de facto* recognition.

The regulation of 25 July 1999 in effect means that the SRSG can make or change any law he wants to simply through the issuance of regulations.[60] These regulations which, once signed by the SRSG, become legally binding, not only because of the authority invested in the SRSG by 1244, but also because all regulations issued take precedence over the applicable law in Kosovo.[61]

Thus, with the stroke of a pen, the international community has effectively wrenched sovereign jurisdiction from Yugoslavia, particularly since the federal (Yugoslav) government exercises no control over, or has no say in the running of Kosovo. The signs can be seen everywhere – from Regulation 1999/4 making the German Mark the official currency of Kosovo rather than the Yugoslav dinar[62] to the proposed Kosovo criminal procedural code, which in every other country is a federal rather than a territorial statute.

Other indications that the international community views Kosovo as an entity separate from that of Yugoslavia include the fact that UNMIK patrols the borders to other parts of Yugoslavia (Serbia and Montenegro) as well as controlling those borders crossing international boundaries (the Former Yugoslav Republic of Macedonia and Albania). UNMIK also has the ability to collect customs and excise taxes (Regulation 1999/3).[63] One recent statistic indicated that the income accrued to UNMIK from the Pristina airport, was approximately 500,000 DM ($250,000 USD) per month.[64]

Although the federal Yugoslav government has an 'office' in Kosovo, based in Pristina, from which Kosovars can obtain passports, often in one day, UNMIK provides 'travel documents' to Kosovars wishing to travel abroad.[65] As well, UNMIK allows countries to establish diplomatic missions in Kosovo, under the term of 'liaison offices'.

Even more important is the new power sharing arrangement with the Kosovar Albanian community currently being implemented (at the time of this writing, Kosovar Serbs are continuing to boycott the international community's efforts to include them in the UNMIK administrative structures until such issues as the return of Serbs to the province, and their safety and security, etc. can be guaranteed). In order to negate the authority of the parallel power structures in place in Kosovo even before the conflict, with the KLA (Kosovo Liberation Army) controlling the majority of the local self-styled administrative councils in the territory, the SRSG negotiated a power-sharing agreement with Kosovar Albanian leaders in December 1999.

Termed the Joint Interim Administrative Structures (JIAS), the new power sharing arrangement calls for the setting up of 20 'departments' to be run jointly with an international and a Kosovar co-head,[66] the most controversial of which is the 'Department of Non-Resident Affairs'. Some view this 'department' as a stand-in for a foreign affairs ministry. Other departments which give Kosovars a substantial amount of power usually associated with a federal government include the Department of Central Fiscal Authority, the Department of Democratic Governance and Civil Society Support, the Department of Justice and the Department Civil Security and Emergency Preparedness.[67]

Ultimately though, according to international community-think, the legitimacy for the current involvement of the international community in Kosovo in place of a functioning indigenous government, is the failure of the Rambouillet Agreement.

The failed Agreement, negotiated in March 1999, which the Serbs refused to sign, in effect reduced the territorial sovereignty of Yugoslavia over Kosovo to name only. The Agreement, which is referred to in 1244,[68] lays out in detail substantial autonomy for Kosovo. While there is nothing particularly problematic about the concept of autonomy for Kosovo, especially given that the Yugoslav Constitution of 1974 granted autonomy for Kosovo until Milosevic removed it in 1989 and oppressively clamped down on the Albanian majority population, leading to the current crisis; Rambouillet, both in spirit and letter, goes further than any other nation-state would have most likely accepted as terms of a compromise for a secessionist-minded territory.

For example, the military part of the Agreement would have allowed "NATO to establish and deploy a force operating under the authority and subject to the direction and political control of the North Atlantic Council (NAC) through the NATO chain of command."[69] While a Joint Implementation Commission would have been established to address all military matters and complaints which would have included Yugoslav military commanders, ultimately it was envisioned that "the final authority to interpret the provisions of Chapter 7 would rest with the KFOR [international] commander."[70]

As well,

> [all Yugoslav army] units, other than 1,500 members of a lightly armed border guard battalion deployed close to the border would have had to be withdrawn from Kosovo... The border guards would have been limited to patrolling the [internal] border zone and their travel through Kosovo would have been subjected to significant restrictions. Moreover, the air defense system in Kosovo would be dismantled and associated forces withdrawn, as would other [Yugoslav] or Serb forces, including the Ministry of the Interior Police (MUP).[71]

While these military provisions are understandable in the context of what was then happening in Kosovo, particularly with regard to the MUP, whose repressive presence and human rights violations were notorious, it still remains that the complete removal of a federal defense, border and internal security force and their power handed over to international organs to control a territory *internal* to a nation-state (as opposed to a defeated Germany after WWII for example) would be regarded by that nation-state as an extreme threat to its sovereignty.

Furthermore, Rambouillet gave ultimate control of the civilian implementation of the Agreement to the international community.

> [O]perating under a Chief of the Implementation Mission (CIM) . . . [the international community] would have had final authority to interpret the provisions of the agreement in relation to civilian implementation.[72]

Again, while understandable in terms of what had been happening in Kosovo for the previous decade and the Milosevic regime's well-known tactics as previously played out in Croatia and Bosnia-Hercegovina; in terms of the concept of territorial sovereignty, these would be considered unacceptable terms. Yet, this is what is currently happening with the present international presence in Kosovo. Mandated under 1244, UNMIK and the international military force (KFOR) operates under a broader and wider framework of control than was originally envisioned with Rambouillet.

While the dilemma of Kosovo (often referred to as the 'powderkeg of the Balkans') can be squarely laid at Milosevic' feet, from the time of removal of Kosovo's autonomy in 1989 to the ten year reign of oppression over Kosovo's 90% Albanian population, the international community must absorb some responsibility as well.

With the revoking of autonomy and the beginning of the repression of Kosovo's Albanian population, the international community weighed in their disapproval with words and paper. Yet much of it was muted, for many of the reasons outlined above. And with the wars raging in other parts of Yugoslavia, the international community's concern in Yugoslavia lay elsewhere within the splintering country.

As well, the international community missed an opportunity to deal with the Kosovo question during the Dayton Agreement negotiations. Leaving Kosovo outside of the framework for negotiations ensured that the problems in Bosnia-Hercegovina would next be carried over to Kosovo, particularly since the pattern of Milosevic to gather the troops and prepare the 'nation' for war under the guise of nationalism in order to tighten his grip on authority would have had a broad appeal given that Kosovo has long been considered the 'cradle of Serb civilisation' by the Serbs.

It was an opportunity lost, not only to deal with the 'Kosovo question' within a more soluble and permanent framework, but also because up until at least 1995, when Dayton was negotiated, the Kosovar Albanians would have settled for a return of their autonomy within the borders of Yugoslavia. As it stands now, the idea of autonomy within Yugoslavia is anathema to most Kosovar Albanians; only full independence will do. This is understandable on an emotional level given the brutality and repression the Kosovar Albanians experienced for ten years.

And as it currently stands, this is pretty much what the international community has given Kosovo, at least *de facto*. While publicly claiming to recognise the sovereignty of Yugoslavia, UNMIK has created for itself an international protectorate in which it may have no choice but to move ahead and accord formal recognition, since it is unlikely to cede federal legislative and executive power for Kosovo back to Yugoslavia. Ultimately the international community may find it paying the price in the precedence it has set for its *de facto* recognition of Kosovo, giving licence for other 'Kosovos' in the world to surface.

CONCLUSION

The tortured and protracted dissolution of Yugoslavia, which continues today with Kosovo and Montenegro, has to date, cost hundreds of thousands of lives and the displacement of millions of people. To what extent recognition aided and abetted this it cannot be conclusively determined as it is hard to conjecture other courses of outcomes, given all the other variables that came into play in the dissolution of Yugoslavia. However, as has been argued above, it can none-the-less be assumed that the act of recognition (whether constitutive, de facto or declaratory) had major political consequences for the break-up of Yugoslavia.

> The attempt by the [international community] to contain the dissolution of [Yugoslavia], or at least force the parties to accept certain legal rules governing their internal and external affairs after the dissolution, may have contributed significantly to the seriousness of the crisis. The principal weapon used by the international community in seeking to influence events was [the act of] recognition [both in their initial attempts to deny it and their later attempts to grant it]; yet the effect of recognition was much misunderstood in this crisis.[73]

What conclusions can be drawn from the observations above? Given what we have witnessed of the role that recognition played in the messy dissolution of Yugoslavia, there is a case to be made for taking away the act of recognition from the political interests of individual states and instead, to create appropriate international channels that would accord or deny recognition on the basis of whether or not an entity fulfils the requirements of statehood as determined by its meaning in the evolution of international law.

While realistically this is unlikely to happen, the Yugoslav crisis has shown that there needs to be a fundamental re-examination around questions underlying such issues as the right to self-determination, minority rights, the exercise and limits of sovereignty and especially the question of outside intervention.

Yet, one thing does remain clear from the wreckage that was Yugoslavia. If the international community is to intervene in any conflict, in the hope of affecting the outcome, whether it is through the mechanisms of recognition or through other political tools; it must be prepared to actively enforce a united commitment to the principles that underlie their intervention.

Given that the international community is now pouring billions of dollars of aid into the former Yugoslavia to hold up a fragile peace; effective commitment to principles they undertook to uphold in the first place, would have undoubtedly been the least costly alternative;[74] both in terms of the lives lost and displaced, and in terms of the human, financial and material resources committed. However, since it is in the nature of the relations between states in the international arena to take a blinkered and shortsighted approach to flashpoints that upset the political order of things, it is an expensive lesson that the international community has yet to absorb.

NOTES

1. J. Gow, *Triumph of the Lack of Will: International Diplomacy and the Yugoslav War.* C. Hurst & Co. Publishers, 1997. p. 67
2. Ibid. pp. 2-3

3. M. N. Shaw, *International Law* (4th ed.). Cambridge University Press, 1997. p. 295

4. Ibid.

5. S. Economides & P. Taylor, Former Yugoslavia. In: J. Mayall, (Ed.), *The New Interventionism 1991 – 1994: United Nations Experiences in Cambodia, Former Yugoslavia and Somalia*. Cambridge University Press, 1996. p. 62

6. Ibid. p. 65

7. M. N. Shaw, *Op. cit.* pp. 296–7

8. D. J. Harris, *Cases and Materials in International Law* (4th ed.). Sweet & Maxwell, 1997. P.144

9. M. N. Shaw, *Op. cit.* p. 298

10. D. J. Harris, *Op. cit.* p. 147

11. Ibid p. 147

12. K. N. Schake, The Breakup of Yugoslavia. In R. von Lipsey (Ed.), *Breaking the Cycle: A Framework for Conflict Intervention*. MacMillan Press, 1997. p. 98

13. J. B. Steinberg, International Involvement in the Yugoslav Conflict. In: L. F. Damrosch, (Ed.), *Enforcing Restraint: Collective Intervention in Internal Conflicts*. Council on Foreign Relations, 1993. p. 34

14. S. Economides & P. Taylor, *Op. cit.* p. 63

15. Ibid.

16. P. Weller, The International Response to the Dissolution of the Socialist Federal Republic of Yugoslavia. *American Journal of International Law*. Vol. 86, 1992. p. 574

17. Ibid. pp. 34-5

18. Ibid. p. 55

19. J. Gow, *Op. cit.* p. 30

20. J. B. Steinberg, *Op. cit.* p. 56

21. J. Gow, *Op. cit.* p. 7

22. J. B. Steinberg, *Op. cit.* p. 66

23. Ibid.

24. Ibid. p. 37

25. S. Economides & P. Taylor, *Op. cit.* p. 91

26. J. B. Steinberg, *Op. cit.* p. 37

27. M. Weller, *Op. cit.* p. 586

28. Ibid.

29. J. B. Steinberg, *Op. cit.* p. 70

30. J. Gow, *Op. cit.* p. 63

31. J. B. Steinberg, *Op. cit.* p. 37

32. S. Economides & P. Taylor, *Op cit.* p. 21

33. J. B. Steinberg, *Op. cit.* p. 70

34. M. Weller, *Op. cit.* p. 587

35. European Policy Council Press Release, 128/91, December 16, 1991. *European Journal of International Law, 4* EJIL 72, 1993

36. M. Weller, *Op. cit.* p. 588

37. Ibid.

38. Ibid.

39. Ibid.

40. R. von Lipsey, *Op. cit.* p. 99

41. J. Gow, *Op. cit.* p. 64

42. J. B. Steinberg, *Op. cit.* p. 35

43. Ibid.

44. R. von Lipsey, *Op. cit.* p. 96

45. M. Weller, *Op. cit.* p. 593

46. Ibid.

47. D. J. Harris, *Op. cit.* p. 123

48. Ibid. p. 121

49. Decision in Kadic v. Karadzic (1995) *International Legal Materials* 34 ILM 1592 (1995) p. 1599

50. Article I of the 1933 Montevideo Convention on the Rights and Duties of a State, defines the state as one having: 'a permanent population; a defined territory; government; capacity to enter into relations with other States'. D. J. Harris, *Op. cit.* p. 101

51. Decision in Kadic v. Karadzic (1995) *International Legal Materials* 34 ILM 1592 (1995) pp. 606–7

52. Ibid. p. 1599

53. Ibid. p. 28

54. J. Gow, *Op. cit.* p. 1

55. Security Council Resolution 1244 states: 'Determining that the situation in the region continues to constitute a threat to international peace and security . . . and acting for these purposes under Chapter VII of the Charter of the United Nations. . .'

56. United Nations Security Council Resolution 1244

57. Security Council Resolution 1244, "[reaffirms] the commitment of all Member States to the sovereignty and territorial integrity of the Federal Republic of Yugoslavia"

58. Security Council Resolution 1244, "[r]equests the Secretary-General to appoint, in consultation with the Security Council, a Special Representative to control the implementation of the international civil presence. . . ."

59. UNMIK/REG/1999/1, 25 July 1999 – ON THE AUTHORITY OF THE INTERIM ADMINISTRATION IN KOSOVO

60. "In the performance of the duties entrusted to the interim administration under United Nations Security Council resolution 1244 (1999), UNMIK will, as necessary, issue legislative acts in the form of regulations. Such regulations will remain in force until repealed by UNMIK or superseded by such rules as are subsequently issued . . ." UNMIK/REG/1999/1

61. "The laws applicable in the territory of Kosovo prior to 24 March 1999 shall continue to apply in Kosovo insofar as they do not conflict with . . . the fulfilment of the mandate given to UNMIK under United Nations Security Council resolution 1244 (1999) . . ." UNMIK/REG/1999/1

62. UNMIK/REG/1999/4, 2 September 1999 – ON THE CURRENCY PERMITTED TO BE USED IN KOSOVO.

63. UNMIK/REG/1999/3, 31 August 1999 – ON THE ESTABLISHMENT OF THE CUSTOMS AND OTHER RELATED SERVICES IN KOSOVO.

64. OSCE Daily Briefing: 18 April 2000

65. REGULATION NO. 2000/18, 2 March 2000 – ON TRAVEL DOCUMENTS.

66. UNMIK/REG/2000/1, 14 January 2000 – ON THE KOSOVO JOINT INTERIM ADMINISTRATIVE STRUCTURE.

67. The complete list of Departments under JIAS are: Sport; Culture; Youth; Democratic Governance and Civil Society Support; Environment; Transport and

Infrastructure; Agriculture; Labour and Employment; Trade and Industry; Civil Security and Emergency Preparedness; Local Administration; Health and Social Welfare; Justice; Public Services; Post and Telecom; Central Fiscal Authority; Education and Science; Non-Resident Affairs; Utilities; and Recontruction. UNMIK/REG/2000/1, 14 January 2000

68. Article 11 (a) of Resolution 1244 states that the international community will "[promote] the establishment, pending a final settlement, of substantial autonomy and self-government in Kosovo, taking full account of annex 2 and of the Rambouillet accords."

69. M. Weller, *The Crisis in Kosovo 1989–1999. From the Dissolution of Yugoslavia to Rambouillet and the Outbreak of Hostilities.* Documents and Analysis Publishing Ltd. 1999. p. 410

70. Ibid. p. 411

71. Ibid.

72. Ibid.

73. M. Weller, *Op. cit.* Footnote 16 p. 604

74. R. von Lipsey, *Op. cit.* p. 100

REFERENCES

Danchev, A., & Halverson, T. (Eds) (1996). *International perspectives on the Yugoslav Conflict.* MacMillan Press Ltd.

Economides, S., & Taylor, P. (1996). Former Yugoslavia in Mayall, James (ed.) *The New Interventionism 1991 – 1994: United Nations Experiences in Cambodia, Former Yugoslavia and Somalia.* Cambridge University Press.

Gow, J. (1997). *Triumph of the Lack of Will: International Diplomacy and the Yugoslav War.* C. Hurst & Co. Publishers.

Harris, D. J. (1997). *Cases and Materials in International Law* (4th ed.). Sweet & Maxwell.

International Legal Materials. Decision in Kadic v. Karadzic (1995). 34 ILM 1592 (1995).

International Legal Materials. Decision in Prosecutor v. Tadic (1995). 35 ILM 32 (1996).

Schake, K. N. (1997). The Breakup of Yugoslavia. In: R. von Lipsey, (Ed.). *Breaking the Cycle: A Framework for Conflict Intervention.* MacMillan Press.

Shaw, M. N. (1997). International Law (4th ed.). Cambridge University Press.

Steinberg, J. B. (1993). International Involvement in the Yugoslav Conflict. In: L. F. Damrosch (Ed.), *Enforcing Restraint: Collective Intervention in Internal Conflicts.* Council on Foreign Relations.

Weller, M. (1999). *The Crisis in Kosovo 1989–1999. From the Dissolution of Yugoslavia to Rambouillet and the Outbreak of Hostilities.* Documents and Analysis Publications Ltd.

Weller, P. (1992). The International Response to the Dissolution of the Socialist Federal Republic of Yugoslavia. *American Journal of International Law, 86.*

7. THE INTERNATIONAL RESPONSE TO THE CRISIS IN YUGOSLAVIA*

Darko Silovic

This paper represents an effort to briefly summarize and evaluate the international response, i.e. the foreign factor in the origins, eruption, unfolding and current situation of the crisis and wars on the territory of former Yugoslavia. Unfortunately, even after the war in Kosovo, the crisis is not over and it is therefore difficult to make a final assessment.

Before even trying to do so it is important to underline that the break-up of Yugoslavia was basically of internal origin, although some sides in former Yugoslavia, in order to deflect their own culpability, and certain outside observers and analysts for various reasons promote the opposite argument. The conditions for this crisis and the war that followed were created by factors inside Yugoslavia. One can go even further and directly blame the political oligarchies of the six constituent republics of the then Socialist Federal Republic of Yugoslavia, the leadership of the League of Communists of Yugoslavia (the only party in power), as well as, and maybe even especially, the collective State Presidency consisting of one representative from each Republic and one from each of the two Autonomous Provinces, Vojvodina and Kosovo (which are part of Serbia).

The Constitution of 1974 elevated the republics and, almost to the same extent, the two autonomous provinces, to statehood. This system could function, although with greater and greater difficulty, so long as President Josip Broz Tito was alive and had the personal authority to act as final arbiter. Following Tito's death in 1980, at an advanced age of 88, the system that was presumably established to prevent a struggle for succession proved to be inadequate. Decisions in

*This article was written early in 1997.

Research on Russia and Eastern Europe, Volume 3, pages 145–157.
2000 by Elsevier Science Inc.
ISBN: 0-7623-0280-1

the State and in the Party presidencies were supposed to be reached by consensus, which enabled any republic and/or autonomous province to block it. Since real power rested in the republics, their respective representatives in the highest federal institutions were, more often than not, delegated on the basis of negative selection and were to a great extent elderly, conservative, frightened politicians reluctant to 'rock the boat' and eager to avoid any of the long overdue reforms in the sterile, stagnant political and economic system. Rather, they opted for the 'ostrich strategy' and refused to understand that everything inside and around Yugoslavia was rapidly changing. The leaderships of the republics, on the other hand, while maintaining and strengthening their grip on power, differed more and more on the concept of development and the future of the country. Some of the differences derived from objective causes, such as particularly disproportionate levels of development and geographic positions. More devastating differences were those arising from irreconcilable concepts concerning the socio-economic system, the proper contribution to and distribution of federal assets, the roles of the Party and the Army. There was one thing, though, they had in common – each guarded jealously its own power on its own turf. Thus, Yugoslavia had practically six parties within a one-party system and its crisis was created at the top and not from the bottom.

SERBIAN AND CROATIAN RELATIONS

It should also be emphasized that Yugoslavia was created (in 1918), it existed, suffered, prospered, and finally broke up on the basis of relations between Serbia and Croatia. The other republics, especially Slovenia, were important but not crucial. Bosnia was always claimed by both Serbia and Croatia and finally became their ultimate victim. The Muslims in Bosnia, as a result of its historical heritage and the salt and pepper of its specific mentality, individuality, and culture, are Slavs who speak the same language used by both the Serbs and the Croats (although today this language is called differently by each one of them and every effort is being made, especially in Croatia, to accentuate, impose and even create artificial differences). The Muslims had been glorified and vilified, depending on momentary needs or alliances, and they were the biggest losers and victims of the war in Bosnia. During the several years that have elapsed since the Dayton Accords were adopted, it is clear that the fate of Bosnia is still not decided, that renewed conflicts are possible or even probable, and that this too will depend on further developments within and between Serbia and Croatia. While Milosevic and Tudjman are in power and support their factions in what has become their parts of Bosnia and Herzegovina, there will be no peace, reconciliation nor stability. The ethnic group that suffered the second-most are the Serbs, particularly the Serbs in Croatia. However, long-term adverse consequences of the Yugoslav crisis are and will be felt also by their brethren in Bosnia and Serbia proper, whose political and, shamefully enough, intellectual élite initiated and caused this tragedy and bears the biggest responsibility for it. It is arguably a self-inflicted wound, but no ethnic group should be punished for the deeds of its leadership. The Serbs took up arms inflamed by Milosevic's regime and nationalistic oratory. The victory of President Tudjman's nationalistic party in Croatia and its pre-election slogans, openly anti-Serbian and reminiscent of the genocidal fascist – Nazi puppet – Ustasha Croatia from 1941–1945, only further strengthened Milosevic's arguments. What began as an assertion of the need of all Serbs to live in one country (as if they did not!?) led to a bloody war and the expelling of most of the Serbs from Croatia, as well as from many parts of Bosnia. Of course, Muslims were also forced from their ancestral homes in Eastern Bosnia and Croatian-held parts of Herzegovina.

THE INTERNATIONAL PERSPECTIVE

While this crisis was still brewing the international community was surprised, maybe worried, but on the whole rather indifferent. Later on, of course, the international forces, which had their own interests and points of view, certainly did try to appease, but also to make the most out of the situation. The international response was too late in arriving and when it started was inadequate, unfocused, and based on a faulty understanding of the real causes and the consequences of the crisis. Also, there were differences among the various international actors who were intervening, and their disagreements made a solution more difficult.

It is important to recall at this point also that for many years Yugoslavia had occupied a unique position in the world. Between the break-up with Stalin in 1948 and the end of the Cold War it was a buffer zone, a socialist country independent of Moscow and as such supported by the West. Yugoslavia was seen from the West as an example for other Eastern European countries – a possible model for their eventual detachment from Moscow's influence.

Located politically and geographically between NATO and the Warsaw Pact countries, Yugoslavia also provided a barrier against Russian access to the Mediterranean and the warm seas. And as a founder, active member, and leader of the Movement of Non-Aligned Countries it was an important player on the international stage and a partner both to the East and the West. It was able to play a special role during the paralysis that was created in international institutions by the Cold War.

Still, the West had an ambivalent attitude towards Yugoslavia, which it supported politically and economically in its independent course from Moscow, while it was nevertheless suspicious and uncomfortable with it as a communist country. There was constant debate as to what side Yugoslavia would take if war between the blocs were to erupt. Because the answer was uncertain, the West also supported some groups and individuals who were opponents of the regime.

During World War II, there were various groups inside Yugoslavia which sided with the occupying forces and left the country afterwards. In the 1990s, while many countries were re-examining their conscience some fifty years after the war, there were new discussions about the Nazi gold in Swiss banks, about the not-so-neutral behavior of Portugal and Sweden, and so on. The same applies to those countries which harbored World War II war criminals and their supporters. Many collaborators of the Nazi regimes from various European countries, as well as from Yugoslavia, were accepted in Canada, in the United States, in Argentina, and elsewhere. In the Cold War atmosphere of ideological confrontation their younger followers were also welcomed and allowed to organize. During the wars in Croatia and Bosnia, the Minister of Defence of Croatia, for instance, was a Canadian citizen whose ideological convictions were well known. He is not the only such figure. These linkages must be understood as part of the West's ambivalence toward Yugoslavia.

The crisis in Yugoslavia not only had internal roots, but it began long before its outward manifestations were becoming obvious. One also has to understand the international conditions in which these manifestations became visible. Major international changes were taking place during that period. In 1989 the Berlin wall fell and the Cold War ended. Germany unified much faster than anyone expected and many countries, both in the East and the West, had reservations about this speed. But the process could not be stopped and it completely changed the geo-strategic position of Germany and the political balance in Europe. Instead of being an Atlantic country, which it had been until unification, it suddenly became a Central European power, very much oriented towards its Eastern neighbors, which so recently had been located behind the Iron Curtain. There were the Gulf crisis; the invasion of Kuwait by Iraq; the Desert Storm which followed; and the Maastricht process, the final phases of which coincided with the outbreak of the open Yugoslavia crisis. The last six months of the Maastricht negotiations took place during the first six months of the manifest crisis and the beginning of the war in Yugoslavia. And finally, there was the collapse of the Soviet Union.

All these events fundamentally transformed the political map of the world and the relations which had existed for decades. Although the Cold War was cold, it was still a war. It had lasted a long time and it ended in a way that nobody could have anticipated even a short period before. The end of the Cold

War was greeted with enthusiasm. A 'new world order' was immediately declared, though of course, one could immediately see that there was, if anything, more disorder in the world. This period was even called 'the end of history', though anyone can understand that history does not end, or else everybody would be dead. After a major earthquake there are many smaller quakes, and the world is going through that process at the turn of the millennium, as evidenced not only in Yugoslavia but in many other places as well.

The immediate and direct consequence of all these gigantic changes for Yugoslavia was that it has almost disappeared from the political map of the world. It lost its previous geo-political and strategic importance and therefore the outside interest in its stability considerably waned. The Yugoslav authorities were even told to expect that. It was clearly explained that "from now on you can't count on our support and you have to put your house in order."

Unfortunately, many political leaders in the post-Tito era of stagnant politics just could not grasp the fact that things were different and changing. They believed that Yugoslavia had earned its position by its own merit and forever. Everybody likes to have a good opinion of himself, but one also has to be able to make judgments. That judgment was missing. Not only had Yugoslavia lost its importance, but it was seen as one of the last remnants of communism at a time when the communist system was crumbling everywhere. This is why there was no enthusiasm to rush in and start helping. On the contrary, the failure of the communist system was welcome. Later, for a brief time the West tried to prevent the dismemberment of Yugoslavia because of concern about developments in the Soviet Union. Yugoslavia was seen as a precedent, an example, of what might happen in the Soviet Union. Western Europe and North America considered it, at the time, better to keep the status quo in the Soviet Union than to encourage changes that might lead – who knows where?

Thus, there was an effort for some time to keep Yugoslavia together, rather than to entice or help the various parts to leave it. Enough has been said in that context about the weak, rather verbal support for the program of economic reforms of the last truly Yugoslav Prime Minister Ante Markovic, which would by its nature have introduced also major political transformations, but that was never to happen.

A CRISIS ARISING

Unfortunately, by that time nationalism was already rampant in Serbia. A few years before by a Bolshevik type coup inside the Serbian communist party, Slobodan Milosevic had taken over the reigns of power and was already getting ready to start his pressure on the rest of Yugoslavia. What followed were the so-called constitutional reforms which, through political pressure and blackmail, fully subordinated the two autonomous provinces to Belgrade in order to switch the balance at the federal level in Serbia's favor. Kosovo and Vojvodina were stripped of their autonomy in 1989.

The first multi-party elections were held in all the republics between April and December of 1990 in an already divided country. The first of these were held in Slovenia and the last in Serbia.

The communists lost the elections in all republics except in Serbia. But, since there were no elections at the federal level – neither multi-party nor a referendum on the fate of the Federation – no possibility existed for the whole population to pronounce itself on the future of Yugoslavia. It was all done within the republics in a situation in which nationalism had already arisen and where people were switching from one ideology to ethnicity in a political vacuum and confusion. This vote was seen as a popular plebiscite for the break-up of the country; it was facilitated by the fact that there were no elections on the federal level prior to those in the republics and it legitimized the push for independence by the new leaderships in Slovenia and Croatia. On the other hand there could have been no elections on the federal level since there was no agreement on the basis for such elections. So it turned into a vicious circle, the net result being that the push for independence was legitimized.

It was at that moment that the West realized that events were evolving in a dangerous direction. Finally, on June 25 1991 when Slovenia and Croatia declared independence, the West started to respond, but even that belated effort was half-hearted and did not address the essence of the problem.

Four days before Slovenian and Croatian declaration of independence U.S. Secretary of State James Baker visited Belgrade for one day. He arrived fully preoccupied by the Gulf crisis and held a series of meetings with representatives of all the republics, the federal government and the presidency. He left his interlocutors as confused as he was himself with what he found there. His overall goal was to persuade all the Yugoslavs to make peace and stay together, having in mind primarily that the break-up could be a bad precedent for the USSR. However, the U.S. was ready to use neither a big stick nor a fat carrot to prevail on the republican leaders, which left the latter with the belief, if not conviction, that they each had the support for their respective and conflicting agendas. At least they understood that the U.S. would not intervene, no matter what happened. Each of Baker's Yugoslav interlocutors concluded that he had the support of the United States for his own program. Milosevic and the army thought: we have a free hand now to intervene because the United States says they don't want the country to break-up. The others only heard that the United States opposed the use of force, so they concluded they could proceed with their process toward independence because the army would not be able to use force.

The European Community also started to get more directly involved after Baker's visit. Yugoslavia had a collective presidency whose president rotated on annual basis in a legally established order. The representative of each republic and/or autonomous province would be president for one year. On May 15, 1991 it was the turn of Croatia. Milosevic controlled half of the presidency by then – Serbia, its two autonomous provinces (Kosovo and Vojvodina), and Montenegro. He blocked the election of the Croatian representative Stjepan Mesic to the function of president of the presidency, which actually paralyzed the supreme decision-making body and chief military commander of the State.

The European Community's (later to become European Union) troika of foreign ministers consisting of the current, preceding, and next in succession presiding country, forced the Yugoslav presidency to elect Stjepan Mesic as its President. The EC considered this to be a major success and confirmation of their growing influence, although by then the system did not work anyway.

The foreign ministers and their governments did not and could not understand that the main protagonists of the Yugoslav drama, namely Serbia, Croatia and Slovenia (since Bosnia and Macedonia were at the time genuinely interested in an agreed solution, for they did not see their survival without Yugoslavia) acquiesced to their pressure and formally accepted Stjepan Mesic as the new president, while in fact they were only buying time needed for the final clash.

After Croatia and Slovenia declared independence on June 25, 1991 the same troika of European Union foreign ministers brokered on July 7 a three-month suspension of the declaration of independence, the so-called Brioni Agreement. It was actually only a postponement of independence for three months and it did not rescind independence.

Slovenia and Croatia were happy with this arrangement because they knew that in three months their independence would be recognized. Serbia used the three months to prepare for war. Actually, immediately after the Brioni Agreement serious fighting started in Croatia. This was when Vukovar was bombed and razed, and when Dubrovnik was shelled and its immediate hinterland burned and pillaged. This is a period in which 30% of the territory of Croatia was taken over by the local Serb population with the direct assistance of the so-called Yugoslav National Army (JNA).

In August 1991 the European Union convened the first Conference on Yugoslavia under the chairmanship of Lord Carrington in The Hague. As the basis for the discussion at this Conference, Lord Carrington first proposed an overall settlement of all the problems and relations between the republics, as a precondition for further discussion and possible recognition of independence for all republics that might wish it. The Conference also established the Badinter Commission, an arbitration body named after its Chairman, a well known international legal figure from France. That Commission declared, at the end of November 1991, that Yugoslavia was in the process of dissolution – an assessment that differed from that of Lord Carrington. That immediately created a crack which then started widening. By early December, the Badinter Commission even stated that Yugoslavia was in 'full disintegration', which actually meant that the country was dissolving and that the way was open to independence of various republics.

Lord Carrington at this point changed his approach somewhat, so that his second proposal included discussions on independence, which was then rejected by Serbia. The Conference was blocked and at that time the European Union realized that they alone did not have the necessary clout to solve the problems of Yugoslavia. So they turned to the United Nations.

Everybody at the U.N. was aware that there was a serious situation developing in Yugoslavia and many countries were willing to help. The non-aligned partners were worried that one of their founding members was in crisis; the neighbors were worried; the European Union likewise. Informal consultations and private discussions about the Yugoslav crisis ensued.

It should be noted that there was also serious opposition in the United Nations to dealing with the Yugoslav crisis. It was an internal matter and many countries did not want to create a precedent and open the door for the Security Council to intervene in their own internal affairs. The federal authorities of Yugoslavia, by now largely controlled by the Serbs, were also opposed. Yugoslavia was at the time Chairman of the Non-Aligned Movement and its rump leadership tried to misuse this position for its goals.

Yugoslavia helped the confusion since, from the beginning of the crisis, it was sending mixed messages to the outside world. While fighting continued on the ground in Croatia and a political tug of war was intensifying at the highest levels, the official message from Belgrade was trying to convince everybody not to worry: "Things will be settled soon, this is an internal, family affair, everything is under control." The diplomatic service was also instructed to do likewise. On the other hand, a completely different signal was coming from the republics, particularly Slovenia and Croatia, but also Macedonia and Bosnia. By now many individuals at various levels of government had sided with the opposing and irreconcilable positions of their respective republican leaderships and this cacophony of confusing signals made it even more difficult for the world to understand what was really at stake.

Neither the foreign ministers or the governments of the prosperous and stable Western democracies who tried to reach an agreed solution among the Yugoslav parties, nor their counterparts from the non-aligned world, who looked at Yugoslavia as a pillar of the Movement and as a model of a multi-ethnic and multi-religious society, could grasp the simple truth that there were people and groups in that country for whom war was a deliberate choice and that they were actually preparing for it while posturing at negotiations and unscrupulously lying into the faces of their interlocutors, making them look almost naïve. This is the bottom line and the reason why foreign response did not succeed. Europe and the U.S. did not understand that for the leaders of Serbia, first of all, but also of Croatia, the use of force to promote their ambitions was an option present from the very beginning. Even more, it was not perceived that ethnic cleansing, executed by rape, torture, burning of villages and every other violent method imaginable was not the consequence, but the means by which the final goal of this war was to be achieved. And that goal was the creation of nationally homogeneous states.

Or maybe Western politicians were not ready to recognize this fact publicly because that would leave them with only one option – the threat of force and readiness to use it. That, however, for internal political reasons in Western democracies and for institutional and procedural shortcomings within the U.N. and NATO, was not possible. Whatever the true reason, the international response in Yugoslavia was a series of blunders and half measures which only prolonged the agony. The result after five years of international involvement and intervention was much weaker and more costly in human loss and material damage than could have been done in the very early stages with a determined and straightforward approach, coupled of course with the determination to follow it up. As was proven later on different occasions, all the warring sides and particularly the Serbs, reversed when confronted with force.

In the beginning the main obstacle at the U.N. was to officially introduce the Yugoslav conflict on the Agenda of the Security Council because it certainly started as an internal problem. The U.N. Charter clearly states in Article 2, para 7: "Nothing contained in the present Charter shall authorize the United Nations to intervene in matters which are essentially within the domestic jurisdiction of any state or shall require the Members to submit such matters to settlement under the present Charter. But this principle shall not prejudice the application of enforcement measures under Chapter VII."

The Belgrade regime, which was winning the war in Croatia but losing credibility in the world, was still recognized as the official Government. It acquiesced to the U.N. involvement only under tremendous pressure. It also had, by then, secured more than its military goals and understood that the recognition of Slovenia and Croatia was imminent, leading to the internationalization of the problem in any case. The presence of the U.N. was therefore seen by Belgrade also as a way to secure the military gains and freeze the situation on the ground.

So the Security Council finally discussed Yugoslavia for the first time at one of the rare meetings at ministerial level in late September 1991. Resolution 713 was adopted on that occasion, imposing an arms embargo, although it was clearly favoring Serbia which controlled most, if not all, JNA resources.

Cyrus Vance was soon after appointed U.N. Secretary-General's Special Envoy and the international mediation now acquired two tracks since the EC Conference in The Hague continued. Vance succeeded in reaching an agreement on the U.N. peacekeeping operation in Croatia.

By the time the troops were actually deployed Slovenia and Croatia were recognized as independent states by the EC and soon by most other countries. This move was and still remains controversial. It is often cited as an example of discord and inconsistency on the part of European powers and is deemed by many to have exacerbated the conflict. The U.N. Secretary-General had an exchange of unusually strong letters with the German Foreign Minister on this issue and the U.S. Administration was also very reserved. At the same time, it left Bosnia and Herzegovina and Macedonia vulnerable, with the only alternative to opt for independence. As foreseen, war erupted in Bosnia immediately.

It is true that the EC did not fully follow the legal opinion of its own Arbitration commission as far as recognition of the various republics goes. That Commission advised that Croatia should first fulfill certain conditions concerning human and minority rights, amend its Constitution accordingly, and respect these in practice. Regarding Macedonia, on the other hand, the Commission recommended its recognition. The EC did the opposite, clearly guided by political and not legal considerations. But, recognition of states is always a political act with, of course, legal consequences. Recognition of Croatia came after the war there stopped and it did not lead to a widespread violation of the earlier signed cease-fire, as was feared by some and even threatened by the Serbian side. But that recognition did weaken the capacity of the Western countries to exert pressure on Croatia and force it to comply with its own laws and with international standards in its treatment of the Serb minority and in abiding more by the rule of democracy. This problem continues and its consequences are still felt in the late 1990s.

The main argument for recognizing Bosnia (as was also the case with Croatia) was to prevent Serbian aggression. That did not succeed and para-military groups together with the Bosnian Serb Army, which was already organized, prepared and supplied by Belgrade, immediately started open warfare, eventually taking over 70% of the territory and causing movement of more than half of the Bosnian population, with up to a quarter-million presumed killed. But if this was done in spite of the recognition, it would have happened without it even more. This way at least it enabled the international community to identify the aggressor and recognize the conflict as international. Without recognition Bosnia would have remained part of a rump Yugoslavia and the Serbian and JNA leadership would have been given a free hand in subduing it completely.

After difficult negotiations, particularly with Milosevic and the Bosnian Serbs, Lord Owen (who succeeded Lord Carrington as EU negotiator) and Vance came out with a plan for Bosnia in January 1993. To everybody's surprise, the new Clinton Administration turned the plan down and opened the door for continued fighting. In spite of that and its election rhetoric, the U.S. refused to get involved, while Europe was unable to unite nor act decisively on its own.

The U.N. mounted a major peacekeeping operation in Bosnia, but remained hostage to its own mandate, limited resources, different interpretations of rules of engagement and the stubborn opposition of the warring parties. There was no peace to keep, and no will – therefore, no mandate – to enforce it.

The war in Bosnia took on a new dimension in May 1993 when the Croatian side, openly supported by Croatian regular military units, attacked their up-to-then allies, the Bosniaks, who for more than a year had to fight for their survival on two fronts. It brought into the open the real aims of Croatia's ruling top – to divide Bosnia with the Serbs and create 'Greater Croatia'. Many atrocities were committed by the Croats in this conflict, including concentration camps for Muslims and the devastation of the Eastern part of the city of Mostar and of the beautiful 400-year-old Turkish stone arch-bridge from which the city got its name. The fighting stopped under tremendous U.S. pressure and an agreement was signed in Washington establishing a Croat-Muslim Federation in Bosnia and its confederation with Croatia. The first has never functioned (still does not) and the second is not even mentioned any more. However, that agreement and the use of military force against the Serbs eventually led to the Dayton Peace Accord, which was to a great extent also the result of U.S. pressure and of internal, electoral needs of the Clinton Administration.

DAYTON

The major flaw of the Dayton Accords was that it promoted Milosevic – who is the biggest culprit for all these developments and has even been indicted as a war criminal – into a peace-maker on whose cooperation the deal was to depend. Secondly, it ignored the acute and brewing crisis in Kosovo, which obviously had to be solved or else would erupt sooner or later. The arms embargo was violated with the knowledge and acquiescence of those who should have imposed it. The same goes for economic sanctions against rump Yugoslavia, so that Macedonia suffered as much as those against whom sanctions were directed. The non-flying ban over Bosnia was only monitored and not enforced. A double control by the United Nations and NATO for the use of force was elaborated so each side had an excuse for not acting. Humanitarian assistance, however important, was too often an excuse not to use force. Impartiality was interpreted as neutrality in a situation where one side was clearly the main culprit and even officially declared as such by the Security Council.

An International Tribunal for War Crimes in former Yugoslavia was established but was for a long time denied financial and political support. Bringing to trial those already indicted, but also those who masterminded and incited war and war crimes (war itself being the ultimate of all crimes) is a necessary precondition for reconciliation and peace. The postwar leaders of Serbia, Croatia, and all sides in Bosnia must be forced to deliver those allegedly responsible from their own ranks and not be allowed off the hook until they fully co-operate with the Tribunal. NATO should also enforce its mandate and arrest people who daily pass their check-points and/or whose whereabouts are well known, as well as the warrants for their arrest.

One of the less noted, but very effective measures against Milosevic's rump Yugoslavia was its suspension from the work of the U.N. General Assembly and all its subsidiary bodies, and the refusal by the Security Council to recognize it as the successor of former Yugoslavia. Economic sanctions, and particularly the refusal of access by Yugoslavia to international financial institutions, were an important instrument in forcing it to compromise on a number of issues. Economic and political isolation proved to be important tools in changing the minds of the leadership in Belgrade in spite of their anti-Western rhetoric. Pro-democracy demonstrations followed in Belgrade and were a visible manifestation of the effectiveness of these measures.

However, the decisive measure which brought about the Dayton Agreements was the resolute use of force. The reluctance to exercise it earlier was the main reason which enabled these protracted wars to take place, with all their consequences.

After the Dayton Accords were adopted, the situation remains tense and far from stabilized. Bosnia remained, and is, if anything, even more divided today than ever. The danger of new conflicts and population movements continue. Refugees and displaced people have yet to return to their homes. Unfortunately, instead of pushing for full and consistent implementation of the Dayton and other agreements, the world has been getting tired of the problem. The Muslim-Croat Federation, one of the pillars of the Dayton Agreement and of Bosnian statehood, has not been working and it will not do so until Zagreb is forced to impose it on its proxies in Herzeg Bosna. Yet this might prove to be the fuse of a new explosion. At the end of the decade, Eastern Slavonia has still to be resolved peacefully. Democracy has yet to take root and develop in Serbia, Croatia, and Bosnia. The rights of minorities, especially of the two largest and most persecuted groups – the Serbs in Croatia and Albanians in Serbia – have not been recognized or respected, as became apparent in 1999 with the catastrophic war over Kosovo.

CONCLUSION

What are the main lessons to be learnt from the Yugoslav crisis and the international response? One, that nationalism is latent, dangerous, and explosive. It can easily be inflamed and should therefore be confronted immediately and decisively everywhere. Second, and equally important, that international intervention in local conflicts cannot be effective without credible threat of force and resolve to use it. The United Nations cannot be blamed for that, because these decisions rest with the governments who for their part often vote for toothless and sometimes even contradictory resolutions at the United Nations. Finally, that today and for the immediate future at least, no international decisive action which includes the use of force is possible without the United States.

The overall conclusion is that the lack of decisiveness, or the rash, pragmatic and selfish political behavior of international actors often has long-term, costly, and dangerous consequences in many countries and regions. Having said that, there is no doubt that the Yugoslav crisis was of internal making and that the responsibility rests with the local oligarchies who provoked it for selfish reasons and struggle for power. The victims of this man-made catastrophe are some twenty million people.

With the possible exception of Slovenia, the country is very much in ruins, the economy in very bad shape, the standard of life in a free fall. A small number of people, those close to the ruling echelons (as well as the rulers themselves) have enriched themselves through smuggling, graft, corruption and privatization schemes, while the vast majority of the population is clearly worse off than before.

PART II

WARS IN SLOVENIA, CROATIA, AND BOSNIA

8. COMPARING THE IMPORTANCE OF INTERNATIONAL AND DOMESTIC FACTORS

Margarita Papandreou

At what point in time do we start in analyzing the factors that contributed to the outbreak of war in Yugoslavia? My starting point – perhaps arbitrary – is the Yalta Agreement; then to the Cold War; to the subtle and sometimes open competition between the two great powers in attempts to seduce Yugoslavia over to 'their side'; to the death of Tito; to the ending of the Cold War which was the glue that helped keep the country intact; to the more intensive efforts on the part of the West to create the conditions for a capitalist transformation in that country; and finally, to the drop that caused the cup to spill over – the secession and quick recognition by Germany of Slovenia and Croatia. These are broadly the main events and developments. This traces a long path in an often manipulated direction.

One goal was the restoration of a neo-colonial market system in the Balkans hegemony, stability, control of energy routes. An overreaching strategy has to do with what is called 'pipeline politics', the geopolitical correlation of various strategic factors that affect pipeline routing and the transportation of oil or natural gas from the oil fields to the consuming markets.

The preconditions for the crisis were there. Unemployment, inflation, government austerity programs. The federal government in the 1980s was pushing for radical economic reform and more centralized power under pressure from the IMF, while confronting the assertions of the republican governments to their rights of sovereignty. On the political front this took the form of constitutional proposals for change and bargaining among the republics.

Research on Russia and Eastern Europe, Volume 3, pages 161–164.
2000 by Elsevier Science Inc.
ISBN: 0-7623-0280-1

The Serbs and the Croats fought for the first time in their history in 1941. Then, as now, it was not the outcome of blood hatreds, but the result of events and political decisions. Following the war, the Yalta Agreement proclaimed 50% of Yugoslavia as belonging to the Eastern camp and 50% to the Western, making for a struggle from both camps to win over Yugoslavia. Tito joined the non-aligned movement and began setting up a socialist society. A number of policy makers supported the notion of an independent neutral socialist country as a 'buffer' between East and West, and a model to Eastern European countries controlled by the Soviet Union.

A 1982 declassified U.S. document entitled, 'United States Policy Toward Eastern Europe' and one in March 1984, entitled 'U.S. Policy Toward Yugoslavia' reveal the primary long-term goal "to facilitate the eventual reintegration of Yugoslavia into the European community of nations." The objective for Yugoslavia held for Bulgaria, Czechoslovakia, Hungary, Poland, the German Democratic Republic, and Romania: a capitalist transformation.

To carry out this policy, the West promoted de-industrialization and dependence. Threats and pressure were used, covert arms shipments were made by nations wanting to empower ethnic entities within the Yugoslav republics, loans were given as a method of control. In return, U.S. planners, trying to advance the capitalist transformation, continued demanding reforms by the governments. Before long the country's currency was weakened, serious balance of payments problems developed, and inflation set in. As expected, the IMF appeared on the scene, forcing an economic policy of austerity. The conditions were set for the aggravation of ethnic tensions.

The path to secession, deliberately or by momentum, had begun. The Bush government wanted reforms to move faster. Then came a significant move in the breakup of Yugoslavia – a U.S. law. On November 5, 1990, the U.S. Congress passed the 1991 Operations Appropriations Law 101-513. A section of this law, suddenly and without previous warning, cut off all aid, trade, credits, and loan from the United States to Yugoslavia within six months. It made continuation of aid contingent on the approval of election procedures and results, which the State Department demanded be held in six republics.

The law caused a crisis in the Yugoslav federal government. It could not pay the huge foreign debt interest or arrange for the purchase of raw materials for industry. U.S. personnel in international financial institutions were required to enforce this cut-off policy. The provision that only 'democratic forces' would be given funding meant that right wing nationalist parties would be favored over socialist parties.

Internal events in Croatia added flame to the warming climate toward war. Franjo Tudjman was elected president in April 1990, with the strong support of committed Croatian nationalists. After his victory he dismissed thousands of Serbian workers, suppressed opposition new media, and spread harsh nationalist rhetoric in the government television and radio stations.

The Yugoslav question was the first serious test of a united European foreign policy. There turned out to be almost as many policy solutions as there were countries in the European Community. And those differed from the policy proposals of the United States. The European Union failed the test, proving to be unable to set up a post-Cold War strategy for the East or to be an equal partner with the United States. One could almost argue that the intervention of the European Union turned a manageable conflict into fratricide. In addition to that, the United Nations seemed impotent to deal with the situation.

On the government side, between April 1992 and April 1994 Berlin transferred $320 million in arms to Zagreb, while Moscow sent $390 million of arms to Belgrade.

In the policy of intervention in Yugoslavia, the focus was on Serbia, which, as the heart of the federal government, had the fourth largest army in Europe, maintained an armaments industry, and possessed modern aircraft and weapons. It was also the most reluctant to accede to Western demands. The West needed groups that were anti-communist. As it happened, most of these groups were also ultranationalist. In Bosnia Herzegovina, the United States and Germany selected anti-Yugoslav pan-Islamists loyal to Alija Izetbegovic as their collaborator. This required at some stage, because of the Bosnian Serb claim to an independent republic within Bosnia to be joined with Serbia, Western help to Izetbegovic's government for retraining and rearming its forces.

By now the United States, Germany, the European Community, the United Nations, and NATO were deeply mired in the whole Yugoslav war, and the Clinton administration, with an election coming up, decided to put an end to the hostilities and force a solution on a weary population. The carrot and stick strategy was used – force and negotiations. The Dayton Agreement terminated, for the time being, armed conflict in former Yugoslavia.

The most common explanation of internal forces at work in the creation of the war in Yugoslavia was 'ethnic conflict'. It is not a real explanation, at least not a sufficient one. Ethnic differences can be used to divide and rule, by external elements – and they were. They can also be used by leaders fighting for economic resources, territory, or the enhancement of their power, and they definitely were.

Yugoslavia had become in the 1980s a buoyant economy as compared to the other Eastern and Central European nations. The majority was not unhappy with its one-party system, social ownership, and other attributes of socialist self-management. There was an intense xenophobia accompanied by a conviction of the strength of their system and way of life.

This may account for the Serbs' woeful record of publicity and public relations in promoting themselves in the global community through use of the mass media. One could say that the propaganda war was not lost by the Serbs – it was never even waged.

Suffice it to say that the media loomed large in the history of the conflict in Yugoslavia. Both externally and internally. On the home front an intense propaganda campaign was waged to change the attitudes of people who essentially all their lives had lived peacefully with their compatriots. Public opinion in the key republics of Croatia, Serbia, and Bosnia was a crucial factor. In Serbia, as compared to the attempt to build public opinion externally by the Croats, Milosovic's nationalist propaganda was turned almost exclusively inward or toward the Serbs of the diaspora. At both levels it was ultimately effective. In the diaspora there was an already existing nationalism.

Western intervention had been going on for some time in attempts to steer Yugoslavia toward a market economy and to integrate it more completely in the European geopolitical space. When there was accelerated movement toward war in the 1989-90 period of time, and attempts were made by the West to manage the crisis, it provided the turning point toward fanatic nationalism and war.

The entire war was waged according to nineteenth century rationale – possession of territory and military control of space. By accepting the principle of self-determination for the independence of states, the West made the fight for the acquisition of territory the natural outcome. Since there were not many differences in economic resources among the republics, the fight for territory became based on ethnic habitation and nationalist revolution. Because ethnic groups were dispersed throughout ex-Yugoslavia, the battling became disastrous.

There were alternative proposals, quite a few developed by non governmental organizations. The ideas were not utopian, but workable – certainly more so than the Dayton Agreement division. The Yugoslavs themselves were never given the chance to decide about their own destiny. When are we going to learn, as Vojin Dimitrijevic said to me, "that 'the well-being of people mostly depends on economic development, good organization, and the ability of democratic institutions and decision-making to fulfill human needs?" And when is the international community going to make the human rights of the people the guiding principle in any intervention?

9. DISMANTLING FORMER YUGOSLAVIA, RE-COLONIZING BOSNIA

Michel Chossudovsky

The press and politicians alike portray Western intervention in the former Yugoslavia as a noble, if agonizingly belated, response to an outbreak of ethnic massacres and human rights violations. After the November 1995 Dayton Peace Accords, the West was eager to touch up its self-portrait as savior of the Southern Slavs and get on with 'the work of rebuilding' the newly sovereign states.

But following a pattern set since the onslaught of the civil war, Western public opinion has been misled. The conventional wisdom, exemplified by the writings of former U.S. Ambassador to Yugoslavia Warren Zimmerman, is that the plight of the Balkans is the outcome of an 'aggressive nationalism', the inevitable result of deep-seated ethnic and religious tensions rooted in history.[1] Likewise, much has been made of the 'Balkans power-play' and the clash of political personalities.[2]

Drowned in the self-serving analyses are the economic and social causes of the conflict. The deep-seated economic crisis which preceded the civil war has long been forgotten. The strategic interests of Germany and the United States in laying the groundwork for the disintegration of Yugoslavia go unmentioned, as does the role of external creditors and international financial institutions. In the eyes of the global media, Western powers bear no responsibility for the impoverishment and destruction of a nation of 24 million people.

But through their domination of the global financial system, the Western powers, pursuing their collective and individual 'strategic interests' helped from

Research on Russia and Eastern Europe, Volume 3, pages 165–177.
2000 by Elsevier Science Inc.
ISBN: 0-7623-0280-1

the beginning of the 1980s bring the Yugoslav economy to its knees, contributing to stirring simmering ethnic and social conflicts. Now, the efforts of the international financial community are channeled toward "helping Yugoslavia's war-ravaged successor states." As the world focuses on troop movements and cease-fires, the international financial institutions are collecting former Yugoslavia's external debt from its remnant states, while transforming the Balkans into a safe haven for free enterprise.

ECONOMIC REFORMS FROM THE 1980s

Adopted in several stages since the early 1980s, the reforms imposed by Belgrade's creditors wreaked economic and political havoc leading to disintegration of the industrial sector and the piece-meal dismantling of the Yugoslav Welfare State. Despite Belgrade's political non-alignment and extensive trading relations with the U.S. and the European Community, the Reagan administration had targeted the Yugoslav economy in a 'Secret Sensitive' 1984 National Security Decision Directive (NSDD 133) entitled 'United States Policy towards Yugoslavia'. A censored version of this document declassified in 1990 largely conformed to a previous National Security Decision Directive (NSDD 54) on Eastern Europe issued in 1982. Its objectives included "expanded efforts to promote a 'quiet revolution' to overthrow Communist governments and parties" while re-integrating the countries of Eastern Europe into the orbit of the World market.[3]

Secessionist tendencies feeding on social and ethnic divisions, gained impetus precisely during a period of brutal impoverishment of the Yugoslav population. The first phase of macro-economic reform initiated in 1980 shortly before the death of Marshall Tito "wreaked economic and political havoc.... Slower growth, the accumulation of foreign debt and especially the cost of servicing it as well as devaluation led to a fall in the standard of living of the average Yugoslav.... The economic crisis threatened political stability ... it also threatened to aggravate simmering ethnic tensions."[4] These reforms accompanied by the signing of debt restructuring agreements with the official and commercial creditors also served to weaken the institutions of the federal state, creating political divisions between Belgrade and the governments of the republics and autonomous provinces. "The Prime Minister, Milka Planinc, who was supposed to carry out the program, had to promise the IMF an immediate increase of the discount rates and much more for the Reaganomics arsenal of measures ..."[5]

Following the initial phase of macro-economic reform in 1980, industrial growth plummeted to 2.8% in the 1980–87 period, plunging to zero in 1987–88 and to −10.6% in 1990.[6] The economic reforms reached their climax under the pro-U.S. government of Prime Minister Ante Markovic. In the Autumn of 1989 just prior to the collapse of the Berlin Wall, the federal Premier had traveled to Washington to meet President George Bush. A 'financial aid package' had been promised in exchange for sweeping economic reforms including a new devalued currency, the freeze of wages, a drastic curtailment of government expenditure and the abrogation of the socially owned enterprises under self-management.[7] The 'economic therapy' (launched in January 1990) contributed to crippling the federal state system. State revenues which should have gone as transfer payments to the republics and autonomous provinces were instead funneled towards servicing Belgrade's debt with the Paris and London clubs. The republics were largely left to their own devices, thereby exacerbating the process of political fracturing. In one fell swoop, the reformers had engineered the demise of the federal fiscal structure and mortally wounded its federal political institutions. The IMF-induced budgetary crisis created an economic 'fait accompli' which in part paved the way for Croatia's and Slovenia's formal secession in June 1991.

THE IMF AGREEMENT AND THE ENTERPRISE REFORMS

The economic package was launched in January 1990 under an IMF Stand-by Arrangement (SBA) and a World Bank Structural Adjustment Loan (SAL II). The budget cuts requiring the redirection of federal revenues towards debt servicing, were conducive to the suspension of transfer payments by Belgrade to the governments of the Republics and Autonomous Provinces, thereby fueling the process of political Balkanization and secessionism. The government of Serbia rejected Markovic's austerity program outright, leading to a walk-out protest of some 650,000 Serbian workers directed against the Federal government.[8] The Trade Union movement was united in this struggle: "worker resistance crossed ethnic lines, as Serbs, Croats, Bosnians and Slovenians mobilized ... shoulder to shoulder with their fellow workers."[9]

The 1989 enterprise reforms adopted under Premier Ante Markovic played a central role in steering the industrial sector into bankruptcy. By 1990, the annual rate of growth of GDP had collapsed to –7.5%.[10] In 1991, GDP declined by a further 15% and industrial output collapsed by 21%.[11] The restructuring program demanded by Belgrade's creditors was intended to abrogate the system of socially-owned enterprises. The Enterprise Law of 1989 required abolishing the 'Basic Organizations of Associated Labor (BAOL)'.[12] The latter were socially-owned productive units under self-management with the Workers' Council constituting the main decision making body. The 1989 Enterprise Law

required the transformation of the BOALs into private capitalist enterprises with
the Worker's Council replaced by a so-called 'Social Board' under the control
of the enterprise's owners including its creditors.[13] "The objective was to subject
the Yugoslav economy to massive privatization and the dismantling of the public
sector. Who was to carry it out? The Communist Party bureaucracy, most notably
its military and intelligence sector, was canvassed specifically and offered political and economic backing on the condition that wholesale scuttling of social
protections for Yugoslavia's work force was imposed . . ."[14]

Overhauling The Legal Framework

Several supporting pieces of legislation were put in place in a hurry with the
assistance of Western lawyers and consultants. A new Banking Law was enacted
with a view to triggering the liquidation of the socially owned 'Associated
Banks'. More than half the country's banks were dismantled, the emphasis was
on the formation of 'independent profit oriented institutions'.[15] By 1990, the
entire 'three-tier banking system' consisting of the National Bank of Yugoslavia,
the national banks of the eight Republics and autonomous provinces and the
commercial banks had been dismantled under the guidance of the World Bank.[16]
A World Bank Financial Sector Adjustment Loan was being negotiated in 1990.
It was to be adopted by the Belgrade government in 1991.

Industrial enterprises had been carefully categorized. Under the IMF-World
Bank sponsored reforms, credit to the industrial sector had been frozen with a
view to speeding up the bankruptcy process. So-called 'exit mechanisms' had
been established under the provisions of the 1989 Financial Operations Act.[17]
The latter stipulated that if an enterprise were to remain insolvent for 30 days
running, or for 30 days within a 45-day period, it must hold a meeting within
the next 15 days with its creditors in view of arriving at a settlement. This
mechanism allowed creditors (including national and foreign banks) to routinely
convert their loans into a controlling equity in the insolvent enterprise. Under
the Act, the government was not authorized to intervene. In case a settlement
was not reached, bankruptcy procedures would be initiated in which case workers
would not normally receive severance payments.[18]

In 1989, according to official sources, 248 firms were steered into bankruptcy
or were liquidated and 89,400 workers had been laid off.[19] During the first nine
months of 1990 directly following the adoption of the IMF program, another
889 enterprises with a combined work-force of 525,000 workers were subjected
to bankruptcy procedures.[20] In other words, in less than two years 'the trigger
mechanism' (under the Financial Operations Act) had led to the lay-off of more
than 600,000 workers (out of a total industrial work force of the order of 2.7

million). The largest concentrations of bankrupt firms and lay-offs were in Serbia, Bosnia-Herzegovina, Macedonia and Kosovo.[21]

Many socially owned enterprises attempted to avoid bankruptcy through the non-payment of wages. Half a million workers representing some 20% of the industrial labor force were not paid during the early months of 1990, in order to meet the demands of creditors under the 'settlement' procedures stipulated in the Law on Financial Organizations. Real earnings were in a free fall, social programs had collapsed, with the bankruptcies of industrial enterprises, unemployment had become rampant, creating within the population an atmosphere of social despair and hopelessness. "When Mr. Markovic finally started his 'programmed privatization', the republican oligarchies, who all had visions of a 'national renaissance' of their own, instead of choosing between a genuine Yugoslav market and hyper-inflation, opted for war, which would disguise the real causes of the economic catastrophe."[22]

The January 1990 IMF-sponsored package contributed unequivocally to increasing enterprise losses while precipitating many of the large electric, petroleum refinery, machinery, engineering, and chemical enterprises into bankruptcy. Moreover, with the deregulation of the trade regime in January 1990, a flood of imported commodities contributed to further destabilizing domestic production. These imports were financed with borrowed money granted under the IMF package (i.e. the various 'quick disbursing loans' granted by the IMF, the World Bank and bilateral donors in support of the economic reforms). While the import bonanza was fueling the build-up of Yugoslavia's external debt, the abrupt hikes in interest rates and input prices imposed on national enterprises had expedited the displacement and exclusion of domestic producers from their own national market.

The situation prevailing in the months preceding the secession of Croatia and Slovenia (June 1991) (confirmed by the 1989–90 bankruptcy figures) points to the sheer magnitude and brutality of the process of industrial dismantling. The figures, however, provide but a partial picture, depicting the situation at the outset of the 'bankruptcy program'. The latter has continued unabated throughout the period of the civil war and its aftermath. Similar industrial restructuring programs were imposed by external creditors on Yugoslavia's successor states.

The World Bank had estimated that there were still in September 1990, 2,435 'loss-making' enterprises out of a remaining total of 7,531.[23] In other words, these 2,435 firms with a combined work-force of more than 1.3 million workers had been categorized as 'insolvent' under the provisions of the Financial Operations Act, requiring the immediate implementation of bankruptcy procedures. Bearing in mind that 600,000 workers had already been laid off by bankrupt firms prior to September 1990, these figures suggest that some 1.9 million workers (out of a total of 2.7 million) had been classified as 'redundant'. The 'insolvent' firms concentrated in the energy, heavy industry, metal processing, forestry and textiles sectors were among the largest industrial enterprises in the country representing (in September 1990) 49.7% of the total (remaining and employed) industrial work-force.[24]

POLITICAL DISINTEGRATION AND POSTWAR RECONSTRUCTION

Supporting broad strategic interests, the austerity measures had laid the basis for 'the re-colonization' of the Balkans. In the multi-party elections in 1990, economic policy was at the centre of the political debate, the separatist coalitions ousted the Communists in Croatia, Bosnia-Herzegovina and Slovenia.

Following the decisive victory in Croatia of the rightist Democratic Union in May 1990 under the leadership of Franjo Tudjman, the separation of Croatia received the formal assent of the German Foreign Minister Mr. Hans Dietrich Genscher who was in almost daily contact with his Croatian counterpart in Zagreb.[25] Germany not only favored secession, it was also "forcing the pace of international diplomacy" and pressuring its Western allies to grant recognition to Slovenia and Croatia. The borders of Yugoslavia are reminiscent of World War II when Croatia (including the territories of Bosnia-Herzegovina) was an Axis satellite under the fascist Ustasha regime: "German expansion has been accompanied by a rising tide of nationalism and xenophobia. . . . Germany has been seeking a free hand among its allies to pursue economic dominance in the whole of Mitteleuropa."[26] Washington on the other hand, favored "a loose unity while encouraging democratic development . . . [the US Secretary of State] Baker told [Croatia's President] Franjo Tudjman and [Slovenia's President] Milan Kucan that the United States would not encourage or support unilateral secession . . . but if they had to leave, he urged them to leave by a negotiated agreement."[27]

The economic reforms being imposed on the 'successor states' are a natural extension and continuation of those previously implemented in federal Yugoslavia. In the tragic aftermath of a brutal and destructive war, the prospects for rebuilding the newly independent republics appear bleak. Despite a virtual press blackout on the subject, debt rescheduling is an integral part of the peace process. The former Yugoslavia has been carved up under the close scrutiny of its external creditors, its foreign debt has been carefully divided and allocated to the republics. The privatization programs implemented under the supervision of the donors, have contributed to a further stage of economic dislocation and impoverishment of the population. GDP had declined by as much as 50% in four years (1990–93).[28]

Moreover, the leaders of the newly sovereign states have fully collaborated with the creditors: 'All the current leaders of the former Yugoslav republics were Communist Party functionaries and each in turn vied to meet the demands of the World Bank and the International Monetary Fund, the better to qualify for investment loans and substantial perks for the leadership.... State industry and machinery were looted by functionaries. Equipment showed up in 'private companies' run by family members of the nomenklatura.'[29]

Even as the fighting raged, Croatia, Slovenia and Macedonia had entered into separate loan negotiations with the Bretton Woods institutions. In Croatia, the government of President Franjo Tudjman signed in 1993, an agreement with the IMF. Massive budget cuts mandated under the agreement thwarted Croatia's efforts to mobilize its own productive resources, thus jeopardizing post-war reconstruction. The cost of rebuilding Croatia's war-torn economy was estimated at some $23 billion, requiring an influx of fresh foreign loans. In the absence of 'debt forgiveness,' Zagreb's debt burden will be fueled well into the 21st Century.

In return for foreign loans, the government of President Franjo Tudjman had agreed to reform measures conducive to further plant closures and bankruptcies, driving wages to abysmally low levels. The official unemployment rate increased from 15.5% in 1991 to 19.1% in 1994.[30]

Zagreb also instituted a far more stringent bankruptcy law, together with procedures for 'the dismemberment' of large state-owned public utility companies. According to its 'Letter of Intent' to the Bretton Woods institutions, the Croatian government had promised to restructure and fully privatize the banking sector with the assistance of the European Bank for Reconstruction and Development (EBRD) and the World Bank. The latter have also demanded a Croatian capital market structured to heighten the penetration of Western institutional investors and brokerage firms.

Under the agreement signed in 1993 with the IMF, the Zagreb government was not permitted to mobilize its own productive resources through fiscal and monetary policy. The latter were firmly under the control of its external creditors. The massive budget cuts demanded under the agreement had forestalled the possibility of post-war reconstruction. The latter could only be carried out through the granting of fresh foreign loans, a process which would fuel Croatia's external debt well into the 21st Century. The cost of rebuilding Croatia's war-torn economy was estimated at some 23 billion dollars.

Macedonia also followed a similar economic path. In December 1993, the Skopje government agreed to compress real wages and freeze credit in order to obtain a loan under the IMF's Systemic Transformation Facility (STF). In an unusual twist, multi-billionaire business tycoon George Soros participated in the International Support Group composed of the government of the Netherlands and the Basle-based Bank of International Settlements. The money provided by the Support Group, however, was not intended for 'reconstruction' but rather to enable Skopje to pay back debt arrears owed the World Bank.[31]

Moreover, in return for debt rescheduling, the government of Macedonian Prime Minister Branko Crvenkovski had to agree to the liquidation of remaining 'insolvent' enterprises and the lay off of 'redundant' workers – which included the employees of half the industrial enterprises in the country. As Deputy Finance Minister Hari Kostov soberly noted, with interest rates at astronomical levels because of donor-sponsored banking reforms, "it was literally impossible to find a company in the country which would be able to . . . to cover [its] costs . . ."[32]

Overall, the IMF economic therapy for Macedonia constitutes a continuation of the 'bankruptcy program' launched in 1989 under federal Yugoslavia. The most profitable assets are now on sale on the Macedonian stock market, but this auction of socially owned enterprises has led to industrial collapse and rampant unemployment.

Yet despite the decimation of the economy and the disintegration of schools and health centers under the austerity measures, Finance Minister Ljube Trpevski proudly informed the press that "the World Bank and the IMF place Macedonia among the most successful countries in regard to current transition reforms." The head of the IMF mission to Macedonia, Mr. Paul Thomsen, concurs that "the results of the stabilization program [under the STF] were impressive," giving particular credit and appreciation to 'the efficient wages policy' adopted by the Skopje government.[33]

REBUILDING BOSNIA AND HERZEGOVINA

With a Bosnian peace settlement holding under NATO guns, the West unveiled a 'reconstruction' program which fully stripped Bosnia-Herzegovina of its economic and political sovereignty. This program largely consists in developing Bosnia-Herzegovina as a divided territory under NATO military occupation and Western administration.

Resting on the November 1995 Dayton Accords, the U.S. and the European Union installed a full-fledged colonial administration in Bosnia. At its head is their appointed High Representative (HR). Initially this was Mr. Carl Bildt, a former Swedish Prime Minister and European Representative in the Bosnian Peace negotiations, who was succeeded by Carlos Westendorp of Spain. The HR has full executive powers in all civilian matters, with the right to overrule the governments of both the Bosnian Federation and the Bosnian-Serb Republika Srpska. The HR is to act in close liaison with the IFOR (later SFOR) Military

High Command as well with donors agencies. To make the point crystal clear, the Dayton Accord spell out that "The High Representative is the final authority in theater regarding interpretation of the agreements" (Dayton Accord, 'Agreement on High Representative', articles I and II).

An international civilian police force is under the custody of an expatriate Commissioner appointed by United Nations Secretary – General Boutros Boutros Ghali. Initially the Irish police official Peter Fitzgerald, with UN policing experience in Namibia, El Salvador, and Cambodia, presided over some 1,700 police from 15 countries, most of whom had never set foot in the Balkans. The police were dispatched to Bosnia after a five-day training program in Zagreb.

While the West has underscored its support to democracy, the Parliamentary Assembly set up under the 'Constitution' finalized under the Dayton Accords, largely acts as a 'rubber stamp'. Behind the democratic facade, actual political power rests in the hands of a 'parallel government' headed by the High Representative and staffed by expatriate advisors.

Moreover, the Constitution adopted in Dayton handed over the reins of economic policy to the Bretton Woods institutions and the London-based European Bank for Reconstruction and Development (EBRD). Article VII stipulates that the first Governor of the Central Bank of Bosnia and Herzegovina is to be appointed by the IMF and "shall not be a citizen of Bosnia and Herzegovina or a neighboring State . . ."

Just as the Governor of the Central Bank is an IMF appointee, the Central Bank was not allowed under the Constitution to function as a Central Bank: "For the first six years . . . it may not extend credit by creating money, operating in this respect as a currency board" (Article VII). Neither was the new 'sovereign' successor State allowed to have its own currency (issuing paper money only when there is full foreign exchange backing), nor permitted to mobilize its internal resources. As in the other successor republics, its ability to self-finance its reconstruction (without massively increasing its external debt) is blunted from the outset.

The tasks of managing the Bosnian economy were carefully divided among donor agencies: while the Central Bank is under IMF custody, the European Bank for Reconstruction and Development (EBRD) heads the Commission on Public Corporations which supervises operations of all public sector enterprises including energy, water, postal services, roads, railways, etc. The President of the EBRD appoints the Chairman of the Commission which also oversees public sector restructuring, meaning primarily the sell-off of State and socially owned assets and the procurement of long term investment funds.

One cannot sidestep a fundamental question: is the Bosnian Constitution formally agreed between heads of state at Dayton really a constitution? A somber and dangerous precedent was set in the history of international relations: Western creditors have embedded their interests in a Constitution hastily written on their behalf, executive positions within the Bosnian State system are to be held by non-citizens who are appointees of Western financial institutions. No constitutional assembly, no consultations with citizens' organizations in Bosnia and Herzegovina, no 'constitutional amendments'.

The Bosnian government estimated that reconstruction costs would reach $47 billion. Western donors pledged $3 billion in reconstruction loans, yet only a meager $518 million dollars were granted in December 1995, part of which was tagged (under the terms of the Dayton Peace Accords) to finance some of the local civilian costs of the Implementation Force's (IFOR) military deployment as well as repay debt arrears with international creditors.

In a familiar twist, 'fresh loans' were devised to pay back 'old debt'. The Central Bank of the Netherlands has generously provided 'bridge financing' of $37 million. The money, however, was earmarked to allow Bosnia to pay back its arrears with the IMF, a condition without which the IMF would not lend it fresh money.[34] But it is a cruel and absurd paradox: the sought-after loan from the IMF's 'Emergency Window' for so-called 'post-conflict countries' is not used for post-war reconstruction. Instead it was applied to reimburse the Central Bank of the Netherlands which had coughed up the money to settle IMF arrears in the first place. . . . While debt is building up, no new financial resources are flowing into Bosnia to rebuild its war-torn economy.

Western governments and corporations show greater interest in gaining access to potential strategic natural resources than committing resources for rebuilding Bosnia. Documents in the hands of Croatia and the Bosnian Serbs indicate that coal and oil deposits have been identified on the eastern slope of the Dinarides Thrust, a region retaken from rebel Bosnian Krajina Serbs by the Croatian army in the final offensives before the Dayton Peace accords. Bosnian officials report that Chicago-based Amoco was among several foreign firms that subsequently initiated exploratory surveys in Bosnia. The West is anxious to develop these regions: "The World Bank – and the multinationals that conducted operations – are [August 1995] reluctant to divulge their latest exploration reports to the combatant governments while the war continues."[35] Moreover, there are also "substantial petroleum fields in the Serb-held part of Croatia just across the Sava River from the Tuzla region."[36] The latter under the Dayton Agreement, is part of the U.S. Military Division with headquarters in Tuzla.

The territorial partition of Bosnia between the Federation of Bosnia-Herzegovina and the Bosnian-Serb Republika Srpska under the Dayton Accords thus takes on strategic importance. The 60,000 NATO troops on hand to 'enforce the peace' are administering the territorial partition of Bosnia-Herzegovina in accordance with Western economic interests.

National sovereignty is derogated, the future of Bosnia decided upon in Washington, Bonn and Brussels rather than in Sarajevo. The process of 'reconstruction' based on debt rescheduling is more likely to plunge Bosnia-Herzegovina (as well as the other remnant republics of former Yugoslavia) into the status of a Third World country.

While local leaders and Western interests share the spoils of the former Yugoslav economy, the fragmentation of the national territory and the entrenching of socio-ethnic divisions in the structure of partition serve as a bulwark blocking a united resistance of Yugoslavs of all ethnic origins against the re-colonization of their homeland.

CONCLUDING REMARKS

Macro-economic restructuring applied in Yugoslavia under the neoliberal policy agenda has unequivocally contributed to the destruction of an entire country. Yet since the onset of war in 1991, the central role of macro-economic reform has been carefully overlooked and denied by the global media. The free market has been presented as the solution, the basis for rebuilding a war-shattered economy. A detailed diary of the war and of the 'peace-making' process has been presented by the mainstream press. The social and political impact of economic restructuring in Yugoslavia has been carefully erased from our social consciousness and collective understanding of what actually happened. Cultural, ethnic and religious divisions are highlighted, presented dogmatically as the sole cause of the crisis when in reality they are the consequence of a much deeper process of economic and political fracturing.

This 'false consciousness' has invaded all spheres of critical debate and discussion. It not only masks the truth, it also prevents us from acknowledging precise historical occurrences. Ultimately it distorts the true sources of social conflict. The unity, solidarity and identity of the Southern Slavs have their foundation in history, yet this identity has been thwarted, manipulated and destroyed.

The ruin of an economic system, including the take-over of productive assets, the extension of markets and 'the scramble for territory' in the Balkans constitute the real cause of conflict. What is at stake in Yugoslavia are the lives of millions of people. Macro-economic reform destroys their livelihood, derogates their right to work, their food and shelter, their culture and national identity. Borders are redefined, the entire legal system is overhauled, the socially owned enterprises are steered into bankruptcy, the financial and banking system is dismantled, social programs and institutions are torn down. In retrospect, it is worth recalling Yugoslavia's economic and social achievements in the post-war period (prior to 1980): the growth of GDP was on average 6.1 per annum over a twenty year period (1960–1980), there was free medical care with one doctor per 550 population, the literacy rate was of the order of 91%, life expectancy was 72 years.[37] But after a decade of Western economic ministrations and five years of disintegration, war, boycott, and embargo, the economies of the former Yugoslavia are prostrate, their industrial sectors dismantled.

Yugoslavia is a 'mirror' of similar economic restructuring programs applied not only in the developing World but also in recent years in the United States, Canada and Western Europe. 'Strong economic medicine' is the answer, throughout the world, people are led to believe that there is no other solution: enterprises must be closed down, workers must be laid off and social programs must be slashed. It is in the foregoing context that the economic crisis in Yugoslavia should be understood. Pushed to the extreme, the reforms in Yugoslavia are the cruel reflection of a destructive 'economic model' imposed under the neoliberal agenda on national societies throughout the World.

NOTES

1. See the account of Warren Zimmermann (former US Ambassador to Yugoslavia), 'The Last Ambassador, A Memoir of the Collapse of Yugoslavia', *Foreign Affairs, 74*(2), 1995.

2. Milos Vasic et al., 'War Against Bosnia', *Vreme News Digest Agency, 29*(13), April 1992.

3. Sean Gervasi, 'Germany, US and the Yugoslav Crisis', *Covert Action Quarterly*, No. 43, Winter 1992–93.

4. Ibid

5. Dimitrije Boarov, 'A Brief Review of Anti-inflation Programs, the Curse of Dead Programs', *Vreme New Digest Agency, 29*(13) April 1992.

6. World Bank, 'Industrial Restructuring Study, Overview, Issues and Strategy for Restructuring', Washington DC, June 1991, p. 10 and 14.

7. Sean Gervasi, op cit.

8. Ibid.

9. Ralph Schoenman, 'Divide and Rule Schemes in The Balkans', *The Organiser*, 11 September 1995.

10. World Bank, op cit., p. 10. The term GDP is used for simplicity, yet the concept used in Yugoslavia and Eastern Europe to measure national product is not equivalent to the GDP concept under the (Western) system of national accounts.

11. See Judit Kiss, Debt Management in Eastern Europe, *Eastern European Economics*, May-June 1994, p. 59.

12. World Bank, op cit.

13. Ibid, p. viii.

14. Ralph Schoenman, 'Divide and Rule Schemes in The Balkans', *The Organiser*, 11 September 1995.

15. For further details see World Bank, *Yugoslavia, Industrial Restructuring*, p. 38.

16.

17. Ibid., p. 33.

18. Ibid., p. 33

19. Ibid, p. 34. Data of the Federal Secretariat for Industry and Energy, Of the total number of firms, 222 went bankrupt and 26 were liquidated.

20. Ibid., p. 33. These figures include bankruptcy and liquidation.

21. Ibid, p. 34.

22. Dimitrije Boarov, *op. cit.*

23. World Bank, *Industrial Restructuring*, p. 13. Annex 1, p. 1.

24. 'Surplus labour' in industry had been assessed by the World Bank mission to be of the order of 20% of the total labour force of 8.9 million – ie. approximately 1.8 million. This figure seems, however, to grossly underestimate the actual number of redundant workers based on the categorisation of 'insolvent' enterprises. Solely in the industrial sector, there were 1.9 million workers (September 1990) out of 2.7 million employed in enterprises classified as insolvent. See World Bank, *Yugoslavia, Industrial Restructuring*, Annex 1.

25. Sean Gervasi, *op. cit.*, p. 65

26. Ibid., p. 45

27. Zimmermann, *op. cit.*

28. Figure for *Macedonia, Enterprise, Banking and Social Safety Net*, World Bank Public Information Center, 28 November 1994.

29. Ralph Schoenman, 'Divide and Rule Schemes in The Balkans', *The Organiser*, 11 September 1995.

30. 'Zagreb's About Turn,' *The Banker,* January 1995, p. 38.

31. See World Bank, *Macedonia Financial and Enterprise Sector*, Public Information Department, November 28, 1995.

32. Statement of Macedonia's Deputy Minister of Finance Mr. Hari Kostov, reported in MAK News, April 18, 1995.

33. Macedonian Information and Liaison Service, MILS News, 11 April 1995.

34. Frank Viviano and Kenneth Howe, Bosnia Leaders Say Nation Sit Atop Oil Fields, *The San Francisco Chronicle*, 28 August 1995. See also Scott Cooper, 'Western Aims in Ex-Yugoslavia Unmasked', *The Organizer*, 24 September 1995.

35. Viviano and Howe, op cit.

36. World Bank, *World Development Report* 1991, Statistical Annex, tables 1 and 2, Washington DC, 1991.

37. World Bank, *World Development Report* 1991, Statistical Annex, tables 1 and 2, Washington DC, 1991.

10. SAFE AREAS AND THE DILEMMA OF CIVILIAN PROTECTION: LESSONS FROM YUGOSLAVIA

Major David Last

This paper is an effort to summarize some of the lessons learned from Yugoslavia about the protection of civilians in civil wars.[1] It concludes that there are at least three models for civilian protection, and that these are probably appropriate under different circumstances. First, there are military means of protecting civilians, but the requirements for successful military protection are stringent, and we have frequently failed. Second, there are alternatives to purely military protection which rely on a combination of limited military and civilian resources; as the reliance on non-military elements increases, so does the importance of host-nation co-operation and the rule of law. Finally there are purely civilian-based strategies that have good prospects when threats are diffuse, influence over the parties is strong, and laws are observed. Yugoslavia provides several examples of conditions under which various tactics do not work. The most important factors in determining how civilians can be protected are the nature of the threats against them, and the resources available to devote to their protection. The protection of international civilian staff in a UN mission is a separate problem that will not be addressed in this paper.

Definitions. According to a handbook on the law of armed conflict, "civilian persons are those who do not belong to the armed forces."[2] However this cannot simply mean those out of uniform. Civil wars bring whole communities into conflict. Men of military age from 16 to 60 may legitimately be considered as military manpower, unless they can be proven otherwise. Even women and children may be combatants. An important aspect of protecting civilians is to

Research on Russia and Eastern Europe, Volume 3, pages 179–196.
2000 by Elsevier Science Inc.
ISBN: 0-7623-0280-1

identify non-combatants clearly and unequivocally, and ensure that they do not take part in the conflict.[3] 'Protection' means the prevention of deliberately inflicted harm. It cannot mean the prevention of all suffering nor the prevention of fear. These are inevitable by-products of violence. '*Safe areas*' are areas in which civilians are free from reasonable fear of abuse or attack, and have basic needs provided (or the opportunity to provide for their own basic needs).

This study draws on military sources to describe some aspects of the efforts to protect civilians in Croatia (Western Slavonia and the Krajina), and Bosnia (Srebrenica, Zepa, and Gorazde). It uses the limited experiences of Human Rights Assessment Teams in the Krajina in August 1995, and a 'protection working group' in the Prijedor area in 1996 to assess the effectiveness of non-military strategies.

FACTORS IN THE PROTECTION OF CIVILIANS

Protecting civilians is not simply a matter of keeping them out of the way of fighting between formed bodies of troops. Both incidental and deliberate harm to civilians occurs as a result of deliberate displacement, starvation, disease, abuse and murder. The UN High Commissioner for Refugees, Mme Sadako Ogata, referred to the UNHCR's three-fold strategy of prevention, preparedness and solutions. This entails complementing provision for asylum abroad with prevention and solutions within the country of origin.[4]

The most extreme form of deliberate harm to civilians is genocide. Genocide might be seen as the final stage of a three-part deterioration in social relationships.[5] In the first stage, there are some violations of individual rights, some discrimination and probably some segregation. Amongst some individuals, intimidation becomes routine. Prejudice is consciously or unconsciously fostered in individual attitudes, possibly through the mass media. The second stage is characterized by sporadic, often cyclical outbursts of unplanned violence. Shop-smashing, looting, arson and riots are typical incidents. The third stage is entered with the active participation of the state, the leadership of individual politicians, the use of police, armed forces and possibly the misuse of social welfare records to identify groups. Although key leaders can be identified, this stage implies substantial public acquiescence or consent.

Nature of the Conflict. If the warring factions are two regular armies, their leaders can be held accountable for the actions of soldiers and units. The units can be expected to behave according to the laws of armed conflict, and intervening forces can deploy in order to counter conventional military tactics. The less central control and the more irregular the forces, the greater the risk of rogue individuals inflicting harm on civilians, individual cases, at night, in remote locations, and random acts of violence. Disciplined armies under firm control may also be used to attack civilians directly, as the Croats did in Western Slavonia in May 1995 and in the Krajina in August 1995, and as the Serbs did in Srebrenica and Zepa in July 1995. In general, attacks on civilians will be more consistent in irregular warfare than warfare between formed armies. In the latter, the laws of war are likely to be either respected or systematically violated.

Threat to Civilians. Irregular, sporadic violence, typical of the second stage of genocide described by Acheson is characterized by threats to civilians generated from the 'bottom up'. Individual decisions or small-group dynamics often lie behind the midnight disappearances, the barn-burnings and personal attacks. It is hard to provide a blanket of protection throughout a society that is experiencing this sort of violence. Some will feel more threatened and others less so. People will use friends and connections to weather the storm. Some will be successful. When the violence becomes systematic and threats emerge from the organized actions of the state, then the stage has been set for genocide. Even this has seldom been sufficient to trigger intervention with sufficient resources to protect civilians.

Relations between Communities. Good inter-communal relations at a personal level can reduce the risk to individuals posed by extremists, as they did in Teslic in 1992. There Serb extremists tried to rouse mob violence against Muslims repeatedly in March and April 1992 at the same time as extremism was taking hold in Prijedor. In Teslic, however, the major factory refused to fire Muslims, neighbors would not identify Muslim houses to outsiders, and a mixed police force continued to behave professionally until it was purged in late summer of 1992.[6] If there is residual goodwill between individuals in separate communities, if some long-term relationships survive, if there are members of the aggressor community willing to take social and political risks, if the communities are intermarried or intermingled, and if it is difficult to distinguish between groups, then many might be spared even if genocide is attempted.

Resources Available to Monitor. The three critical variables for monitoring attacks on civilians are ability to witness, ability to comprehend, and ability to communicate. Coverage depends on the number of reliable witnesses and their ability to move freely by day and night. The ability of witnesses to understand through observations, interviews, and media monitoring cannot be taken for granted. Monitors must also have a secure means of communicating without compromising themselves or their witnesses. Monitoring is a necessary but not a sufficient condition for the protection of civilians.

Resources Available to 'Protect' Civilians. Civilians can be protected by removing them from danger or by preventing danger from approaching them.

Traditionally this has meant assisting vulnerable individuals to seek asylum in other countries. Protection officers to screen applicants, secure transport, safe houses, evacuation corridors, and countries willing to offer asylum are the resources needed for this sort of protection through removal. There is often an upper limit on the number of individual cases that can be effectively screened and protected. The limits of asylum underlie the UNHCR's recent emphasis on prevention. It is better for all concerned if those at risk can be afforded protection without giving up their homeland. Providing a 'safe area' within a country in civil war might be an effective alternative to moving large numbers outside the country, only to resettle them after the war. A safe area requires sufficient military resources to demilitarize an area and defend it from anticipated attacks. In 1995 international military staff estimated that 35,000 troops with armor, artillery and aviation support would be needed to protect the safe areas of the Zepa, Srebrenica, and Gorazde.[7]

Influence over the parties. The fate of the so-called safe areas in Bosnia emphasises a final factor that will influence the ability of outside parties to protect civilians. The greater the external influence over the parties, the less likely full scale combat will be necessary to protect the civilians. This proposition has also been suggested in the form of corollaries to deterrence theory. During most of the exchanges between the Bosnian Serbs and the international community, neither the conditions for compellence nor those for deterrence were met.[8]

Knowledge of these factors is useful in planning to protect civilians. The most important, which others influence to various degrees, are the nature of the threats against civilians and the resources available to protect them. Most of the factors varied predictably as the conflict in the former Yugoslavia progressed. Populations became physically segregated early in the conflict. The degree of central control over violence increased with time, so that local authorities could resist the spreading attacks in 1992 in Teslic, but could not by 1993.[9] As people moved with the tides of civil war, relationships that protected friends and neighbors in the early stages were broken. As communities closed ranks to fight it was increasingly difficult to monitor and to extricate vulnerable individuals.[10]

THE CRUCIBLE OF OPERATIONS

To generalize from the play of these factors during the conflict in Croatia and Bosnia, one might suggest three sets of tactics for the protection of civilians, according to the balance of threats and available resources. When threats are low and the rule of law is intact, limited resources are adequate to protect civilians. Civilian defence and human rights monitoring will probably be adequate. When threats are high and laws are unobserved, significant military resources are needed to protect civilians; mere monitoring is not enough. The greatest dilemmas arose in the middle range, where threats were rising and resources were insufficient to protect civilians. In this dangerous middle ground, illustrated in Fig. 1, our Yugoslav experience offers two suggestions: islands of supervision and safe areas. These are imperfect solutions, the limitations of which should be understood before they are advocated.

Everyone has heard of the failure of the so-called safe areas of Srebrenica and Zepa. The military details of that failure are less well known. There are also other examples of efforts to protect civilians from which we can draw important lessons.

Operation 'Backstop' was conducted successfully by the Argentinean General Zaballa in Sector West in January 1993 to prevent Croatian incursions into the United Nations Protected Area (UNPA) in Western Slavonia. During the Medak pocket operation in May 1993, French and Canadian troops were unable to intervene while Croatian soldiers 'ethnically cleansed' an area. Half a day later, Croats were pushed out of the area through a combination of firepower and negotiation. By May 1995, Operation 'Backstop' was no longer viable, and Croatian troops easily overran the UNPA in Western Slavonia.

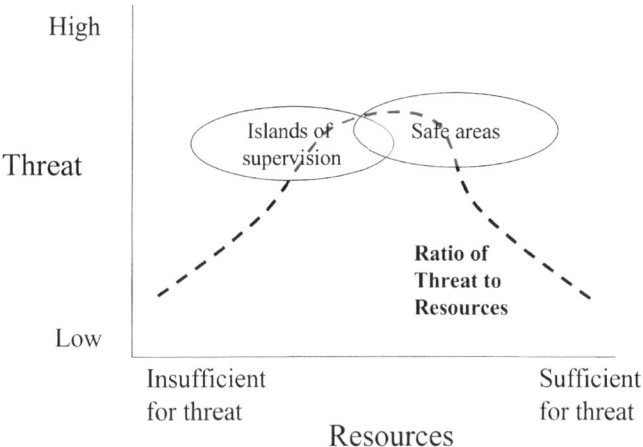

Fig. 1. Protection of Civilians.

From the establishment of Bosnian safe areas in 1992 to their fall in 1995, diplomatic pressure and limited military presence had combined to deter successive threats and attacks. Diplomatic leverage was significantly reduced by the UN's failure to prevent the fall of Western Slavonia and by the persistent use of the safe areas by Muslim forces. On several occasions, the local UN commanders had to make difficult compromises on the spot in an effort to preserve their inherently untenable positions. In 1994 and 1995 the tightening siege by the Bosnian Serb Army weakened the UN military presence in the safe areas. Units were forced to remain long past their scheduled rotations. Fuel and equipment were stolen in transit. Weapon-systems passed to successive replacement units deteriorated from lack of maintenance and spare parts. When Srebrenica was finally overrun in July 1995, the Dutch troops were deficient in food, fuel, ammunition, vehicles, radios, anti-tank weapons, and even protective vests and helmets. They faced superior forces in terrain that made air support almost irrelevant, even when weather permitted it. Maneuver was impeded by tens of thousands of refugees, packed in an indefensible area. The protectors had become helpless witnesses.[11]

Operation 'Active Presence' was an attempt in July 1995 to re-deploy troops to deter a Croatian attack in the Krajina. The troops established strongly fortified observation posts along the main avenues of attack into the Serb-held Krajina. Again, the deterrence gesture failed and about 170,000 Serbs fled advancing Croatian troops in August 1995. Despite deliberate shelling of towns and documented attacks on UN troops,[12] there were several cases of effective protection. About 750 civilians sheltered in the UN compound in Knin. More than 2000 temporarily encamped around the UN compound in Topusko before safe passage was negotiated. Outside the Danish compound at Dvor on the Bosnian border, however, several elderly handicapped people were beaten and shot by troops in sight of the UN positions.

As Srebrenica fell to Bosnian Serb troops and Zepa and Gorazde were under attack in July 1995, UNPF Headquarters looked for alternatives to provide for protection of civilians. Metta Spencer of the University of Toronto helped by contacting organizations in Europe and America, which faxed suggestions to UNPF headquarters. At that late stage, even confirmed believers in civilian-based and non-violent defence had reservations about advocating it. Nevertheless there were suggestions about the ways in which protection of civilians could be enhanced. Some of these techniques are theoretical, and others have been tried under various circumstances. All deserve consideration in those desperate circumstances when threats to civilians are increasing, but resources are inadequate to provide guarantees of protection.

LIMITS TO MILITARY PROTECTION OF AN AREA

Military protection of limited areas in the former Yugoslavia began with Security Council Resolution 762 (30 June 1992). This resolution initiated a familiar sequence of military actions: supervision of a cease-fire, simultaneous withdrawal by both sides of heavy weapons to a distance of 30 kilometers, withdrawal of the JNA from Croatia, and withdrawal of Croatian forces from the UNPAs.[13] Preventing the militarization of the UNPAs, and defining the boundaries of the UNPAs (the 'pink zone' dilemma) remained problems for as long as the UN operated protected areas. Despite the familiarity of the tasks, lack of co-operation by the parties to an ongoing civil war led to an unstable and deteriorating cease-fire, and frequent withdrawal of the limited and conditional consent of the parties to the mandate of the force. Expulsions, coercion and intimidation within the UNPAs, for example, were a problem from the outset, but were linked to similar events outside the UNPAs over which the UN had no jurisdiction:

> From the very outset, UNPROFOR has been faced with a situation in the UNPAs where terrorist methods such as physical abuse, coercion, harassment and even killings, have been used to force non-Serb families to leave their homes . . .
>
> UNPROFOR believes that there is a direct link between this situation and the presence of large numbers of refugees in these areas. The influx into the UNPAs of Serb refugees from other parts of Croatia, and more recently from Bosnia and Herzegovina, continues unabated. The refugees from Croatia, some of whom appear to foresee no possibility of return to their homes . . . . have reportedly occupied houses left vacant by the departure of their non-Serb occupants.[14]

When UN forces were first deployed for the protection of the UNPAs, they often operated from a position of strength. Argentinean and Canadian troops deployed with armored vehicles and heavy weapons. In general they faced lightly armed and poorly organized irregular troops on both sides of the confrontation line. There was considerable diplomatic influence over both Croatia and Serbia, though less over local elements that may have been manipulated by the principal parties to the conflict. More importantly, the UN military forces were the lead element in what was evolving into an international trusteeship over part of Croatia's territory in order to protect civilians from a spreading internal conflict. There were grave reservations about the responsibilities that UNPROFOR was beginning to assume:

> The Force Commander's latest recommendations relating to the UNPAs illustrate the extent to which the evolution of the situation in what was the Socialist Federal Republic of Yugoslavia is drawing UNPROFOR into quasi-governmental functions which go beyond normal peace-keeping practice, have major resource implications and may stimulate demands for yet deeper UN involvement in this troubled region.[15]

Already by July of 1992 it was apparent that adequate resources were not available for the tasks assigned to the UN forces. In the following three years, tasks in Croatia, Bosnia and Macedonia were enlarged and amended by more than 70 Security Council Resolutions and Presidential Statements. Lack of co-operation by the parties, a poor framework for operations, confused rules of engagement (ROE), and inadequate military resources combined to reduce the extent to which military protection could be offered to civilians in Croatia and Bosnia.

Lack of co-operation. The consent of the combatants both to a viable cease-fire and to the arrangements for international monitoring has long been the single most important pre-condition for a peacekeeping mission. It became increasingly evident to soldiers on the ground that many of the parties to the conflict did not in fact want them there. Consent is never absolute for a mission. It is always contingent upon the scope and mandate of the mission, and co-operation varies according to the degree of consent. The successes as well as the limitations of long-standing missions in the Golan, Cyprus and Sinai are a direct result of limited consent to a constrained but clearly defined mandate.[16] Co-operation and consent are the major factors that determine how use of force by a party will influence the legitimacy of the mission. If consent and co-operation are high then any attempt at coercion by the peacekeepers will undermine the legitimacy of the mission. Force should be used only in self-defence and as a last resort; Soldiers go to Cyprus expecting not to fire their weapons in anger. If consent and co-operation are low, then the belligerents, who normally outnumber and out-gun the peacekeepers, will surpass any show of force by the peacekeepers.

The lesson already expressed in that first report of July 1992 is that if the international community is going to meddle in unsettled civil wars with limited consent, it should "go big, or stay home."[17] Forces need to be sufficient to deter or compel, depending on the tactical situation. In January 1993, one batallion provided sufficient deterrence to keep the UNPA safe. This was a function of diplomatic pressure, negotiation at all levels, efforts to demilitarize the UNPAs, and the balance of forces. By August 1995, five battalions and the (empty) threat of air strikes were insufficient to deter Croatian forces from over-running the Krajina. American training, European loans and equipment, OSCE dithering, Serb intransigence and aggression all undermined the co-operation needed to preserve the UNPAs. Understanding how much co-operation there is (in the region and the international community as from the belligerents) is crucial to judging how much force is enough.

Poor framework for Operations. When it is likely that force will have to be used to protect civilians, there must be a clear and legal framework within which they can operate. The absence of such a framework for operations was one of the greatest barriers to effective use of the forces that were available. It can be traced directly to successive Security Council Resolutions, which shaped the evolving UNPROFOR mandate. In Security Council Resolution 836 (1993), for example, operative paragraph 9 provides a Chapter Six mandate for ground forces, authorizing them to fire in self-defence to deter attacks on the safe areas. Paragraph 10 provides a Chapter Seven mandate, allowing all necessary measures to be used by national air forces in support of the UN. As air strikes were introduced to bring pressure on the Bosnian Serbs in 1993 and 1994, the Serbs found ways of returning and escalating the pressure. This culminated with large-scale hostage taking in May 1995. The problem was compounded by the Bosnian government, which made every effort to manipulate incidents demanding a forceful response, in order to bring air power to play on their side.

There were three requirements for a framework within which force could be used for the protection of civilians. First, vulnerable elements subject to being taken hostage had to be withdrawn. This included UN Military Observers as well as all units remaining in Safe Areas or on Serb-held territory. Second, means had to be found to defend lightly armed UN troops from harassment by indirect fire and road blockades. Third, a consistent and effective means of using air power to put pressure on the Serbs had to be devised. The first condition was met with the fall of Srebrenica and Zepa, the evacuation of Gorazde, and subsequent withdrawal of UNMOs and other vulnerable elements. The second condition was met with the deployment of the Rapid Reaction Force (RRF) in July 1995. The RRF was made up of NATO troops (British, French, and Dutch), with armor, artillery, target acquisition and attack helicopters. It had a coherent operational staff drawn from NATO's Allied Rapid Reaction Corps (ARRC) and was trained and equipped to carry out high-intensity operations. Notwithstanding this, it was part of UNPROFOR and operated under exactly the same rules of engagement and mandate as the rest of the force. The third condition was met when NATO and the UN signed a memorandum of understanding about the use of air strikes, and NATO's Southern Command began to prepare a comprehensive air-ground campaign in conjunction with the commanders of UNPROFOR (General Rupert Smith) and UNPF (General Bernard Janvier).

The impact of this new framework of operations can be seen in Sarajevo in August 1995. Before the deployment of the RRF, in June 1995, there had been more than 2000 firing incidents on an average day. When the artillery and target acquisition assets of the RRF deployed, the UN could respond to heavy weapons and artillery fire in kind. By 5 August, average firing incidents were down to fewer than 500 a day. The air campaign waged by NATO in August was designed to secure withdrawal of Serb heavy weapons from around Sarajevo, thus protecting the civilians of the conflict's 'poster city'.

Confused rules of engagement. Even during this final application of force in the last months of UNPROFOR, there was a great deal of debate about the 'rules of engagement' or the conditions under which force could be used, both for self-defence and defence of civilians. There were at least two major issues that were never fully resolved. First, who were protected persons, and when could UN troops use deadly force to protect them? Second, what means could be used, and under what circumstances? The first question concerned the identification of protected groups; this the mission never did satisfactorily. The second question revolved around the use of close-air support, heavy weapons and pre-emptive engagement. While it was generally accepted that soldiers did not always have to wait to be shot at before they fired their small arms in self-defence, the same was not true of the RRF's artillery, at least from New York's vantage point.[18]

Most aspects of the mission were carried out under a Chapter Six mandate, which limited soldiers to the use of force in self-defence. Some national contingents had further restrictions on the use of force. Canadians, for example, could not use deadly force in protection of property, while Danish troops could not interpret threat to a non-UN person as a justification for use of force.[19] There were some tasks that allowed them wider authority, such as the protection of humanitarian aid, supported by 'all necessary measures'. The wording of Resolution 836 concerning safe areas created some confusion amongst the units assigned. Were all civilians in protected areas 'protected persons', even if they were men of military age on leave from the front? In Croatia, legal staff prepared a statement which was to be read to persons seeking protection in UN compounds in 1995: they had to surrender weapons, submit to a search, and might be surrendered to authorities with jurisdiction for criminal activity. A prolonged standoff outside the UN compound in Knin resulted from the unwillingness of the UN commander to surrender all Serb civilians alleged to be sought for criminal charges.

Inadequate military resources. To defend Srebrenica, Zepa and Gorazde effectively could have required up to 35,000 troops (according to a DPKO staff estimate). The scale of the manpower requirement alone, let alone the equipment and logistics support for what would amount to an army of occupation, was always beyond the capacity of UNPROFOR. Even without the unrealistic objective of defending remote and isolated territories from a superior enemy, UNPROFOR had serious military deficiencies for a mission facing low consent, lack of co-operation and frequent hostile acts (by all three factions). Lack of effective target acquisition made it difficult to identify the source of indirect fire harassment of protected areas and UN forces.[20] When artillery was deployed with the RRF, its use had an immediate impact on the number of cease-fire violations. Similarly, more tanks and the demonstrated will to use them quickly opened up the road to Gorazde for food supplies in August.

Most of the limitations to military protection of civilians revolve around factors that limit the military effectiveness of UN formations in complex peace-keeping missions. Many of these limitations were overcome by the deployment of a cohesive NATO force (the Rapid Reaction Force) and a military planning staff which operated without many of the normal constraints of a UN mission headquarters. (The ARRC planning staff operated as the military planning head-quarters for General Rupert Smith, both improving UNPROFOR effectiveness and preparing the way for IFOR).

Given sufficient co-operation by the parties and non-military means of influ-encing the parties, it is plausible that limited military forces (short of an army of occupation) can help to deter harm to civilians, if those forces are properly constituted and effectively used. However even when military forces are deployed with full diplomatic backing and co-operation of the host government, they might fall short of affording protection due to local lack of co-operation, a flawed mandate, confused rules of engagement, and inadequate resources.

CIVILIAN-BASED DEFENCE AND NON-VIOLENT RESISTANCE

When the severe limitations of armed protection are apparent through experience, civilian-based defence and non-violent resistance seem more plausible. Civilian-based defence is a "policy which uses non-military forms of struggle, either as a supplement to military means or as a full alternative to them to deter and defend a society against attacks."[21] Non-violent resistance is "a technique of conducting protests, resistance, and intervention without physical violence . . ."[22] An array of techniques might enhance the security of civilians under different circum-stances. In their responses to the situation in Bosnia in July 1995, several experts in the field expressed reservations about the use of these techniques. Lack of genocidal intent seems to be a basic requirement. Some techniques require the intact social network that exists prior to a violent conflict, while others might be equally applicable in the wake of violence. Advocating non-violent resistance during murderous violence, however, might be irresponsible.

Inventory of potential techniques. The Civilian Based Defence Association, amongst other organizations has advocated civilian-based defence as a basis for national security. As a comprehensive national policy, developed with all the means at the disposal of a national government, it has considerable potential to resist physical occupation by a hostile power. The non-violent actions associated with civilian-based defence include acts of omission (strikes, non-co-operation, etc.) and commission (demonstrations, sit-ins, etc.).[23] Civilian-based defence as a national policy is theoretical. When the techniques of non-violence are applied by groups without the backing of a state apparatus, they often focus on gaining broader support or recognition for a cause, and mobilizing third parties to influence the opponent. Media strategies, particularly targeting international audiences, are often central. Advocates argue that the tactics must be shaped for specific circumstances. For example, specific individuals under threat might be hidden from authorities, moving them from house to house. Neighborhoods might establish intervention squads to react quickly to mob threats against particular households. International organizations might provide witnesses who enjoy the protection of their passports to accompany persons at risk. Civilians might use video cameras to deter acts of violence. In the context of protecting civilians in the so-called safe areas, an urgent plea in July 1995 elicited many practical suggestions, but also some skepticism that non-violent resistance was appropriate. Table 1 groups some of the suggestions that were made.[24]

Community requirements and caveats. All these tactics involve risks, which increase with hostility of the state apparatus or decline in the rule of law. They also require cohesive social relationships, which quickly break down as a result of inter-communal violence. For example, non-co-operation which worked in Teslic in 1992 was no longer practical by 1993 – too many people had moved, too many were afraid, too few would accept the increasing risks. Many of those who responded to the July 1995 plea emphasised the need to plan and prepare for non-violent resistance. No military commander would contemplate mounting a complex defensive operation with untrained units on less than 48 hours notice. Effective non-violent resistance requires a disciplined, organised and well-motivated civil population with well-developed communications and leadership infrastructure. Standing up to physical intimidation, to say nothing of organised rape and murder takes courage and confidence that cannot be instilled at short notice.

Attempting to apply techniques during violence. This raises the problem of trying to develop and use non-violent resistance and civilian-based defence in the face of genocidal violence. Few experts were optimistic. The fact that military protection under the UN had signally failed did not justify experimentation which could easily lead to even more death and suffering. Bruce Jenkins of the Albert Einstein Institute wrote, "I do not think civilian resistance is a very good emergency response mechanism in the midst of a near-genocidal war."[25] Michael Randle reflected on the practical problems: "I imagine that concerted civilian action would be well-nigh impossible while shelling or street fighting is actually

Table 1. Non-violent Techniques to Enhance Civilian Protection

Organization and support measures
- Form a non-violent defence council
- Keep the community together
- Maintain inter-generation and inter-gender solidarity
- Integrate unarmed military observers with civilians trained in non-violent resistance
- Set up international support groups, hotlines, working groups outside the zone of conflict
- Use children to get information out of the conflict zone

Observation and witness measures to increase international pressure
- Get an international media presence full-time
- Arrange VIP visits with accompanying media
- Send human rights monitors with a clear mandate, and make it clear that violations will be prosecuted
- Appeal to church leaders and others who might have influence with the perpetrators

Resistance measures
- Try to prevent physical separation, maybe with UN help
- Refuse any dialogue with aggressors, or open dialogue and negotiate (there was conflicting advice on this topic)
- Collectively resist movement; link arms, sit down, refuse co-operation. This may slow deportation but risks increasing brutality (Jenkins)
- Make the territories less 'politically digestible'

General cautions and suggestions
- It is dangerous to mix violence with non-violence
- If UN soldiers are expected to resist non-violently without training, they should have good role models
- Dramatic demonstrations of courage help dissuade soldiers from brutality

taking place for control of a town, and extremely difficult immediately after a town has fallen in battle."[26] This was also the assessment of several military planners, who were skeptical about the ability of refugee groups to organize under more favorable circumstances. Divisions within the Muslim leadership in Srebrenica made any organization difficult. Separate groups had been involved in guerrilla raids, black-market operations and people-smuggling out of the enclaves. One group of Muslims negotiated safe passage to Tuzla, while another group started out and was hunted down.[27] Metta Spencer summed up many comments: "the place where non-violent struggle is weakest is in the prevention of genocide."[28] Far from being well placed to use non-violent techniques, the civilians in safe areas were geographically concentrated, physically weakened by prolonged siege, cut off from outside communications and international observation, and subjected to psychological pressure that left them vulnerable.[29] All these factors made any alternative to military defence with political backing problematical.

Applying before or after violence. Without having attempted civilian-based defence or non-violent resistance, it is hard to be sure that it would not work, or to say where the cut-off points are for attempting it. The techniques depend very much on preparation, organization, social cohesion, and grass-roots leadership. All of these will suffer in a threatened community as social conflict escalates; all will deteriorate during the course of a violent conflict; and all will take time to recover after a conflict. It is likely that these techniques are most useful in what Acheson describes as the first stage of social deterioration leading to genocide. They might have made their most significant contribution in the former Yugoslavia from 1990 to 1992, which underscores the importance of early warning and preventive measures. In January 1996, it was not immediately possible to introduce any of these measures for the protection of minorities or returnees in Bosnia. They remain impractical to this day in Croatia. It was not until March 1996 that enough human rights observers and civilian police had been deployed to provide a framework for international supervision. Protection working groups with representation from UN Civil Affairs, OSCE Human Rights Observers, UNHCR protection officers, and the International Police Task Force (IPTF) were established in several parts of Republika Srpska by April. More ambitious programs like the 'neighborhood facilitators' depend on international participation as much as local co-operation and consent of the authorities for success. The Neighborhood Facilitators Project is reported in detail in David M. Last *From Peacekeeping to Peacebuilding: Theory, Cases, Experiments and Solutions.* Royal Military College of Canada, Department of Politics and Economics, Working Paper 99/01. (Kingston: Royal Military College, 1999). In the meantime, evictions and beatings continue to be a risk for refugees returning to any community.

Non-violent resistance and civilian-based defence may have little or no capacity to protect civilians from genocide or ethnic cleansing. However, there might be scope to combine some techniques with a military presence that is inadequate to provide protection by itself.

CREATING 'ISLANDS OF SUPERVISION'

Normand Beaudet, of *Le Centre de Ressources sur la Non-violence* came closest to advocating a combination of military and non-military techniques. He suggested that UN troops might become "witnesses who refused to leave." Many of the techniques in Table 1 depend on a high-profile international presence.

The problem is that celebrities, church leaders and media personalities may be unwilling to take the risks, or, if willing, might simply be physically removed or killed by aggressors to avoid embarrassment. A carefully defended 'island of supervision' might be an effective way to use limited military resources to maximize the impact of non-violent resistance techniques. This is more than an idea; there are several examples of this occurring in Croatia and Bosnia.

Examples. Operation 'Active Presence' involved establishing military strong points on the main avenues of Croatian advance into the Krajina in August 1995. Advancing Croatian forces did not seize most of these strong points. It was easier to by-pass them, because the real objectives were civilian population centers. After an extensive psychological campaign including radio and print media, advancing Croatian forces shelled towns and then occupied them. In Knin and Topusko, headquarters compounds were located in urban areas. When the shelling stopped and Croatian troops were advancing into the town, civilians made their way to the compounds for protection. In Topusko they camped outside the compound, while in Knin, about 750 (mainly women and children) were admitted. In both cases, international media were present. The UN compounds were sufficiently well defended to prevent them from being over-run or intimidated (as happened to the Ukrainian Company in Zepa in July 1995). The behavior of Croatian commanders attempting to detain civilians was reported on major European news channels. Eventually a negotiated settlement allowed safe passage of refugees to Serbia.[30]

Limitations. Like a medieval castle, an 'island of supervision' has many limitations. If too many are admitted to the compound, or if they are not screened before entry, the compound can become a dangerous trap. If the UN has not stocked its compounds adequately or cannot re-supply them, protection is flawed. It is unlikely that most of the population at risk will find space within the walls, and they will have to camp outside the walls. Here, the view of the 'defenders' is limited. Faced with a militarily superior foe determined to commit genocide, a compound full of international witnesses may be unable to stop the deportation or even witness the events. People may be taken away or attacked at night, or the witnesses may be able to do no more than record a massacre.

CONCLUSIONS: LESSONS FROM YUGOSLAVIA ABOUT PROTECTION OF CIVILIANS

The first lesson to emphasize from the former Yugoslavia is that within all groups in the conflict, there were innocent civilian victims whose plight warranted protection. In many cases, the international community was unable to protect them. We can learn from that failure.

Military protection. The military limitations of the UN can be overcome by regional organizations with more coherent military forces, although NATO may be the only example. All the other instruments of national and international influence need to be put in the balance before a decision is made to take the risk of affording military protection. If the parties are weak and international will, diplomatic pressure, and economic leverage are strong, then light forces may be a sufficient deterrent. This appears to have been the case at several points during the life of UNPROFOR. As co-operation and consent of the parties declines, or as the balance of other factors shifts, it may be necessary to increase military forces rapidly to avoid a disaster. This assumes that it is possible to increase forces without resorting to an 'army of occupation', which may not be the case. When the military resources available are insufficient to provide protection, given the level of consent and co-operation of the parties, providing islands of supervision might be the only alternative role, however unsatisfactory this may be. Understanding this in Croatia in July of 1995 might have meant a shift of emphasis for Operation 'Active Presence' – occupying population centres rather than military approaches.

Civilian-Based Defence. Despite the range of techniques that might be attempted, neither civilian-based defence nor non-violent resistance is appropriate when genocide is threatened. However, there are phases in the deterioration of social relationships when it is not obvious that genocide is likely, yet serious threats are posed to individuals and groups. Introducing effective means of mobilizing civilians for non-violent resistance and social action early might be the best means of mitigating or perhaps even preventing later violence. The less the leverage of the groups involved the greater the importance of international supervision. Media presence alone, however, is not a reliable means of augmenting non-violent resistance, because of its short attention span and its propensity to be manipulated by the parties for their own ends. Effective monitors with a good understanding of local language and society are vital if international pressure is to be brought to bear effectively, with or without a military presence. As the situation becomes more dangerous, the problem of dead witnesses (both domestic and international) becomes more acute, and the need for an international military presence also increases.

Hybrid 'islands of supervision'. When there is an increasing threat to civilians, but monitoring and protection resources are inadequate, one solution might be to create protected islands of supervision. These afford less of a guarantee than a safe area, which can be militarily defended. But when military resources are inadequate to defend a large number of people in a large area, they might be used to guarantee the presence and safety of international witnesses and the direct passage of information to the outside world. Meanwhile, strengthened by the knowledge that their risks are witnessed, civilians might engage in bolder forms of resistance and non-co-operation, increasing the costs to an aggressor. This is not an ideal solution. It may not work in many circumstances. Faced with the risk of genocide and inadequate resources to deter violence or defend innocent victims, it might be attempted as a means of using limited military resources and civil leadership to the best effect. In this sense, it deserves to be considered early in the contingency planning process.

The experience of trying to protect civilians in the former Yugoslavia has not been very positive. Where it has worked, it has been a result of effective use of all the means of protection at the disposal of the threatened group and the international community. Military protection and social action are not perfect substitutes for each other; both should be used.

This chapter has addressed protecting civilians by removing them from danger, or keeping the danger at bay. Changing the behavior of the perpetrators is another approach altogether.

NOTES

1. This work does not represent the views of the Department of National Defence, Canadian Armed Forces, or Royal Military College, but only the views of the author.

2. Frederic de Mulinen, *Handbook on the Law of War for Armed Forces*, (Geneva: ICRC, 1987), Summary for Commanders, p. 2.

3. See definitions included in draft protocol I to the Geneva Conventions.

4. Sadako Ogata, 'The Interface Between Peacekeeping and Humanitarian Action', in *New Dimensions of Peacekeeping*, edited by Daniel Warner (London: Martinus Nijhoff, 1995), p. 120.

5. Donald Acheson, 'Preventing Genocide: Episodes must be Exposed, Documented and Published', *British Medical Journal*, Vol 313, 7 December 1996. There are more sophisticated descriptions of genocidal processes, But Acheson's simple model is adequate for my purposes.

6. Interviews with former residents of Teslic at Banja Luka in May 1996.

7. UN documents

8. Frank Harvey, "Rigor Mortis or Rigor More Tests: Deterrence and Compellence in Protracted Ethnic Crises." Presented to the International Studies Association annual meeting, in San Diego 2000 (the paper is forthcoming in International Studies Notes).

9. Interview with former residents of Teslic in Banja Luka,May 1996.

10. Interviews with UNHCR protection officers and UNCA representatives in Zagreb, August 1995 and in Banja Luka, April 1996.

11. There are numerous accounts of the fall of Srebrenica, including the comprehensive Dutch official Board of Enquiry and several media versions. I watched events from Zagreb and was subsequently involved in an enquiry for the Force Commander, UNPF.

12. The attacks included direct shelling and tank-fire against clearly identified UN positions, taking of UN hostages who were made to walk in front of advancing Croatian troops, and grenades thrown into UN bunkers.

13. *Report of the Secretary-General Pursuant to Security Council Resolution 762 (1992)* S/24353, 27 July 1992.

14. S/24353, pp 1-2.

15. S/24353, p. 9.

16. David Last, 'Peacekeeping In Divided Societies: Limits To Success', *Low Intensity Conflict and Law Enforcement*, 6:1 (Winter 1997).

17. I am indebted to Dana Eyre of the Naval Post-graduate School for this expression of the relationship. It is also discussed by Charles Dobbie,

18. An exchange of cables and phone calls between UN DPKO and UNPF headquarters in August 1995 made this clear. DPKO was uncomfortable with the concept of pre-emptive counter-battery fire (i.e. firing at a rocket launcher when it was loaded, elevating and traversing).

19. UNSHIRBRIG conference, Wiener Neustadt, November 1995.

20. Counter-gun/counter-mortar radar were deployed in Bosnia before the RRF, but were shot at by at least two sides and often lacked spare parts. The US embargo on parts for Pakistan meant that the UN could not repair the AN-TPQ 36 radars deployed near Tuzla.

21. Gene Sharp, *National Security Through Civilian-Based Defense* (Omaha: Association for Transarmament Studies, 1985), p. 47. The field of civilian-based defence, non-violent action, transarmament, and related ideas is well developed, and I cannot do justice to a literature review in this context. The Albert Einstein Institute is a good place to start for further information.

22. Sharp, 1985, p. 51.

23. Sharp, 1985, pp. 51-52.

24. These are drawn from correspondence forwarded to UNPF headquarters in July 1995 by Metta Spencer. Some of the contributors were: Normand Beaudet, *Le Centre de Ressources sur la Non-violence*; Christine Schweitzer, Balkan Peace Team; Curtis Doebbler, London School of Economics; Graeme MacQueen ; Hans Sinn (Perth); Bruce Jenkins, Albert Einstein Institute; Michael Randle (Bradford); Michael Beer (Non-violence International); and A. Ross Wilcock.

25. Bruce Jenkins, fax to UNPF Headquarters, dated 17 July 1997.

26. Michael Randle, e-mail dated 17 July 1995.

27. Interviews with UNMOs and debriefing documents, UNPF, August-September 1995.

28. Metta Spencer, fax dated 17 July 1995, summarizing the advice of many of the people who provided responses.

29. Fax from Normand Beaudet, *Le Centre de Ressources sur la Non-violence*, 18 July 1995.

30. This was uneventful, with the exception of attacks by civilians in Sisak and delays at the border with Serbia.

11. PEACEMAKING FROM BELOW – THE NGOs[1] LESSONS LEARNT, LESSONS TAUGHT

Jan Øberg[2]

Lessons-learnt studies[3] is a fast-growing industry. It is easy to *state* the need for 'learning lessons' about international conflict-management and peacekeeping. It usually expresses the wish to apply gained knowledge and experiences to tomorrow's new situation and emergency so as to optimise the handling of it, i.e. securing that actors take fewer counterproductive and more productive steps tomorrow than yesterday towards peace. It is more difficult to *implement* lessons learned and to *do* things right.

First of all, lessons learned exercises frequently depart from the simplifying, tacit assumption that all actors who intervened with the aim to help solve a conflict did so without other motives in mind. It presupposes basically noble or altruistic motives, which are not the most frequent ones in international politics.

Second, learning a lesson comes close to apportioning blame. It's a kinder way of saying that conflict-managing actors should behave differently in the future, and that "it's not our fault that things went wrong." Lessons-learned studies (whether by institutions, scholars and diplomats) can pass the buck regarding responsibility for conflict-management failures.

Third, does 'learning' imply mere recognition of a lesson or also the remedying of the underlying problem and the training of people to ensure that the same mistakes will not be repeated?

Fourth, lessons-learned discussions often proceed from the assumption that "we know what would have happened and how the crisis could have been

Research on Russia and Eastern Europe, Volume 3, pages 197–213.
2000 by Elsevier Science Inc.
ISBN: 0-7623-0280-1

solved had something else been done in that situation." This resembles the historian's 'contra-factual hypothesis'. It is not possible to know that things would have been different, if only. . .

How can it be known that the UN would have succeeded in Bosnia if only it had been given the resources needed to fulfil its mandate? Who *knows* that "if only we'd bombed the Serbs earlier, the war would have stopped?" Nor can we know, for instance, that if NGOs had played a bigger role, prospects for peace would look better today.

Finally, powerful conflict-managers teach lessons to supposedly more primitive conflict actors. Political body language, if not explicit statements, communicate messages such as "if you don't comply with our proposal, sanctions will follow." If conflict management initiatives (mediation, peacekeeping, humanitarian aid, threats, crimes tribunals, etc.) don't yield the desired results conflict-managers assume that it is *their* fault.

Discussions about lessons learned in international conflict management can be useful as a heuristic tool. When we see that things don't work, we are challenged to come up with constructive new policies. One simple thing we know from conflict-resolution psychology is that it is smarter to address the problem than to attack people. So, too, with lessons learned. It is less important to say *who* should learn than to ask, *what* can be learnt?

Besides, do we want to learn lessons useful for power politics or for conflict-management and conflict-resolution? There seem to be no commonly accepted criteria for evaluating what is bad conflict-management. The Yugoslav experience since 1990 has not even generated a debate about how to analyse conflicts, intervene in them, and what it means to solve – or just manage – a conflict.

In the case of ex-Yugoslavia, we all have something to learn – not only about those 23 million people and their complex circumstances, but also about ourselves as conflict-interventionists and mediators. that things would have been different, if only. . .

POLITICS AS USUAL, CONFLICT-MANAGEMENT AS UNUSUAL

The power model of conflict-management

The most widespread, common perception of conflict can be described in the following manner: a conflict is an unwanted disturbance of social order, usually about two parties who cannot agree and who both struggle to win their way. Conflicts can be deterred, or otherwise prevented, by one or more of the parties or by a third party (e.g. judges, mediators, prisons, military). Should this fail, all conflicts can be contained, provided sufficient coercion is employed by stronger external actors.

According to this school of thought, settlement is at hand when one party who was 'right' wins (is rewarded somehow) and the other(s) who was wrong, loses (is punished somehow).

We shall call this the *power* model. Its first question is: who is guilty? It rests on implicit guiding assumptions, such as *that* it is more important to stop maladaptive behaviours than understanding their root causes, *that* their one actor who has a 'higher' right to judge and employ force for the sake of society's order, or *that* conflict-solution is essentially a competitive activity wherein the conflict will go away the moment the guilty side is punished.

The terms 'conflict' and 'violence' are often used indiscriminately. If advocates of this model of conflict are asked about the causes of conflict, they are likely to answer that human evil, animal-like[4] instincts and aggression, primitivism, or low level of development and education, or cultural-ideological inferiority, accounts for most of the world's conflicts. In other words, human beings with a particular set-up of characteristics are supposedly the root causes – which is why control or punishment is the preferred method of conflict-resolution.[5]

The problem-solving model of conflict-management

Another view negates much of this and can be described in this way: Conflicts are unavoidable and 'normal' in all human activity and any social order is bound to create conflict; conflicts are therefore *generically* rooted in the structures of human relations and institutions (such as the state, but not only there).[6] On close inspection, therefore, conflicts usually have more than two actors and more than one issue. In addition, within each actor – be it an individual or a group – there are inner conflicts that influence conflict behaviour vis-à-vis others. In short, conflicts are inherently about human needs. When *needs* are not satisfied or even addressed, conflictual behaviour can be expected.

To be sure, what we often see is people struggling to triumph over – and even humiliate – each other as a method of settling their differences. However, this is not inevitable for all conflicts, but expressive of frustrations from certain unsatisfied needs, which may result only from lack of knowledge concerning other potential options.

Conflicts cannot always, perhaps not even usually, be solved by force – either through a showdown between the parties themselves or through enforcement by third parties. This view considers that power, force and violence mostly aggravate the conflict and make a long-term solution more difficult to find.

Indeed, violence harms the doer as well the object that can never be repaired. Killing and most other forms of violence can never be undone.

Table 1. Conflict management styles by 'third parties'.

	POWER	PROBLEM-SOLVING
Philosophy	change people's behavior, punish	attack the problem; it's more important to support the good guys (the many) than to punish the bad guys (the few).
Competence	any diplomat, military or lawyer can be a mediator, no special knowledge or training needed	helping others solve their conflicts is a science and an art; it can be learnt but one cannot do it right away
Conflict	it was theirs originally but they were not capable of solving it, so it's now ours ('conflict snatching') and they must live with our solution	since they suffer and since they must live with the solution, third parties can only assist, suggest, facilitate; it's their conflict and the solution must be theirs to be viable ('conflict mitigation')
Attitude	people in conflict must be told a few things and taught lessons	people in conflict need empathy and a fair hearing; people don't kill and destroy for fun
The intervenor's role	judge: 'you are wrong to conduct war, we are right in telling you to stop, and how'	doctor: 'we are here to help you on all sides, to stop this madness'
Violence	= human evil	= structures and complexities leading to frustration; the result of unresolved problems
Actors addressed	elites, GO, IGOs	civil society and elites
Activities	judgment, blame, threats, deals	diagnosis, prognosis, therapy
Conflict-resolution	quick fix, cease-fire forced, if necessary	conflict-transformation, long-term cooperative with conflicting parties
Peace	non-war, must trickle down from peace centers	rooted in society, in the human dimensions, ground-up
Peacemaking	linear	circular
Civil society	destroyed and helpless nothing to build on, must be democratized later, and from outside	is always there, potentially at least, must be empowered

A solution is not, therefore, arrived at by employing force against actors although that may be necessary under specific circumstances to *regulate* a conflict process. For instance, we feel a natural urge to prevent someone from committing suicide, a child from running out in front of a car, and a leadership from carrying out a genocide. Again, our violent intervention does not address the underlying causes of why these acts were in the making.

A solution requires that we address the root causes, the sources in social structures and processes that provide a fertile ground for conflict, and the incompatibilities of needs, values, perceptions, and expectations. It is not always possible or desirable to analyze who is right and who is wrong; in fact, all the conflicting parties may be both. They each may have a 'story' to tell with which the analyst can empathize. Conflicts are problems to be solved rather than opportunities for apportioning blame, guilt, and punishment.

We shall call this the *problem-solving* model. The first question it poses is: What is the problem? Its second question is: How can we care for the victims and help the wrongdoer to be more constructive?

This model, too, rests on a series of assumptions – a belief that all human beings have good and evil within themselves; that their various potentials can be activated for good and bad under specific circumstances; and – most importantly – that an evil deed does not prove that a human being is exclusively evil or non-human. Seeking to understand conflict behaviour and violence is crucial to conflict-resolution, to the reduction of violence worldwide, and to the realisation of potentials for the common good.

Mediators of this 'problem-solving school' of conflict-resolution tend to foster the joint exploration of future options (preferably win-win) and appeal to the highest level of creativity that the participants can bring. They tend not to see a conflict situation as a win-lose battle.

The present author sympathises with the problem model. He maintains that the power model is one reason why many of the world's ongoing problems remain unsolved. The Dayton peace deal is a clear example of the power model and it is my hypothesis that it will not succeed in conflict-resolution.

Advocates of the power model sometimes criticize the problem-solving model for being blind to individual responsibility and unable to deter future crimes. Advocates of the problem model criticise practitioners of the power model for setting themselves up as moral judges that are blind to structural causes and unable to get to the roots of conflicts. This quarrel is not productive, even when it is friendly. No doubt, the power mode *can* also address underlying problems and no doubt the problem-solving model *can* also deal with punishment.

The power model is likely to prevail in the foreseeable future since it is so strongly institutionalised and rooted in the nation-state, the free market, the capitalist economy – indeed, in the Occidental cultural program that dominates global society.

It remains in each case to see which of these methods can be applied.

What is absolutely essential, in my view, is that the problem-solving model must be strengthened worldwide, and become increasingly practised by states or by other trans-state actors. The problem model has a clear advantage in domestic conflict bordering on the existential, whereas the power model – due to the anarchical structure of the international system – still works best in the international context.

The case of former Yugoslavia

In the Yugoslav case, one can speak of a *remarkably conflictual approach to conflict management.* The Yugoslav crisis undermined the international system's ability to handle future complex conflicts elsewhere. It also seriously weakened the legitimacy of 'post-modern' security concepts in the European Union.

Many – probably too many – actors have been involved, performing roles in several collective organizations with different mandates, capabilities and aims. As in all other political conflicts, one can distinguish between official, explicit objectives on the one hand and real, implicit ones on the other hand. Each actor has performed contradictory roles and each collective organisation has promoted policies that were incompatible with the objectives of others.

For instance, members of the UN Security Council have, in addition to that role *also* played the following roles in the Yugoslav crisis:

(1) worked for some kind of conflict-resolution resolutions or UN mandates;
(2) conducted big power politics reflecting national interests;
(3) supplied arms and ammunition through private dealers and perceived the Balkan quagmire in a geostrategic and political framework;
(4) conducted policies with ultimatums, threats, deterrence, peace enforcement, and finally massive bombings;
(5) provided soldiers for peacekeeping and monitoring on the ground (except China) based upon the UN Charter and, later, on NATO;
(6) served as mediators;
(7) the U.S. has become a military alliance partner to Federation-Bosnia, Croatia and others;
(8) introduced and monitored the sanctions against Serbia and Montenegro and handled their effects on third parties;
(9) brought humanitarian assistance to various parties;

(10) applied pressures through such actions as diplomatic isolation of some states, recognition of others, and suspension of others from membership in important international forums.

Each of these roles conflicts with at least one of the other roles the state has played simultaneously.

The abundance of actors and institutional frameworks within which they have tried to work has prolonged the war and probably rendered more difficult the search for a genuine long-term settlement. The so-called international community has become confused by the combination of the above-mentioned roles. For instance, actors have tried to do impartial mediation in one setting, as in the International Conference on the Former Yugoslavia (ICFY) in Geneva, while simultaneously taking sides with policies of recognition, sanctions, ultimatums, and bombings. From a conflict-resolution point of view this is bound to fail.

It is evident that the attempt to play roles simultaneously in institutional frameworks with widely diverse aims has aggravated the conflict. The Yugoslav crisis has illustrated, par excellence, the international community's unskilled conflict analysis and the lack of appropriate post-Cold War institutions.

The same 'international community' has been able to blame either the Serbs or United Nations or both for the continuous aggravation of the crisis. If a 'peace' initiative was not accepted or did not work, that was automatically blamed on one of more of the parties. The initiative itself was not questioned. I have argued elsewhere that the Dayton Agreements is a conflict-resolution fraud, whatever else its merits, and that if, let's say, the Russians had facilitated it, opinion-makers around the world would have denounced it as misguided, if not immoral.

Evidently no leader had the courage to state the obvious – "Yes, we the international community have made serious mistakes," or "Yes, each of us have national interests; we sometimes said one thing and did something else. We should not have created the false impression that foreign policy is conducted by the principles of the Good Samaritan."

Civil Society in War and Peace

In the literature on "civil society" the term means, by and large, whatever each writer chooses for it to mean. Perhaps 'civil society' is not a set of actors but a sentiment, a political culture, an ever-changing combination of world views, myths, communication structures, and unorganized individuals. It seems to be defined by people who are not – rather than by people who are – members of civil society.

Civil society is hardly a society within society. Indeed, perhaps it is not even a 'society' as defined in sociology textbook. Perhaps it is not exclusively nonviolent; (after all, 'revolutionary' movements sometimes release violent activity).

When social systems fall apart as was the case in the ex-Soviet Union and ex-Yugoslavia, does 'civil society' continue to exist as a kind of 'society of souls' rather than as a political community? Or to turn the question around, is there a 'civil society' today just because authoritarian structures fell apart?

And what, after all, do we do with 'civil society' actors who turned into warlords, local war heroes, media propagandists, private arms dealers, drug traffickers, mafiosi, or Hells Angels? Or, if more than 1000 people are killed per year in the greater Chicago area, does that area nevertheless constitute a 'civil society'? Perhaps civil society is neither a society nor truly civil (civilised, non-military, non-governmental, or civic).

Let's approach it from another angle: 'Civil society' is about *power* or, rather, about a dissenting conceptualisation of power.

Power is, of course, another difficult concept. It is usually defined as the ability of A to make B do what A wants B to do or abstain from doing what A does not want B to do. Thus, power is power over others, the control and command over resources in a zero-sum game. The state is the highest expression of that concept and we invariably relate the state to other concepts such as self-interest, national interest, rule, force, authority/tarianism, domination, warfare, repression, dictatorship – and to welfare, the power to remunerate allegiance and punish disobedience, or simply keep others out (migration and refugees).

However, power has many other definitions. For instance, Robert C. Tucker argues in his fine little book *Politics as Leadership*[7] in favour of a Platonian concept. He says: "Just as medicine is the art of tending to the body with a view to restoring it to a healthy state, so statesmanship is tendance of the soul. . . . The true statesman, possessing knowledge of what is good for man, is a physician of souls. . . . I think it is fair to characterise Plato's dissenting view as one that equates politics with leadership." (1981, p. 2).

He argues that there are politically very important leaders who possess power and occupy high political offices but are non-constituted, such as e.g. Gandhi, King, Schweitzer, Djilas, Monnet, Sakharov and others. Thus, power holders are not necessarily leaders, whereas genuine leaders always have power, however differently defined from the power of the Prince.

Tucker quotes James MacGregor Burns to the effect that *leadership* can be seen as "leaders inducing followers to act for certain goals that represent the values and the motivations – the wants and needs, the aspirations and expectations – *of both leaders and followers.*" Tucker maintains that although we cannot deny that Hitler and Stalin were political leaders, they were predominantly absolute wielders of brutal power.

Genuine leaders, not dictators, flourish in *problem situations* by doing basically three things: (1) diagnosing the situation, (2) prescribing a course that will meet the situation, and (3) mobilise the community and its support for action towards a commonly defined, acceptable solution.

I am attracted to the possibility of defining 'civil society' as social actors and ways of acting that rest on *leadership* – embodying the goals, values and motivations of leaders as well as followers – and which therefore does not rest on force. From a theoretical and ethical viewpoint it is important, it seems to me, *not* to equate power with brute force or the threat of using it.

Any individual, movement, institution that rests essentially on this understanding of power is part of civil society. Thus, for instance, social movements rise, says Tucker (ibid, p. 85), through successful leadership authority of a sort not being undertaken by constituted authority, i.e. through non-constituted leadership. And he predicts, I think correctly, that "the capacity of human civilisation to endure on earth has come to depend upon the development of leadership of and for the whole. The emerging crisis of survival [of the state-dominated global system, JØ] is, along with all else, one of leadership deficiency. The spaceship cannot fly much longer unpiloted."

We are here, in effect, reminded of the definition of power provided by Gene Sharp in *The Politics of Nonviolent Action. Power and Struggle,*[8] one of his comprehensive analyses of Gandhian politics. He states:

> Basically, there appear to be two views of the nature of power. One can see people as dependent upon the goodwill, the decisions and the support of their government or of any other hierarchical system to which they belong. Or, conversely, one can see that government or system dependent upon the people's good will, decisions and support . . . Non-violent action is based on the second of these views: that governments depend on people, that power is pluralistic, and that political power is fragile because it depends on many groups for reinforcement of its power sources. The first view . . . appears to underlie most political violence.'

Thus, rulers are *dependent* upon people who accept them as rulers because of, let's say, the material resources they command, the obedience they require, their psychological and ideological outlook and the sanctions at the ruler's disposal. When 'civil society' members decide to withdraw from that power *source*, the ruler will no longer be able to rule. Hitler could not have fought the battles, burnt the people, and arrested all opponents with his own naked fists.

The Needs and the Resources

What are the needs of civil society in peace, conflict and war? It is common knowledge that highly developed countries bring development assistance and conduct trade and investment policies in so-called developing countries from which the donors benefit more than the recipients. Asymmetric structures create mechanisms that, in turn, reinforce the structure.

Conflict-managing countries usually belong to the centre of the global structure – as do the Security Council permanent members. They intervene in structurally weak, peripheral regions and countries, seldom in other central regions and countries – and, even then, in weaker ones such as Yugoslavia in the periphery of Europe. Conflict management has become isomorphic with the Western-dominated global dominance-dependency structure and serves to maintain that system, the Western control of it and the promotion of Western values and institutions. The countries making the headlines speak for themselves in this perspective – Iraq, Somalia, Rwanda, Haiti, Cambodia, Liberia etc. Like development assistance, conflict intervention is legitimated vis-à-vis the public with reference to humanitarian concerns and the moral need to 'do something'.

This is not to say that stronger, highly developed countries ought not engage in conflict management or – certain types of – humanitarian interventions. But two aspects deserve emphasis: (1) the needs and potentials of the recipient/object, i.e. the conflict-ridden, countries are seldom taken into account and (2) conflict-management and peace proposals have become vital aspects of global order-formation, and an integral part of power politics. To believe that powerful states intervene in Somalia or Yugoslavia exclusively or even predominantly for the sake of humanity, peace, compassion and the sole wish to help these people is simply naive. Had that been the case, early warning, preventive diplomacy and genuine socioeconomic assistance – in Somalia in the 1970s when its new visionary regime asked for it and in Yugoslavia in the 1980s when they needed it – would have been the chosen method. And if this had been the case, issues such as world hunger and poverty, global inequalities and environmental degradation would have been addressed with determination long ago.

Mainstream thinking has it that since countries have fallen into war, they have little or no subjective wish and little or no objective resources for peace-making. Therefore, it is argued, Third Parties intervening from outside must come in, speak with those in the political, military and economic power at the top of society and suggest measures that will lead to cease fire, negotiation, a settlement, reconstruction and post-conflict normalisation. If the party in question is judged to be the guilty, it is less a matter of suggesting as much as dictating the rules as well as the outcome.

This approach ignores a lot. It addresses actors, or people, but not issues. It ignores the potential – found in any civil society – for peacemaking, be it elders, institutions, cultural norms and activities, and social structure itself. It ignores, in short, civil society and addresses only state-based, elite actors who, paradoxically, are the ones responsible in the first place for larger-scale violence.In consequence, what we call NGOs and local society is ignored, undermined. In short, civil society is not only victimised from inside by its own state but, by and large, also by intervening, conflict-managing states from outside.

The counter argument is obvious: but you have to speak with those who have it in their power to stop the fighting in the first place. There is no reason to dispute that, but top leaders are not the only ones and the present balance between *ruler-oriented* conflict-management and *civil society-oriente*d conflict-management is very counterproductive.

NGOs, LIMITATION AND POTENTIALS

Civil society as government

There is a tendency to see civil society as 'good' – as more honest and democratic and less involved in raw power politics and deception. That is of course true but there is not a wall between the two.

Governments often succeed in co-opting civil society, either individual members or virtually all of it, willingly or unwillingly. Within some of the above-mentioned definitions, 'civil society' groups – in the sense of 'non-government' – sometimes do take part in government-induced ethnic cleansing, hoisting of nationalist flags, singing of anthems and voting for constitutions that makes life virtually impossible for minorities. With sufficiently low education and being sufficiently deprived of welfare, or having its expectations toward the future thwarted, civil society can be made quite ready to serve as the cause of human suffering.

The most frequent type of wars today are domestic. Several wars are embodied in one war theatre. There are wars on the mind, wars on perceptions, war in the schools, the media and even between and inside families. Civil wars are also fought by 'civil society' under the monolithic state-based power theory when that has penetrated the minds of a sufficient number of civilians. There is no clear borderline. Young people joining voluntarily the army or paramilitary units in moments of crisis leave behind them 'civil society' and 'leadership' power and join the petty- or state-based monolithic power world.

So, in various ways 'civil society' can be co-opted to provide the sounding board for nationalist or otherwise authoritarian government policies. No government can get away with warfare without having some support by some sections of civil society.

Likewise, civil society increasingly takes over the role of government. Local lords who are de facto leaders in regions (e.g. Fikret Abdic) or have proclaimed their own states with a few followers (e.g. South Ossetia), or set up small economic empires and mafiosi culture (e.g. the Croatian Herceg-Bosna in Mostar) etc. Consequently, ordinary government power is being eroded unless – as is the case in Mostar – such proto-states serve interests of real centres. Governments don't seem to know what to do when faced with these new types of leaders – gangsters, thugs, arms traders, warlords, terrorists, secessionists, or non-violent but constituted leaders such as Dr. Rugova in Kosovo.

When civil society takes over government functions, traditional government and the state lose control as part of a long term fragmentation strategy. It is fragmentation of government power and reintegration in new (but ideologically old) authoritarian structures. New states invariably have bigger 'old' states as role models. The net result is brutalisation.

Civil society organisations (CSOs) in the limited sense related to peace, rights, gender equality, and environment, who espouse peace values are double losers. Their 'civilised' norms are exposed as powerless; civil society is being undermined by the thugs, and ordinary government which was (perhaps) once quite decent becomes more brutal, more primitive – more difficult to influence by peace values.

Civil society has been militarised both in terms of mentality and violent means. Indeed, privatisation and the appropriation of military power (warlordism) is on the rise. The monopoly on the control over the means of violence, which used to be a defining characteristic of the state, is also eroded by arms traders.

Another visible trend – much less harmful, but nonetheless deplorable – is seen in NGOs, private voluntary organisations that are so happy to get power that they gladly water down their aims, goals and values to appear 'realistic' in the corridors of government and other elite settings. As lobbyists they 'go native', lending not only their competent brains but also their hearts to the powers that be – all in the name of promoting their cause. They compete with any other NGO standing in their way – and willingly queue up at any Hilton Hotel reception to market their message on behalf of the damned of the earth.

Such CSOs organise themselves as corporations or governments. They speak the same diplomatic language. Instead of doing what they are supposed to do – such as supporting human rights, bringing humanitarian aid, reporting objectively, promoting culture, or showing empathy and solidarity with the victims of wars – they hold press conferences about who is right and wrong, condemn whole countries, argue for sanctions and war tribunals and military interventions – in short, imitate governments in form and substance. The next step, of course, is to *be* in government or high power – through membership of a party, media corporation or other career-making machinery.

When 'civil society' adopts the methods of governments one way or another they can, of course, no longer serve to restore civil society's values or help those who suffer and need humanitarian assistance. Civil society is not necessarily 'innocent'; it can create violence in a variety of ways.

CSOs in the case of ex-Yugoslavia

This final section is based on some of the above reflections plus the author's experience of having seen NGOs in action one way or another since 1991.

(1) What it means to be a CSO in civil society is less clear cut nowadays with the breakdown of societal and state structures in focus of conflicts and warfare. Civil society is not only 'good' but has proved to be able to take on features and functions which are part of the problem, not the solution.

(2) CSOs are drawn into society-building in new ways for which they are not trained or have much experience. Without training in the much broader, integrated tasks, they are likely to make mistakes from the point of view of conflict-resolution.

(3) CSO and government activity is not necessarily compatible. It is wise to only strive for functional integration between the two when feasible. Otherwise CSOs will increasingly be asked to sweep the floors when governments have walked through conflict areas and be handed over 'missions impossible'.

(4) CSOs should make their own analyses, not rely too much on media and government images, fact-finding, and interpretations.

(5) To facilitate conflict-resolution in a broad sense, CSOs would do wise to follow the problem-solving approach rather than the power approach to people in conflict. Many do, but CSOs with semi-independent status increasingly seem to enjoy playing according to the power model.

(6) CSOs should avoid crowding around certain places and certain tasks because that's where money and fame are. Sarajevo after Dayton is a case in point. Coordination and exchange of information is a helpful first step, but the 'disease' is deeper – a kind of "we must also plant our flag and fax machine there to come into the limelight." Interesting tasks and evident needs elsewhere in former Yugoslavia have been grossly ignored.

(7) CSOs in the business of conflict-resolution training and reconciliation should take care not to become travelling circuses, holding the same seminar on Monday with refugees in Bosnia, Tuesday with McDonalds in Moscow, Wednesday with Siemens top managers in Germany, passing Israeli and Palestinian politicians on Thursday on their way to spending the weekend in Rwanda. Long-term commitment based on field-analyses and knowledge is a must to really help people.

(8) Peace must be built top-down but equally important from the ground-up. The needs and potentials of local society are often ignored. War-torn societies often get what donors have available (be it NATO, OSCE, intelligence agents, macro-economic and infrastructure loans and projects, surplus weapons from another war, standard aid programs and constitutions written above their heads by lawyers in Chicago) rather than what an objective assessment made in partnership with local society would recommend.

(9) CSOs have many comparative advantages. They are able to get into a conflict area early, they don't have military force, they can relate to local society and to all society in ways government diplomats cannot (or at least usually don't). CSO staff people often have a much longer working experience in troubled areas than do government representatives, and CSOs do not – in principle, at least – have national interests. Neither do they have to be reactive; they can be pro-active. One would wish to see these important potentials developed more.

(10) Here are some concrete proposals related to conflict-resolution in the field. A few CSOs are doing these kinds of things, but international and local CSOs could expand their agenda and:

- Intercept conflict 'signals' and serve as facilitators of meetings among adversaries.
- Serve as impartial, fair, and legitimate 'citizens diplomats' and 'conflict doctors.'
- Combat hate speech and propaganda in media and education, establish parallel centres of information, radio and television.
- Train local citizens in civil disobedience (conscription, hiding, escape, breaking laws, sitdowns – but not in war zones, of course), boycott those who threaten war and other emergency situations. Non-violent protests (marches, letters, strikes, happenings – as in Belgrade in the winter 1996–97).
- Alerting, informing and appealing to international bodies, combating reductionist media images of complex conflicts.
- Impartial conflict-journalism rather than war-reporting.
- Build alternative or parallel institutions in support of civil society and local citizens.

- Parallel, alternative international and local negotiations/peace plans.
- Systematic attention to the human dimensions of conflict.
- Dispatch 'White Helmets,' 'Peace Brigades,' retired diplomats and military staff – socio-psychological humanitarian intervention.
- Support the victims/'good guys' rather than punishing perpetrators/'bad guys'
- Brainstorm on peaceful ways out of the emergency.
- Help develop a peace culture among peoples, manifesting concrete activities in defiance of those who promote the catastrophe.
- International monitoring, protection and support for 'dissidents' and others.
- Community building and -maintenance, citizens peacekeeping, peacemaking and peacebuilding. Focus on 'peace pockets' together with the people, help them see alternatives to local warlords and profiteers.
- Experimental Peacebuilding Zones based on negotiated agreement and international protection.
- Using 'windows of opportunity' for peace and normalisation, such as checkpoint meetings, sports, cultural events, competitions, common building projects, etc. among citizens belonging to all sides.
- Courses and seminars with professional groups (teachers, journalists, community leaders etc.) and movements increasing their skills in conflict understanding, negotiation and reconciliation.
- Joint institution-building: police, media, factories, agriculture and infrastructure.
- Mobile teams of citizens diplomats and 'conflict-doctors,' preferably also local teams drawn from the war zone.
- 'White helmets' of NGOs and others operating with Blue Helmets.
- Establish 'Conflict Consortiums' – rapid analytical teams of area experts, conflict-resolution experts, and representatives of humanitarian and other CSOs to do fact-finding, early warning and devise strategies – and do so in time.
- Projects for former, demobilised soldiers.

CSOs can find tasks all along the classical conflict cycle – conflict risk assessment, early warning, violence-prevention, early peacemaking, peacekeeping, ceasefire, mediation, negotiation, peacemaking, conflict-resolution, peace planning, de-militarization, reconstruction, peacebuilding, reconciliation and normalization towards peaceful development – in short, conflict *transformation.*

There are no limits to what can be done by non-violent, peaceful means. But history – also that of ex-Yugoslavia and its dissolution – tells us that there are limits to what can be achieved by violent means. Perhaps there is a lesson to learn?

CONCLUSION

Suggested points for the conference's 'lessons' list

1. The international 'community' used ideas, *institutions and means from the Cold War era* that have little, if any, relevance in the Yugoslav case. We must learn to develop new policies and the necessary institutions to handle conflicts more effectively.
2. *Succesful conflict-resolution has at least four basic elements*: (1) comprehensive diagnosis, (2) accurate prognosis and (3) a cohesive plan for treatment as well as (4) respect for and cooperation with the conflict-ridden society. We can learn these four practices and do better in the future.
3. To intervene in somebody else's conflict with a view to help the parties is a science and an art. It will do more harm than good if 'third parties' ignore simple rules of conflict-resolution or have mixed motives. Only *people with some minimum of training* should be sent as mediators.
4. *The human dimensions of conflict tends to get lost.* No people expert in that, nor any women, were involved in negotiations and peacemaking in ex-Yugoslavia – presumably a major reason for the failure to bring peace.
5. There is *a vast potential in any civil society* for conflict-prevention, peacekeeping, peacemaking and peacebuilding and for post-war reconciliation. We can learn to employ these resources better and work with civil society about solutions.
6. Most of all, perhaps, violence may regulate a conflict, but can seldom solve it. A military 'occupation for peace' a la Dayton solves none of the original problems or those created by the war itself. The richness of the idea of *peace by peaceful means is ignored because it doesn't fit power politics.*

NOTES

1. The terms non-governmental and NGO is unfortunate. We ought to use positive terms for desirable goals such as non-violence, non-proliferation, non-militarization, non-pollution and non-war. In this case it would, as pointed out by Johan Galtung, be as reasonable to call governments NPOs – non-peoples/popular organizations. Thus, we prefer CSO, Civil Society Organisations.

2. Dr Jan Øberg, co-founder and director of TFF, head of its conflict-mitigation team in former Yugoslavia and Georgia. The team has conducted some 1500 interviews at all levels of society and in all parts of ex-Yugoslavia during 24 missions, leading to some twenty publications, some alternative peace proposals and a series of courses and seminars in the region.

3. This paper deals with former Yugoslavia but also with certain aspects of 'lessons learnt' approaches that, if clarified, may help prevent academic lessons learnt discussions from sliding into mudslinging about the 'correct' interpretation and position.

4. The author does not like this term since animals are more peaceful in that they don't engage in planned mass-killing of their own species.

5. Sections 3.1 and 3.2 builds to a considerable extent on inspiration from John Burton's and associates seminal works at George Mason University, see John Burton 1990, *Conflict: Resolution and Provention*; John Burton (ed) 1990, *Conflict: Human Needs Theory*; John Burton and Frank Dukes (ed) 1990, *Conflict: Readings in Management and Resolution*; and John Burton and Frank Dukes (ed) 1990, *Conflict: Practices in Management, Settlement and Resolution,* all St. Martin's Press, New York.

6. Even though we may focus on conflict it is not enough to categorize conflicts according to the ways in which only states are involved as does e.g. Peter Wallensteen 1994 in his *From War to Peace. About Conflict-Resolution in the Global System*, Almqvist & Wiksell (in Swedish), Stockholm, namely: inter-state conflicts, conflicts about state power and conflicts about state-formation. This may be helpful as a categorization tool but not so for analysing why people behave the way they do in these conflicts and in those conflicts in which states are not participating – such as intra-actor conflicts. But, admittedly, Wallensteen departs basically from a power model rather than from a problem-solving model and provention in the Burton sense of the word does not come to the fore.

7. Robert C. Tucker 1981, *Politics as Leadership*, University of Missouri Press, Columbia.

8. Gene Sharp 1973, *The Politics of Non-violent Action. Power and Struggle*, Porter Sargent Publishers, Boston.

12. THE LESSONS OF YUGOSLAVIA: INTERVENTION BY INTERNATIONAL NGO's

Dorie Wilsnack

INTRODUCTION

Since the war began in the former Yugoslavia, there have been hundreds of debates among European and North American peace organizations about how they could effectively intervene to help end the war with non-violent means. Often it was just talk. But sometimes the words led to concrete action. There are lessons to be learned from their numerous mistakes, but there are also programs and campaigns to point to that were effective in important, small ways. Reviewing these can help set guidelines for future work in the region, as well as conflicts in other parts of the world.

AN EARLY MISTAKE

An early mistake that international peace groups made was that they did not listen. They did not heed the voices of concern from Yugoslav peace activists about the impending violence. Marko Hren, a peace activist at the Centre for the Culture of Peace and Non-violence in Ljubljana and an active member of War Resisters' International, wrote to WRI members on June 28, 1991, right after the Yugoslav federal army attacked Slovenia, "For years we tried to warn you about the situation in Yugoslavia; first about the violations of human rights,

Research on Russia and Eastern Europe, Volume 3, pages 215–226
Copyright © 2000 by Elsevier Science Inc.
All rights of reproduction in any form reserved.
ISBN: 0-7623-0280-1

later about state violence in Kosovo, and over the past few years about the threats made by the federal state against Slovenian rights. What everybody feared, happened yesterday."

Part of the problem was that most peace activists found it difficult coming to terms with the complex political questions raised by the war. The War Resisters' International staff and office responded quickly to requests for workshops in non-violent conflict resolution skills and made public statements immediately in support of draft resisters. But for WRI members, there was much international debate. At a WRI Triennial conference in July 1991, Marko Hren posed a tough question: What should be the position of the peace movement concerning borders? Can this discussion be avoided? What is the relation between self-determination and borders? I don't want to discuss wishful thinking or the old pacifist dislike of borders . . . All pacifists deny the legitimacy of borders, since most borders are a result of war, not of peace. But at the moment, the question of borders is posed concretely, since we find ourselves in a time when new borders might be drawn and old ones changed. So what position should we have? Or should we avoid having a position and talk about principles only?

AMERICAN SELF INTEREST

For most Americans, the war was a faraway conflict in an unfamiliar part of Europe. Their awareness was based on television coverage. Their response was passive; they sat and watched and shook their heads to express their frustration and confusion. Fortunately, a few organizations and communities actively concerned themselves with ending the war.

But even among these, unfortunately, their prime motivation was that they saw the war as an opportunity to justify their own particular philosophy or ideology. A related motivation for some activists was self-importance. The war provided them with an opportunity to be the 'people who could help' or 'the people with the answers'. These same organizations held a sincere interest in stopping the violence and protecting the rights of civilians, but too often that seemed to take second place to their political and organizational concerns. Ironically, American activists were behaving no differently than their government representatives.

IDEOLOGICAL LIMITATIONS 1:
THE BOSNIA SOLIDARITY MOVEMENT

The most vocal American protests against the war came from organizations that identified strongly with the Bosnian Muslims, who were the greatest victims of violence and ethnic cleansing. Students Against Genocide (SAGE), Friends of Bosnia, Jews Against Genocide, and the American Committee to Save Bosnia are just a few of the hundreds of organizations that created a nationwide movement in the early 1990s. Their membership came from people who were appalled by what they saw on the television news: the concentration camps, people in Sarajevo running for cover from snipers, the photos of women who reported being raped. They rallied around a campaign to lift the international arms embargo and thus enable the Bosnian government to defend itself against the Serbs. Their rhetoric carried strong attacks against the Serb community, not making a distinction between Serbs who supported the war and those that did not. In communities and on campuses across the country, they held rallies, organized conferences, lobbied their Congressional representatives, and tried to make the war a visible issue to the American public.

These organizations felt strongly that their point of view would help to end the war. They did not see, and probably would not agree, that their perspective and strategies were heavily influenced by their close affiliations with the Bosnian government and the ruling Party of Democratic Action (SDA). They did not work with independent peace and human rights NGOs in Bosnia. They often presented the SDA's definition of the war and political agenda as their own. They invited government representatives to be the star speakers at their rallies.

The limitations from this close relationship began to show as the complexities of the war in Bosnia became more apparent. Their identification with a single party and its ideology left them unable to adapt to new information, such as when Bosnian opposition parties and political leaders like Haris Silajdžić became more outspoken and critical of the party's lack of commitment to a multi-ethnic society. When the SDA became associated with censoring the independent Bosnian media, employing Iraqi mujahdeen as soldiers, and abusing the human rights of Bosnian Serbs and Bosnian Croats still living in Bosnia, few of the American Bosnia solidarity organizations protested. They did not question the growing Muslim nationalism in the SDA.

The enemy attitude they adopted toward Serbs limited them as well. With notable exceptions, such as Nalini Lasawiecz at Bosnia Briefings, Bosnia solidarity groups gave little support for the pro-democracy movement which surfaced in Serbia in late 1996. When street protests took place in Belgrade, there were lively debates on the Students Against Genocide computer chat board. Many SAGE members expressed their distrust of the student protests, convinced that most students in Belgrade were Serbian nationalists who had supported the war.

IDEOLOGICAL LIMITATIONS 2:
THE PACIFIST MOVEMENT

The summer of 1995 was a sad and desperate one for peace organizations in the former Yugoslavia. The cycle of violence was spinning faster. Seventy-eight young Bosnians were killed while sitting at an outdoor cafe in Tuzla; thousands of Bosnian Muslim women and children were driven from their homes in Srebenica while their husbands and sons disappeared; and over 200,000 ethnic Serbs fled their homes in the Krajina region as the Croatian military reclaimed military control of their lands. That fall, an American peace group, the U.S. chapter of Women's International League for Peace and Freedom, posted their first e-mail message on the computer conference, <yugo.antiwar> which would reach peace activists everywhere, including throughout the former Yugoslavia. In their message, no reference was made to the traumatic times. Instead the letter was to call "upon all individuals and organizations who treasure the hope for peace in the former Yugoslavia to contact President Clinton and urge him not to send any U.S. troops to the former Yugoslavia on their peacekeeping mission, unless there is gender parity amongst those discussing and negotiating the peace process for the region." Whatever the merits of this idea, its irrelevancy and insensitivity to the profound sense of loss local peace activists were feeling, paints a picture of American peace activists bent on their own priorities. It was part of an already familiar pattern.

American pacifist and disarmament organizations made strong statements against the war, supporting non-violent solutions ranging from negotiations to possibly, non-violent resistance. The American Friends Service Committee wrote to President Clinton, "We stand for peaceful means to resolve the complex conflict in Bosnia, not military means. We appeal to you to lead the world to exercise moral imagination and creativity in place of the immorality of violence and destruction." The Fellowship of Reconciliation wrote to its members that any military action aimed at relieving the suffering in Bosnia will only exacerbate the crisis and end with more people killed. Their statements covered the same points: war is not the solution; more arms to the region is not the solution; local peace groups should be supported; and the United States should not get militarily involved.

But most of the national peace organizations did not move beyond these pronouncements to work on the creative strategies they envisioned. And local peace groups and individual pacifists took their cues from these organizational leaders. With one exception, no national peace organization developed special projects or hired organizing staff to work on this issue. The exception was the Fellowship of Reconciliation and their Bosnian Student Program, which brought hundreds of Bosnian young people out of war zones and arranged for them to attend U.S. high schools and colleges while living with American families. For the most part, American pacifists were eager to call for non-military solutions but were not able to suggest strategic ideas for how non-violence could be used effectively, nor were they willing to commit the time and expense of working with NGOs in former Yugoslavia on this challenge.

One limitation was Americans' predominant interest in quick solutions. Their protest statements called for immediate cease-fires and rarely mentioned long term strategies. This was a very different approach than that voiced by the Anti-War Campaign in Croatia in 1993, when they wrote "From the very beginning, we have understood that our work can't have fast results. We do not have the social power to change the current global course of events. Consequently, we have decided to develop long term activities in human rights, peace education, conflict resolution."

In 1995, when the Clinton Administration announced its decision to send troops as peacekeepers in Bosnia, opposition from American pacifists and disarmament organizations was loud. A number of groups declared that this was U.S. colonialism under the guise of peacemaking. Anti-war activists in Croatia and Bosnia expressed frustration with this kind of protest. In the face of a war that would not stop, many of them viewed international military intervention as a viable option. They interpreted the American protests as proof that the principle of pacifism was more important than the lives that were being lost daily.

STRATEGIC LIMITATIONS:
MIR SADA AND SJEME MIRA

In 1993, there was an attempt at an international non-violent intervention in Bosnia. The campaign was initiated by an Italian group, Beati i Construttori di Pace. In 1992, they had sponsored a trip of approximately 500 people, mostly Italian, to Sarajevo, to express solidarity with the people of Sarajevo and to call upon all parties to cease military action against each other.

Later in that same year, they aimed for something more ambitious, an international effort with larger numbers, where people would come to work in peace camps in three Bosnian towns, one in Bosnian-controlled Sarajevo, one in Serbian-controlled Ilidža, and one in Croatian-controlled Kiseljak. The camps would operate throughout the summer, working on local humanitarian projects, monitoring human rights, the participants also acting as citizen-ambassadors.

The reality of the war made the peace camp idea impractical, and the project finally took shape as a two week caravan of peace activists into Sarajevo, traveling without military escort and bringing both material aid and messages of support. The project was co-sponsored by Beati i Construttori and a French humanitarian aid organization, Equilibre. This was a self-invited international intervention; it was welcomed by one peace center in Sarajevo, but there were no local collaborative sponsors or partner organizations. The media was told there would be as many as 10,000 participants. In the end, 2000–3000 people started out on the journey, 840 of them traveled into Bosnia and only one busload arrived in Sarajevo.

Mir Sada was besieged with internal problems from the very beginning. Some participants had joined as a holiday adventure and were unprepared for the confusions, the hardships or the dangers. Other activists on board had conflicting political agendas and this led to long tedious decision-making sessions and infighting. As participant Christine Schweitzer summed up in a report, "Not only did we not reach Sarajevo, but after a few days, organizers and participants split hopelessly. The rest of the story was fourteen days of discussions, until almost everybody was close to a nervous breakdown and all returned home with the feeling, 'never more something like this'."

The first major split came between the two co-sponsors. Equilibre and its followers left, taking most of the walkie-talkies and other practical equipment with them. As the group came closer to areas of fighting, it became apparent that most participants were unprepared for the risks, despite the fact that they had been warned about them and asked to sign release forms in advance. Military checkpoints added one more challenge as most guards were unrelenting in their refusal to let the group pass. Eventually, one bus made it to Sarajevo after experiencing some very dangerous moments on the road.

Mir Sada was considered an embarrassing and frustrating fiasco by both the participants and local Croatian and Bosnian peace activists. One exception was the people who welcomed the final group to Sarajevo, quite moved that people had taken such trouble to reach them. But with all the money and energy spent, this effort at non-violent intervention was considered by most a failure.

Some of the participants tried to learn from the mistakes and embark on another intervention, a much smaller one, three months later. Sjeme Mir (Seeds of Peace) involved 20 activists, mostly Americans. They focused on Mostar, a divided city in the middle of the war zone. Their introductory leaflet stated "We come hoping to speak to people on all sides of the conflict here and to help relieve the suffering. We ask all the people in former Yugoslavia to abandon military force as the means by which you struggle for freedom and justice. We ask you to adopt nonviolent resistance as your means of struggle."

They made no arrangements in advance with Croatian or Bosnian peace groups, but upon arrival, they volunteered for humanitarian aid efforts with local agencies. While individual members may have made valuable local contributions during their visit, the visit was of more obvioius benefit to the participants themselves, and their ability to later report back home on their attempt to bring a peaceful influence to Mostar.

ANOTHER APPROACH: THE BALKAN PEACE TEAM

Fortunately, in the later years of the war, there were other attempts at intervention that were more successful. The Balkan Peace Team (BPT), for example, was started in 1994, as a cooperative effort of European and international peace groups, including Brethren Volunteer Service, Bund für Soziale Verteidigung, Eirene Intenational, hCa Geneva, International Fellowship of Reconciliation, Mouvement puor une Alternative Nonviolente, Peace Brigades International, and War Resisters' International.

Balkan Peace Team was built on the idea to send volunteers to areas in the former Yugoslavia where community tensions were high, and where the presence of international volunteers might change the dynamics in such a way as to alter the immediate potential for violence and make it easier for local human rights and peace organizations to carry out their work.

Inspiration for this project came from Peace Brigades International, an organization which sent volunteers to such countries as Guatemala, El Salvador, and Sri Lanka, where they served as non-violent accompaniment to human rights monitors, trade union leaders and community activists whose lives had been threatened. Balkan Peace Team took the concept one step further, preparing volunteers to provide a wide variety of forms of support to local activists, making use of their international presence, their contacts with outside media, their access to an international activist network, and their other organizing skills.

Balkan Peace Team built its initiative on two principles. The first was that a team would only be placed in a country if there was an invitation from local groups. The second was that once a team was in place, its mandate was to be non-partisan, concerned for all ethnic groups and communities affected by the war. This required a delicate balance, being of service to local NGOs and yet remaining independent from them, and it has been a great challenge to the volunteers. Yet these two elements have provided BPT with an integrity that has helped it to build trust among the local groups and population.

Originally, BPT intended to establish its first team in Kosovo. The tensions there between the predominately Albanian population and the Serbian state were considered a potential flashpoint for war, despite a strong non-violent stand taken by the Kosovo Albanian leadership. The Balkan Peace Team planners considered this to be a valuable place for an international team, and began exploring the idea with local NGOs. The exploration moved slowly, however, and in the meantime, peace and human rights activists in Croatia eagerly invited BPT to place a team in their region.

So the first team began working in Zagreb in March 1993, and six months later, a second team was established in the southern Croatian city of Split. In 1996, the Zagreb team relocated to Karlovaç. In late 1994, a team finally began working in Prishtina, Kosovo. A year later, because of police monitoring, the team's office was moved to Belgrade, but its primary focus remained in Kosovo, spending about 10 days there every month. Their presence in Belgrade allowed them to support Serbian groups that were working on issues related to Kosovo.

Each BPT team had two or three volunteers who have made a one to two year commitment. They received room and board and a small monthly stipend. Before becoming volunteers, they went through a lengthy screening process which included a written application, a detailed face-to-face interview, and a four day assessment workshop. Those who were selected after the assessment had a ten-day intensive training before being placed on a team. Once in the field, volunteers who were not already fluent in Serbo-Croatian or Albanian devoted their first months to language lessons.

The two BPT teams in Croatia worked under the name Otvorene Oči (Open Eyes) and their work concentrated in the human rights field, supporting Croatian NGOs that were involved in human rights monitoring and protection. Their work included attendance at trials; being present when during house evictions, which was a frequent form of abuse against minority Serb residents; accompaniment of local human rights groups on visits to isolated villages in the Krajina region; building links between local groups and international journalists; and assisting groups with the preparation of their funding proposals to European and American sources.

The Balkan Peace Team for Kosovo focused on supporting groups in the Albanian and the Serb communities that were working outside traditional political structures and are interested cross-cultural dialogue. BPTs work there included acting as a point of introduction between individuals and groups in the two communities that were interested in dialogue, and when appropriate, providing encouragement and skills workshops for common projects; it also helped local groups to develop links with international journalists, funders, embassy staff, and European activists. In the winter of 1996–96, the team monitored the pro-democracy street protests in Serbia and provided support to NGO's involved in that effort.

The Balkan Peace Team has continued to work in Kosovo at the grassroots level, adapting to the changing political situation. All three teams file regular public reports on the political situations in their regions, on the work of local

NGOs and on their support efforts. Public reports are made available to the media and to international networks through the Internet. Other reports are shared with a more limited audience of local and/or international NGOs who are engaged in related work.

A GOOD MODEL: THE PAKRAĆ RECONSTRUCTION PROJECT/VOLUNTEERS PROJECT PACRAC

Before the war, Pakrać was a town of 6000, a population mix of ethnic Croats and ethnic Serbs, plus some Hungarians and people with other ethnic identifications. It is situated in farming country in central Croatia. With its mixed population, it saw fierce fighting between August 1991 and the beginning of 1992, as the armies of both sides vied for control. Most of the townspeople fled during this period, and almost all the buildings and homes in the area had some physical damage.

When the heaviest warfare stopped, residents returned and Pakrać became a divided community, with the Serbs on one side of town and Croats on the other. The final cease fire, brokered in 1992, created a demilitarized zone between the Serb-controlled region of Western Slavonia and the Croatian government-controlled territory. In Pakrać, this UN cease-fire line ran right through the center of town. It was enforced by the presence of United Nations Protective Forces (UNPROFOR).

The Pakrać Reconstruction Project was initiated in July 1993 by the Croatian Anti-War Campaign (ARK) based in Zagreb. Their idea was to help the townspeople rebuild their community physically and socially by bringing international volunteers to Pakrać for short term (three weeks) or long term periods to help with the labor and to convey a message of support for the townspeople.

At the same point in time, the United Nations Office in Vienna together with the UN Development Program were working together with the UN Civil Affairs office in the Pakrać area to initiate a Social Reconstruction Project and they were looking for a local NGO to work with that would be willing to work in a non-partisan manner, working on both sides of the cease-fire line. The Pakrać Reconstruction Project became the partner, and took responsibility for bringing in international volunteers to help with the labor. The UN agencies supplied the building materials to the townspeople interested and willing to begin the manual reconstruction of their homes. The local city authorities on the Croatian side, where the first work began, organized their local citizens into working brigades. The homes in the worst conditions were to be dealt with first.

The Pakrać volunteers came for three week sessions, approximately 10–15 in a group, and attached themselves to these brigades or applied their skills in other areas where they could be most helpful. Each new group of volunteers went through an orientation session beforehand. Eventually, many volunteers decided to lengthen their stay and a category for long-term volunteers was established.

The international volunteers, as they worked under the coordination of local Croatian activists, brought much more to Pakrać than their physical labor. Their youthfulness, friendliness and creativity led them to establish a number of additional projects to help the townspeople: classes and programs for the school children, cultural events and dances, visits for tea with elderly and homebound residents. Eventually, there was a newsletter and a weekly radio program. This attention provided Pakrać residents with a message, "You have not been forgotten," which was as important for reconstructing the community as the bricks and mortar supplied by the United Nations.

From its beginnings, the Pakrać Reconstruction Project conceived of itself as an effort supporting the needs of both Serbs and Croats in Pakrać, but it was many months before that became a reality. Having begun on the Croatian side of town, the Project organizers learned that to establish work on the Serb side without destroying the trust they were building on the Croat side would require slow but consistent efforts. It was not possible to get permission to live on the Serb-controlled side. For the first six months, volunteers crossed the demilitarized zone on short trips to make home visits to families in need. Eventually, they began delivering messages and mail between former friends and neighbors. In January 1994, the first group of four international long-term volunteers started living and working on the Serb side. These were the first civilian 'outsiders' to take on this kind of work during the war. As their effort was beginning, the Belgrade-based peace group, MOST, was contacted and they developed a program for volunteers coming from Belgrade to work on the Serb side.

In May 1995, when the Croatian military carried out Operation Flash, wresting control of Western Slavonia from the Serbs, the situation in Pakrać changed dramatically. Thousands of Serb residents in the Western Slavonia region fled to Bosnia, Eastern Slavonia, or Serbia, fearful for their lives and believing that there was no longer a future for them, as Serbs, in the region. There was an exodus from Pakrać as well, but there were also notable and visible signs of Serbs who chose to remain, including the mayor of the Serbian side of town. In the years since then, some residents who fled have returned.

Although there are no comparative figures, there are some observers who say that more of Pakrać's residents considered staying than in other communities nearby.

The Pakrać Reconstruction Project formally ended its work at the beginning of 1997. By that time, more than half of its volunteers and staff were local

Croatian activists. Many of its ongoing programs, such as a traveling puppet show, computer classes for school children, and a traveling tool exchange have now been picked up by local NGOs. The town now hosts a human rights center, a women-run micro-enterprise laundry project, a local computer electronic mail project, and a local youth center. In the meantime, the model of the Pakrać Reconstruction Project has been adapted in other parts of former Yugoslavia, such as the Gornji Vakuf Project in central Bosnia.

IMPORTANT ELEMENTS IN COMMON

The Balkan Peace Team and the Pakrać Reconstruction Project contain elements in common which may explain why they have been successful where other efforts at non-violent intervention have not. In both programs, the international volunteers were invited by local NGOs and activists. They came to the region with flexible agendas and without high expectations, so as to be able to work most creatively with the immediate needs of the communities. During their time in the region, both sets of volunteers were non-partisan, working with people from all ethnic groups, and giving their attention to the people on the ground, not the parties in power. Both programs involved long term volunteers, making it more likely that there would be time for the development of trust with members of the local community. Finally, both programs involved their volunteers in an orientation or training before they went to work in the community, and expected long term volunteers to learn the local language.

Within workable models such as these, it is important to note, that mistakes are made as well. For example, the process of orientation and training developed out of lessons learned when volunteers were not appropriately prepared. The involvement with local activists while maintaining an appropriate nonpartisan distance created a number of misunderstandings before the correct balances were reached. But the larger sense of trust that the projects had developed allowed the space for this kind of learning.

IN THE FUTURE

It is important that the lessons which international NGOs have learned in this region about what effective and ineffective forms of intervention be remembered and considered for other conflicts, in other parts of the world. And these lessons remain very relevant and important during the period of rebuilding and refugee return in ex-Yugoslavia. It is troublesome, then, to hear some peace groups in North America and Western Europe express the view in 1997 that they are needed in the former Yugoslavia to bring skills in conflict resolution and non-violent action. This attitude, which mixes sincere concern and enthusiasm with short-sightedness, political ignorance, and arrogance, could mean more mistakes and misused energy.

Those international activists who know the Balkan region and have an understanding of the recent war and its complexities face a creative challenge when they hear such talk coming from activists in their home countries. Without dampening the spirit of it, we must try to channel the activists' deep opposition to war and their eagerness to help, into efforts that are welcomed and truly needed by the local communities.

PART III

AFTER THE BREAKUP AND DAYTON

13. DIVISION AND DEMOCRACY: BOSNIA'S POST-DAYTON ELECTIONS

Timothy Donais

In mid-1995, the Muslim inhabitants of the eastern Bosnian town of Srebrenica became the victims of one of the worst massacres in Europe since the Second World War. After having laid siege to the enclave for over two years, the Bosnian Serb army finally overran Srebrenica – a so-called United Nations 'safe area' – in July of 1995. Serb forces, under the command of indicted war crimes suspect General Ratko Mladic, were undeterred by the small contingent of Dutch UN peacekeepers in Srebrenica, who became helpless bystanders as upwards of 7,000 Muslim men were killed, while the remainder of the town's inhabitants were sent fleeing westward.

While the tragic fate of Srebrenica will long be remembered as one of the most humiliating failures in the history of UN peacekeeping, the town remains a vivid symbol of the daunting obstacles facing those attempting to bring peace and democracy to post-war Bosnia. In September 1997, the same Muslim women whose husbands, brothers and sons now lie in mass graves around Srebrenica helped elect a Muslim majority to the town's new municipal assembly. While virtually every current resident of Srebrenica at the time was a Serb, Muslim candidates were elected to 25 of 45 seats in the new assembly on the strength of absentee ballots from former residents who remained in exile abroad or in Muslim-controlled areas of Bosnia.[1] Predictably, in the aftermath of the elections Srebrenica's municipal government remained in a state of paralysis – a symbol of ongoing ethnic intransigence rather than a beachhead to inter-ethnic

Research on Russia and Eastern Europe, Volume 3, pages 229–257.
2000 by Elsevier Science Inc.
ISBN: 0-7623-0280-1

reconciliation and reintegration – since few Muslims dared venture back into what was now Bosnian Serb territory. As one Serb politician in Srebrenica told *Agence France Presse* prior to the elections: "I don't think it would be possible for any Muslims to sit on the municipal councils. The first night they come back, it means fighting."[2]

Srebrenica's confused political situation is broadly indicative of the awkward political arithmetic – three mutually-antagonistic ethnic groups, two entities, one country – that has characterized Bosnia's post-war political climate. By early 1998, more than two years after the Dayton Peace Accords brought the civil war to an end, Bosnia's Serb, Croat, and Muslim[3] inhabitants were still driving cars with different licence plates, using different currencies, and arguing bitterly over the look of the new national flag. The fact that these issues were ultimately resolved not by Bosnia's political leadership but by the International High Representative to Bosnia, Carlos Westendorp, who imposed solutions with the help of strengthened powers given to him by an increasingly exasperated international community, only underlines the ongoing insecurity and fragility of Bosnian statehood. Similarly, few of the more than two million Bosnians uprooted by the war have been able to return to their pre-war homes, and many of those who have attempted to return have been greeted with hostility or violence. Thus, while the war may have been brought to an end, Bosnians have yet to achieve genuine peace.

National and local elections were to be the centrepiece of the Bosnian peace-building process which began with the Dayton peace agreement, forged at an abandoned airforce base in the American midwest in November 1995 and signed in Paris a month later. Combined with economic reconstruction and the calming presence of heavily-armed NATO peacekeepers, elections were widely seen as the vehicle through which Bosnians of all ethnicities would begin to put the fear and hatred of civil war behind them and start to rebuild their country. Unfortunately, events have shown that the early optimism of Dayton was mis-placed. Rather than promoting ethnic reconciliation and political reconstruction, Bosnia's post-Dayton elections have further divided an already deeply-riven soci-ety, entrenched the power of Bosnia's ruling nationalists, and solidified ethnic-ity as the sole criteria of social organization in post-conflict Bosnia.

This chapter will examine the three rounds of country-wide elections that have taken place in Bosnia in the first three years of Bosnia's post-Dayton existence. It will suggest that given the poisoned political climate, as well as the absence of fundamental democratic institutions such as a free press and a dynamic civil society, it was entirely predictable that the real victors of Bosnia's post-war democratic experiment would be the nationalists of all three sides. This result is also in large part the product of the flawed compromise at the heart of the Dayton peace agreement, which left the central issue of the Bosnian conflict unresolved, thereby guaranteeing that the central issues over which the war was fought would continue to be played out at the level of the ballot box. By the autumn of 1999, with the fourth anniversary of the Dayton Peace Agreement approaching, Bosnia remained suspended between partition and unity, and between war and peace. And despite recent encouraging signs of progress, it remained far from clear whether history will remember Bosnia's Dayton days as a half-time intermission in the country's civil war, or as a historical turning-point which, despite its difficulties, laid the foundations for a lasting and democratic peace.

THEORETICAL PERSPECTIVES ON NATIONALISM AND DEMOCRACY

In the turbulent years of the late 1980s and early 1990s, as the former republics of the Soviet Union and the states of Eastern Europe struggled to re-organize themselves in the aftermath of the sudden collapse of communism, two phenomena in particular emerged to fill the vacuum left by the dramatic implosion of the communist order. On the one hand, nascent institutions of democracy began to emerge, as populations of the former communist states were finally given a voice in choosing their own leaders. On the other hand, nationalism quickly asserted itself (in some cases re-asserted itself) as a formidable political force and a powerful instrument of popular mobilization.[4]

Many scholars have viewed the resurgence of nationalism in Eastern Europe in negative terms, often describing the region's resurgent ethnic nationalism as an irrational return to a pre-modern and insular 'tribalism' of an inevitably anti-democratic and violent nature. As *The Economist* once put it: "The virus of tribalism . . . risks becoming the AIDS of international politics – lying dormant for years, then flaring up to destroy countries."[5]

Writing in the immediate aftermath of the communist collapse, however, the Georgian political philosopher Ghia Nodia argued that far from being mutually-incompatible, nationalism and democracy "are joined in a sort of complicated marriage, unable to live without the other, but co-existing in an almost permanent state of tension."[6] Nodia's central argument is that throughout modern history, nationalism has been central to defining the social and territorial boundaries of political communities. The idea of self-determination has always been a problematic notion within international relations, largely because no natural or objective criteria exist by which to determine the precise social and political boundaries of self-governing entities. Such boundaries are always historically contingent, the product of either conscious political effort or violent military struggle in which nationalism has played a primary role. As Nodia stressed, "whether we like it or not, nationalism is the historical force that has provided the political units for democratic government . . . the political cohesion necessary for democracy cannot be achieved without the people determining themselves to be 'the nation'."[7] Democracy, in this sense, only becomes possible once the rules of the game have been established and when the community of players and the limits of the playing field have been determined.

Writing in *The Atlantic Monthly*, Robert Kaplan has made a similar case that democracy can only emerge after a state has been firmly established, and that efforts to create states through democratic means are doomed to failure. States, he suggests, have never been formed by elections, but rather by geography and the complex and often violent interactions of ethnicity. The evolution to democracy, he argues, comes only after additional social and economic achievements, such as the emergence of a stable middle class and effective bureaucratic institutions. As Kaplan suggests, "because democracy neither forms states nor strengthens them initially, multi-party systems are best suited to nations that already have efficient bureaucracies and a middle class that pays income tax, and where primary issues such as borders and power-sharing have already been resolved, leaving politicians free to bicker about the budget and other secondary matters."[8]

Ultimately, both Kaplan's and Nodia's arguments come down to the contention that a critical pre-condition for democracy is a widely-accepted sense of 'political community'. As David Welch has argued, the notion of a political community implies the existence within a state of "an inclusive code of political understanding, a shared political culture, commonly respected symbols of statehood, and, most critical, a shared view that the outcomes of the political processes (most notably, elections) are legitimate."[9] In other words, if democracy is to succeed, the ties that bind a political community together must be stronger than the divisions pulling it apart.

However, if a certain cohesive dose of nationalism is necessary to foster a sense of political community, and to define not only the 'demos' but also the territorial limits and political arrangements of a democratic state, on another level the idea of nationalism is fundamentally at odds with the principles of liberal democracy. As an ideal-type, liberal democracy combines the principle of popular sovereignty with the privileging of individual human freedom as the foremost political value and the assertion of universal and equal individual rights.[10] This is encapsulated in the phrase 'one person, one vote', which implies a fundamental equality of all individuals within the democratic arena. However, while liberalism emphasizes the individual, nationalism focuses on collective claims rooted in culture, ethnicity or other unique 'national' characteristics. And where liberalism privileges individual freedom and choice, ethnic or national identities are not chosen but inherited. One cannot choose to be a Serb or a

Croat, for example, in the same way that one can choose to become a socialist, a democrat, or even a Catholic.

As ascriptive categories, national and ethnic identities almost invariably imply certain social hierarchies. As Francis Fukuyama has suggested, "if . . . liberalism is about the universal and equal recognition of every citizen's dignity as an autonomous human being, then the introduction of a national principle necessarily introduces distinctions between people."[11] Thus, the Croatian constitution of 1990 declared the Republic of Croatia to be "established as a national state of the Croat nation and the state of members of other nations and minorities."[12] Such wording, which is echoed in the constitutions of other former Yugoslav republics, treats the state as belonging to a specific nation, and relegates minorities to second-class status as 'historical guests'.[13]

The relatively fixed nature of ethnic identities also heightens the relevance of the democratic notion of majority rule, and the accompanying problematic of the 'tyranny of the majority.' In democratic systems with cross-cutting or shifting social and political cleavages, electoral defeat still leaves open the possibility of forging new political coalitions that may carry the day next time. In societies that are deeply divided along ethnic lines, however, demographics are all-important, since minority status may result in permanent exclusion from political power. As Donald Horowitz has written:

> Democracy is about inclusion and exclusion, about access to power, about the privileges that go with inclusion and the penalties that accompany exclusion. In severely divided societies, ethnic identity provides clear lines to determine who will be included and who will be excluded. Since the lines appear unalterable, being in and being out may quickly come to look permanent.[14]

In ethnically-diverse states, ironically, the transition to democratic forms of governance magnifies the political relevance of ethnic difference. As political parties organize themselves for electoral competition for the first time, ethnic identity provides "a convenient core of symbols upon which to mobilize supporters for the competition."[15] The presence of enduring and mutually-exclusive ethnic identities tends to override other potential sources of political cleavage in newly democratizing states, not least because, as Horowitz has noted, "ethnic affiliations provide a sense of security in a divided society, as well as a source of trust, certainty, reciprocal help, and protection against neglect of one's interests by strangers."[16]

Of course, minority fears of permanent majority rule are heightened to extreme levels in states emerging from beneath the wreckage of civil war. In post-conflict Bosnia, Serbs, Muslims, and Croats alike are justifiably anxious about living as minorities in areas dominated by another ethnic group. Consequently, "'majoritization' has become the defining principle of social organization in post-Dayton Bosnia, as the ruling parties of each camp to greater or lesser degrees endeavor to concentrate their own communities geographically."[17]

The challenges of making democracy work in ethnically-diverse societies have long been recognized by political scholars, and there exists a significant body of literature which explores potential institutional mechanisms through which to reduce or alleviate the inherent tensions between nationalism and democracy. One of the foremost theorists of democracy and ethnic accommodation is Arend Lijphart, whose ideas on power-sharing or 'consociational democracy' have gained widespread prominence. Given the durability of ethnicity as a social category and the subsequent futility of attempting to eliminate ethnic groups by forging them into a homogeneous nation, Lijphart has suggested that "the only solutions to the problems of ethnic division or strife that remain are power-sharing and partition or secession."[18] While he argues that partition should never be rejected outright in cases of chronic ethnic conflict, the considerable disadvantages of partition should always render it the solution of last resort. As the seemingly endless map-making exercises of the Bosnian peace negotiations demonstrate, the complex ethnic intermingling that is characteristic of most states makes it virtually impossible to draw clean and just territorial boundaries between ethnic groups.[19] As a result, large-scale population exchanges – with tremendous costs in both human and economic terms – are usually the inevitable consequence of partition. Given the difficulty of arriving at mutually-acceptable and humane terms of partition, Lijphart argues that "it is almost always better to accommodate political influence and groups in the same state with proper guarantees of political influence and autonomy – the power-sharing approach – than to assign them to separate territorial states."[20]

Lijphart's power-sharing approach includes four central characteristics. The first and most critical is the joint exercise of power by relevant groups, particularly at the executive level, in order to ensure that no significant group is completely excluded from power. The constitution of Belgium, which decrees that the Belgian cabinet must comprise an equal number of Dutch-speakers and French-speakers, provides one example of executive power-sharing.[21] Ensuring that all relevant groups have at least some executive power avoids the dangers of winner-take-all elections, where the losers have little incentive to accept peacefully the election results.

Group autonomy is the second major pillar of the power-sharing approach, and rests on the principle that "on all issues of common concern, decisions should be made jointly by the different groups or their representatives; on all other issues, decisions should be left to be made by and for each separate group."[22] Federalism is the most common institutional embodiment of this principle, particularly in cases where major groups are territorially concentrated.

Lijphart's third criteria is proportionality, according to which political appointments, public funds, and political representation should be divided among major groups according to their share of the overall population. At the level of elections, for example, this principle suggests that proportional representation electoral systems are superior to plurality or 'first-past-the-post' systems, which tend to produce unproportional results. The minority veto is the final characteristic of the power-sharing approach. Even if a minority is included within a power-sharing government, it could still be outvoted on all major issues; the minority veto is therefore "the ultimate weapon that minorities need to protect their vital interests."[23]

As a means of ensuring that the interests of all relevant groups are protected within the principles of democratic majority rule, the power-sharing approach is not without its weaknesses. By institutionalizing principles of group representation within democratic systems, power-sharing can implicitly reinforce the very cleavages it seeks to accommodate. There exists a very real danger, for example, that de-centralized political arrangements will accelerate centrifugal forces within a state, with newly-autonomous groups viewing their autonomy as merely a stepping stone to outright independence. Lijphart is correct to argue that in cases where there is no correspondence between the nation and the state, ethnic identity cannot simply be re-engineered to produce such a correspondence. At the same time, however, any power-sharing solution must balance the benefits of de-centralization and group autonomy with the need to produce and continually reproduce a shared sense of 'political community' which legitimizes the ongoing existence of the state. As Donald Horowitz has suggested, devolution can help avert separatism, but only when combined with policies that give the relevant groups a strong stake in the centre.[24]

At the same time, the power-sharing approach can be a recipe for political paralysis, particularly in cases involving a history of animosity among the groups sharing power. Lijphart himself acknowledges the possibility that the overuse or abuse of the minority veto power, for example, could undermine the entire power-sharing system. Certainly, the dangers of power-sharing paralysis are heightened in post-conflict contexts, where former enemies are often expected to collaborate amicably at the executive level. Bosnia's post-war tripartite presidency, for example, comprising representatives of each of Bosnia's three main ethnic groups, exemplifies the potential for power-sharing to lead to deadlock and acrimony.

Clearly, institutional arrangements in and of themselves are insufficient to resolve the inherent tensions between democracy and ethnicity. Political actors themselves are also critical, since power-sharing cannot work in the absence of a moderate and tolerant political leadership willing to make it work. Scholars such as Donald Horowitz, therefore, have suggested that in severely divided societies, the creation of formal institutional mechanisms of governance is often less critical than "the lifesaving goal of making interethnic moderation rewarding."[25]

Constitutional prescriptions aimed at promoting interethnic accommodation – such as institutionalizing minority vetoes or forbidding parties to make ethnic appeals – have largely failed in practice, suggests Horowitz, because they do not take into account the reality that most politicians place their own self-interest above all. Since political leaders will pursue conflict over accommodation if they perceive such a strategy to be in their interests, Horowitz suggests that the key to interethnic accommodation within democratic systems is "to secure the adoption of electoral and governmental structures that give politicians incentives to behave in one way rather than another."[26] In other words, political systems should be structured in ways that reward moderation and discourage extremism and intolerance.

While creating conflict-reducing political systems may require a broad range of strategies, including provisions for federalism, regional autonomy, or minority vetoes, the centrepiece of any such system is an electoral system that promotes interethnic cooperation. As Horowitz has noted, the most reliable way to make political moderation pay, under conditions of democratic elections, "is to make politicians reciprocally dependent on the votes of members of groups other than their own."[27] The most commonly-cited example of this principle in practice is the short-lived Second Nigerian Republic, which existed from 1979 to 1983. In an effort to temper the destabilizing ethnic tensions within Nigerian society, the republic's constitution required political parties to establish branches in a majority of Nigeria's states, while requiring the victorious presidential candidate to garner at least 25 per cent of the vote in at least two-thirds of Nigerian states.[28] Such provisions were designed to ensure that both parties and presidential candidates could not achieve electoral success solely through narrow ethnic appeals, but rather had to attract supporters across ethnic groups. Despite the ultimate collapse of the republic and the ongoing failure of Nigerian democracy, Horowitz has argued that by forcing politicians to moderate their messages to attract interethnic support, Nigeria's constitutional provisions succeeded not in transcending ethnic differences, but in reducing the likelihood that interethnic conflict would tear the country apart.[29]

Comparable electoral solutions have been proposed as a means of encouraging interethnic reconciliation in post-war Bosnia. However, the ongoing dominance of political forces within Bosnia whose own interests would be undermined by such reconciliation has blocked the implementation of such strategies. Rather than searching for electoral solutions that balance the interests of all of Bosnia's ethnic groups within the context of democratic political institutions, Bosnia's political elites have used elections as a means of advancing their wartime objectives through non-military means. As Susan Woodward has pointed out, the Dayton agreement is more of a ceasefire than a political settlement, and consequently, each side in the Bosnian conflict "is still fighting the war for statehood; only their means of securing territory and national survival have changed."[30] In Bosnia, therefore, nationalism and democracy are indeed joined in a complicated marriage, and ensuring that democratic institutions do not become simply vehicles through which virulent and intolerant nationalisms are perpetuated and legitimated remains one of the key challenges facing Bosnia's would-be peacebuilders.

ELECTIONS AND PEACEBUILDING IN BOSNIA

The Dayton Dilemma

One of the central reasons for the ongoing peacebuilding impasse within Bosnia is the failure of the Dayton Accords to settle the central issue over which the war was fought: would Bosnia remain a united, multi-ethnic country or would it be partitioned along ethnic lines? As Chester Crocker and Fen Osler Hampson have argued, "Dayton created the outlines of a political transition toward an ultimate political settlement, but the nature of that settlement remains ambiguous in the extreme."[31] Indeed, the very ambiguity of the Dayton deal was the key to securing the agreement of the three warring sides in the first place. Dayton meant, and continues to mean, fundamentally different things to the different parties. For Bosnia's Muslims, Dayton upheld, and promised to protect with international military force, the ideal of a single, united Bosnia. For the Bosnian Serbs, represented in Dayton by Serbian President Slobodan Milosević, Dayton legitimized the ethnically-pure Serb entity that had been created through force over more than three years of war.[32] At the same time, Dayton left Bosnia's Croats in control of much of Western Bosnia, and therefore strategically well-placed for an eventual territorial union with Croatia proper should the re-constituted Bosnian state ultimately collapse.

The agreement that emerged from Dayton represented a delicate and uneasy balance between maintaining a single, unified Bosnia and carving the country up into ethnically-defined statelets. The result was a single Bosnia which maintained its pre-war borders but which was divided into two relatively autonomous 'entities'. Under Bosnia's new post-Dayton architecture, 51 per cent of Bosnian territory would be governed by the fractious Muslim-Croat *Federation of Bosnia and Herzegovina*, with the other 49% – the entity of *Republika Srpska* – under the control of Bosnia's Serbs. In order to keep the peace along the 1,000-km inter-entity boundary line (IEBL) dividing Bosnia's two halves, Dayton also authorized the creation of a 60,000-strong NATO peacekeeping force.

The political institutions created at Dayton also reflect the agreement's creative ambiguity between unity and partition. A three-member collective presidency for all of Bosnia – comprised of one representative elected from each of Bosnia's Muslim, Serb, and Croat communities – was established, along with a bicameral Parliamentary Assembly in which legislators from both entities would sit. Formed on the basis of power-sharing principles, these central institutions were given primary responsibility for Bosnia's foreign policy, as well as for international and inter-entity aspects of trade, transportation, communications, law enforcement and economic policy. Most other aspects of governance, including the critical issue of defence, were left in the hands of the entity governments.[33] As Janusz Bugajski has argued, real power in the new Bosnia lies not with the weak central government, but rather with the two entities, which "will be able to veto legislation; paralyse important policy initiatives; operate their own economic, military, and security structures; and consolidate their sovereignty and independence."[34]

According to the Dayton blueprint, Bosnia's new federal political structures, as well as municipal assemblies in both entities and 10 cantonal assemblies within the Muslim-Croat Federation, were to be brought into existence through nation-wide elections held no later than nine months after the entry into force of the peace agreement. The extremely narrow time frame between the Dayton agreement and the elections, as well as the fact that the vote was to take place within the context of a fundamentally unresolved conflict, virtually guaranteed that the electoral process would be plagued with problems from the start. As Crocker and Hampson noted in the lead-up to the September 1996 poll:

> The Bosnian election is so charged with controversy because the United States and its Western allies – at the peak of their diplomatic momentum in Dayton – refused to decide between partition and unity. Instead, they persuaded the parties to agree to both. This postponed the war's ultimate outcome, making the election itself a source of guidance in the suspended Bosnian peace process.[35]

Bosnia's electoral process was also complicated from the outset by the fact that the Dayton agreement placed primary responsibility for building peace on the shoulders of the nationalists who led the descent into war in the first place. Despite the fact that Bosnian Serb President Radovan Karadžić was excluded from the Dayton talks, the representatives of the Bosnian parties at Dayton – including Milosević and Croatian President Franjo Tudjman – upheld the interests of the three main nationalist parties who ruled during the war and who would subsequently be contesting the elections. In the words of Bogdan Denitch, the international community decided early on in the Dayton process "that it was easier to deal with the authoritarians in power, no matter how unlovely, than to help build up a democratic and non-nationalist opposition or institutions of civil society like the numerous existing non-governmental organizations."[36] As a result, one of Dayton's key political legacies thus far has been to strengthen the hand of nationalists on all sides and further entrench ethnicity as the sole relevant criteria of social organization in post-war Bosnia. As Jonathan Landay has suggested, the Achilles' Heel of the Dayton Accords is that "they institutionalize and strengthen the power of Bosnia's nationalist parties and their communist-style bureaucracies and militaries and do little to nurture a rebirth of the war-shattered moderate political middle ground."[37]

National Elections – September 1996

Even in the best of circumstances, creating conditions for 'free and fair' elections amidst the rubble of war-torn Bosnia would have been a monumental achievement. As events unfolded, however, the intransigence of the ruling nationalist parties and the unwillingness of the international community to enforce the civilian provisions of the Dayton Accords virtually ensured that the elections would do little more than ratify the continued rule of nationalists on all sides of the conflict.

From the outset, Bosnia's electoral timetable was guided at least as much by external considerations as by the political realities on the ground. The fact that Bosnia's elections were scheduled to take place in the midst of an American presidential election campaign has generated much speculation that the timing of the Bosnian elections had more to do with the Clinton Administration's desire for a foreign policy triumph on the eve of American elections than with a genuine desire to bring democracy to Bosnia.[38] As Michael Williams has argued, the international community, and particularly the United States, appeared determined to push ahead with the Bosnian vote regardless of whether conditions for 'free and fair' elections had been met, and "if the earth needs to be declared flat in the process, so be it."[39]

As a consequence of Dayton's compressed electoral timetable, opposition parties on all sides were effectively denied the opportunity to organize themselves as credible alternatives to the ruling nationalists. At the same time, holding elections so close to the end of the war virtually guaranteed that a traumatized and embittered electorate would vote along ethnic lines. In the words of David Rieff, "in the ruin that Bosnia has become, people are more vulnerable than ever to appeals based on lowest-common-denominator politics: the politics of hatred and revenge."[40]

Similarly, the intransigence of the ruling parties on all three sides of the conflict further complicated the task of creating conditions for free and fair elections in the lead-up to the September vote. The ruling nationalists maintained a virtual media monopoly throughout the campaign, and since the electoral rules did not compel the ruling parties to campaign directly against each other or to seek support across ethnic lines, they used their control over the media both to buttress their own support and to silence or smear their internal opposition. In Republika Srpska, the reporting of the official Bosnian Serb media was so offensive and biased in favour of the ruling SDS party that former International High Representative Carl Bildt accused them of broadcasting propaganda that "even Stalin would be ashamed of."[41] In Croat-controlled areas, both local Bosnian Croat media and the main television network from Croatia proper equated a vote for the ruling HDZ party with a vote for the Croat nation.[42] News reporting in Muslim-controlled areas of the Federation, meanwhile, was generally more even-handed, although coverage by the state-owned television network increasingly favoured the ruling SDA party of President Alija Izetbegović as the campaign progressed.

At the same time as they were denied access to the media, opposition parties across Bosnia were subjected to harassment, intimidation, and outright violence. In the most widely reported incident, former Bosnian Prime Minister Haris Silajdžić, who had left the ruling SDA party to lead the opposition Party for Bosnia and Herzegovina, was beaten up by SDA supporters at a campaign rally in the northern Bosnian town of Cazin. On all three sides of the ethnic divide, opposition party rallies were regularly disrupted by ruling party supporters, while opposition members were often threatened, beaten, or fired from their jobs for their political activities.

No less damaging to efforts to create a neutral political environment across Bosnia was the continued presence on the political scene of indicted war criminals – in particular former Bosnian Serb President Radovan Karadzić. Although the Dayton Accords stipulated that indicted war criminals could neither hold public office in the lead-up to the elections nor run as candidates, Karadzić's image was omnipresent during the election campaign in Republika Srpska. In addition to making a mockery of the international community's commitment to bringing war criminals to justice, Karadzić's presence and his ongoing influence within the ruling SDS party during the campaign were direct challenges to virtually everything that Dayton represented. As Michael Dobbs of *The Washington Post* reported: "Balkan experts in the (U.S.) administration and elsewhere agree that the goal of a self-sustaining peace in Bosnia will remain illusory as long as people like Karadzić stand in the way of even a minimal reintegration of the country."[43]

Freedom of movement was yet another critical component of a 'neutral polit-ical environment' that was not achieved in the lead-up to elections. Despite promises by authorities in both entities to facilitate travel throughout the country and especially across the IEBL, "individuals who ventured into areas or enti-ties not under the control of their own ethnic group were often threatened, subjected to violence, detained, or even murdered."[44] And while conditions were difficult for displaced persons wishing to visit their former homes on the other side of the IEBL, the situation was even worse for refugees and displaced persons wishing to return permanently to homes in areas from which they had been ethnically cleansed. While the right of refugee return was stipulated in the Dayton Accords, by the time of the elections only 200,000 of Bosnia's more than 2.5 million refugees and displaced persons had returned, and of those only a handful had returned to areas in which they would be an ethnic minority.[45] As Jane M.O. Sharp has suggested, in the absence of international protection, "refugees cannot be expected to return to communities where those who murdered and raped their loved ones not only remain free, but in some cases run local police stations."[46] Many of those who did make the effort to return were openly terrorized by members of the majority group; returnees often faced open discrimination, outright violence, and attacks on their homes. Indeed, popu-lation movements in the months leading up to the elections left Bosnia more 'cleansed' than when the war ended, as some 90,000 people living in vulner-able minority situations left their homes and moved into majority areas between the signing of the peace agreement and the September elections.[47]

The voting rights of Bosnia's refugees and displaced persons in fact became a key strategic battleground in the run-up to elections. While the electoral regu-lations stipulated that most Bosnians were expected to vote in their pre-war constituencies (either in person or by absentee ballot), exceptions were made for uprooted persons wishing to live and vote in a new municipality. Authorities in both Republika Srpska and in Serbia proper attempted to take advantage of this exception to engineer Serb majorities in key strategic towns. Since many of the municipalities now comprising Serb-controlled Bosnia had Croat or Muslim majorities before the war, the prospect of those displaced populations electing non-Serbs to municipal councils in Serb-dominated towns represented a serious problem for Bosnian Serb authorities intent on consolidating their territorial gains. Thus, of some 123,000 Bosnian Serb refugees residing in Serbia, some 31,000 were 'assigned' to vote in the disputed town of Brcko, while an additional 20,000 were registered to vote in the formerly Muslim-majority town of Srebrenica. As the International Crisis Group reported, "in practical terms, all formerly Bosniak-majority municipalities were strategically stacked with Serb refugee votes."[48]

Similar strategies were employed within Republika Srpska, where displaced Serbs were systematically pressured into registering to vote in Serb-held municipalities instead of in their pre-war municipalities within the Federation. In the front-line town of Doboj, for example, officials in the SDS-dominated Commission for Refugees and Displaced Persons declared that only those displaced persons who registered to vote in municipalities within Republika Srpska would be eligible to receive housing or humanitarian assistance.[49] Despite the determination of the international community to hold the elections on time, the extent of the electoral engineering, which violated the spirit if not the exact letter of the Dayton agreement, could not be ignored. In late August, less than three weeks before election day, the Organization for Security and Cooperation in Europe (OSCE), the international agency charged with overseeing the electoral process, announced that the municipal component of the elections was being postponed due to "widespread abuse of rules and regulations."[50]

Elections for higher-level political bodies, including Bosnia's tripartite collective presidency, went ahead as scheduled despite the registration imbroglio and despite repeated warnings by international non-governmental agencies that the elections would undermine rather than further the goals of Dayton. Human Rights Watch, for example, declared in an extensive pre-election report that "elections that are conducted under current conditions – where persons indicted for war crimes monopolize the media, using it for their own nationalistic goals; and those who would voice an alternative, multi-ethnic view of Bosnia and Hercegovina are silenced – will only consolidate the power of the extremists."[51]

Despite the turbulent lead-up to the elections, polling day itself unfolded peacefully, if not unproblematically. Inaccurate voters lists disenfranchised many voters, while restrictions on freedom of movement meant that out of some 150,000 voters expected to cross the inter-entity boundary line to cast their ballots in their pre-war communities, only about 15,000 made the journey.[52] And many of those who did make the trip, along with many refugees and displaced persons voting by absentee ballot, received a rude shock when they received their ballot papers. Bosnia's new constitution decreed that the Bosniak and Croat members of the tripartite presidency would be elected from the territory of the Federation, and the Serb member from the territory of Republika Srpska. Consequently, those Muslims and Croats who had been driven from their homes in what was now Republika Srpska but had opted to vote in their pre-war municipalities received presidential ballot papers printed in Cyrillic and containing only the names of Serb candidates. Similarly, Serbs who had chosen to remain within the Federation or those who had been displaced but chose to vote in their pre-war municipalities found themselves unable to vote for a Serb presidential candidate. Predictably, many who found themselves in this situation chose to spoil their ballots. As one embittered Muslim refugee from Banja Luka, voting by absentee ballot from the Croatian city of Rijeka, complained: "How can we vote when the only candidates on the ballot are war criminals?"[53] As it turned out, the winner of the Serb presidential contest, Momcilo Krajisnik, was arrested on war crimes charges by NATO forces on April 3, 2000.

In the end, as most observers had predicted, Bosnia's main nationalist parties sailed to easy victories. In the presidential contest, SDA candidate Alija Izetbegović and HDZ candidate Kresimir Zubak took the Bosniak and Croat presidential slots, each garnering well over 80 per cent of the popular vote. Momcilo Krajisnik of the ruling SDS party won the Serb seat on the three-member presidency with some 67 per cent of the popular vote. Ironically, Krajisnik's less overwhelming margin of victory was largely attributable to anti-SDS absentee votes of Bosniak and Croat refugees and displaced persons.[54] Similarly, Bosnia's three main nationalist parties swept 35 of the 42 seats in Bosnia's new House of Representatives. As Susan Woodward has noted, while the international community had hoped that Bosnia's first post-war elections would cue a triumphant international withdrawal from Bosnia, the results made that scenario seem highly unlikely: "Far from producing a smooth transition and an easy exit for IFOR (the NATO Implementation Force), the election predictably gave the democratic stamp of approval to the three nationalist parties that had waged the war."[55] In the aftermath of the elections, creating a united, peaceful and democratic Bosnia seemed less likely than ever.

Municipal Elections – September 1997

In many ways, the most significant aspect of Bosnia's municipal elections is the fact that they took place at all. Postponed no less than four times, and threatened with boycotts by the major parties up until the last minute, the elections finally took place on September 13–14, 1997, precisely one year after they were originally scheduled.

Despite the delays, Bosnia's political climate had changed little by the time Bosnians returned to the polls to elect local authorities. The national elections of the previous year had indeed entrenched the power of the ruling nationalists, whose mutual animosity continued to hamper progress towards peace and reconciliation. The ruling parties of all three sides also maintained their grip on the major media outlets in areas under their control, severely limiting the ability of non-nationalist opposition forces to make their voices heard. And even though SFOR – the trimmed-down NATO 'Stabilization Force' that replaced IFOR at the end of 1996 – stepped up its efforts to arrest indicted war criminals, most of Bosnia's prominent war crimes suspects remained at large.

The absence of freedom of movement and the glacial rate of refugee return remained serious obstacles to the peacebuilding process. Displaced persons attempting to visit gravesites or former homes on the other side of confrontation lines continued to suffer harassment or worse, while refugees attempting to return to their former homes fared little better. When several hundred Muslims attempted to return to their former homes in Croat-controlled Jajce, for example, they were turned back by an angry mob. When the returnees made a second attempt several days later, land-mines had been laid around several of their houses.[56] The prospects for refugees and displaced persons wishing to return what was now Republika Srpska was even worse. "It's like trying to persuade Hitler to take the Jews back," *The Washington Post* quoted one frustrated Western official as saying. "They (the Bosnian Serbs) think they fought a pretty good war and got rid of all these Muslims and Croats and that it's a preposterous idea to take all of them back."[57]

In spite of the difficult political environment, the OSCE invested considerable institutional resources to overcome the problems that had precipitated the original postponement of the municipal elections. In an effort to produce accurate voters lists and head off the possibility of widespread electoral fraud, the OSCE oversaw a Bosnia-wide voter registration process during the spring of 1997. Some 2.52 million Bosnians were registered during this period, including some 535,000 refugees residing outside of Bosnia.[58] Similarly, the Provisional Election Commission tightened the rules for those who did not wish to vote in their pre-war municipalities. Voters within Bosnia had to provide documentary proof of continuous residence in a new municipality in order to be given the right to vote there. And while refugees retained the right to register to vote in a municipality to which they intended to move, they were required to demonstrate a "pre-existing, legitimate, and non-transitory nexus with the future munici-pality," such as a title to property or an offer of employment in the new municipality. Since refugees were further required to travel to their intended municipality of residence to present their documentation, few ultimately bothered. In the end, less than a thousand refugees registered under the 'future municipality' rule, limiting severely the potential for electoral engineering.

Despite these efforts, however, the OSCE did uncover serious cases of fraud throughout the registration period. In tightly-contested municipalities such as Serb-controlled Brcko, for example, officials were caught issuing fraudulent identity papers to Serb refugees in an effort to boost the number of Serbs voting for the municipality. Such manipulation was not restricted to Serb-held areas; falsified documents also turned up in contested municipalities within the Muslim-Croat Federation, such as in Croat-controlled Žepce where the OSCE de-certified a number of leading HDZ candidates as a result of the ruling Croat party's efforts to manipulate the registration process.

Across Bosnia, the lead-up to the municipal elections saw a process of 'majoritization' similar to that witnessed in the months leading to the 1996 national elections. Given the significant political authority vested in the municipal assemblies, both the Serb and Croat ruling parties viewed electoral victory in municipalities under their control as critical to consolidating war-time territorial gains. Hence the efforts of the Bosnian Serbs to achieve electoral victory at any cost in Brčko, a key strategic town in northeast Bosnia which sits on a narrow land corridor linking the two halves of Serb-controlled Bosnia. The Bosnian Muslims, on the other hand, saw the municipal elections as an opportunity to regain a political foothold in territories, particularly in Eastern Bosnia, from which they had been brutally cleansed during the war.

In the end, therefore, Bosnia's municipal elections were above all a struggle among the country's nationalist parties to consolidate and expand territorial holdings or to reinforce claims over lost territory. Within this context, non-nationalist parties were little more than bit players, ultimately winning only six per cent of council seats across Bosnia.[59] While non-nationalist parties did erode nationalist support in some areas, most notably in the northeastern city of Tuzla, Bosnia's municipal elections did far more to harden the country's ethnic divisions than to overcome them.

While the three ruling parties emerged from the municipal elections with a clear majority of council seats, the most interesting and potentially volatile outcomes were in places like Srebrenica, where candidates representing displaced voters won clear majorities. Such results produced de facto governments-in-exile not only in Srebrenica, but in several towns within the Muslim-Croat Federation where Serb-majority assemblies were elected.

Ensuring the implementation of local election results therefore proved to be as great a challenge as organizing the elections in the first place, particularly since most assemblies had at least some representation from minority ethnic groups. The scope of the challenge became apparent in early 1998, when Srebrenica's Muslim councillors, along with their escort of UN police, were turned back by an angry Serb crowd as they attempted to enter the town for the inaugural meeting of the municipal assembly. By April, Srebrenica was still without a functioning municipal government, and the OSCE appointed an interim executive board charged with overseeing the municipality's affairs while the search continued for a power-sharing agreement acceptable to all parties.

As a result of arduous negotiations, however, by mid-1998 virtually every Bosnian municipal assembly outside of Srebrenica had received final certification. Nevertheless, in many cases the delicate power-sharing arrangements required to sustain functioning local multi-ethnic governments have been strained to the point of collapse. The Croat-controlled town of Drvar provides a particularly dramatic example of these tensions. Absentee votes from displaced Serbs, who had formed the vast majority of Drvar's pre-war population, gave a coalition of Serb parties a majority in the town's municipal assembly and helped ensure that the new mayor of Drvar would be a Serb. This arrangement, which included a Croat deputy mayor, did not survive the first sustained Serb returns to the region. In the aftermath of an April 1998 incident in which two elderly Serb returnees were found murdered in a village near Drvar, the International High Representative dismissed the deputy mayor. Croats in the town reacted by staging a riot in late April, during which offices and vehicles belonging to international organizations were torched and the Serb mayor was badly beaten. By the autumn of 1999, tensions in Drvar remained high and the town still lacked a functioning municipal government.

Despite the drama of the implementation process, in many ways the results of the municipal elections were overshadowed by the power struggle which emerged in mid-1997 between the then-President of the Republika Srpska, Biljana Plavsić, and her predecessor Radovan Karadzić. Supported by the West and more specifically by NATO peacekeepers, who helped her gain control of key television transmitters across Republika Srpska, Plavsić began to progressively weaken the grip of Karadzić and the ruling SDS party in Serb-held Bosnia. Although her Serb National League was formed too late to contest the municipal elections, Plavsić orchestrated a new round of elections for the Bosnian Serb parliament in November 1997. While the President and her supporters failed to win an outright majority in new elections, they did manage to prevent Srpska's hardline nationalist parties from capturing a majority.

More significantly, Republika Srpska's new electoral landscape allowed Milorad Dodik of the Independent Social Democrats to emerge as the Serb entity's new Prime Minister in mid-January 1998. Dodik is widely considered to be a moderate democrat who supports the Dayton Peace Accords and the eventual reintegration of Serb-held Bosnia with the Muslim-Croat Federation. With the international community prepared to offer Dodik's new government significant support in both economic and political terms, his election appeared to represent, in the words of *The Washington Post,* "a significant step forward in the difficult process of restoring peace and building a new nation out of the wreckage of the post-Yugoslavia war."[60]

National Elections – September 1998

The results of Bosnia's second round of post-Dayton national elections – held precisely two years after the first – represented another major setback for international efforts to instill the principles and practices of civic democracy in Bosnia. While there were some bright spots, most notably the defeat of Karadzić loyalist Momcilo Krajisnik in the race for the Serb member of the Bosnian Presidency, more generally the results reflected a continuation of the same political trends which marked the 1996 elections.

At the cantonal level within the Muslim-Croat Federation, for example, an SDA-dominated coalition won absolute majorities within all predominantly Bosniak cantons, while the HDZ retained a stranglehold on power in both Croat-majority cantons.[61] And despite a well-publicized split within the HDZ which saw the Croat member of the joint presidency, Kresimir Zubak, break off to form his own party, both Zubak and his New Croatian Initiative failed to significantly erode HDZ support on polling day. For those looking for signs of a weakening of the forces of nationalism within Bosnia, most disappointing of all was the result of the election for the president of Republika Srpska. Despite the massive influx of international assistance into Republika Srpska which followed Biljana Plavsić's successful campaign against the Pale-based Karadzić clique, both Plavsić and her Western backers were repudiated on election day. In one of the most closely contested races of the campaign, Plavsić was narrowly defeated by the hard-line Nikola Poplašen of the Serb Radical Party. Plavsić's defeat was a major setback for Western policy in Republika Srpska, and a blow to the belief that economic incentives could turn Bosnia's Serbs away from nationalism.

For those in the pro-Dayton camp, the 1998 election results were particularly disappointing in light of the amount of effort expended by the international community to level the political playing field in Bosnia and give non-nationalist parties every opportunity to compete on equal terms with the nationalists. In the lead-up to the elections, the Provisional Election Commission banned paid political advertising on all broadcast media, and required radio and television stations to provide 'fair and equitable' amounts of free broadcast time to all political parties. Similarly, the OSCE opened Political Party Resource Centres across the country whose services, while theoretically available to all political parties, were primarily directed at helping opposition parties organize and mobilize themselves. Despite these efforts, however, with a few notable exceptions the results revealed Bosnia's non-nationalist opposition parties to be fragmented, disorganized, and largely ineffectual at countering the appeal of the main nationalist parties.

The 1998 election results also lent renewed credibility to the argument that the electoral system constructed at Dayton was itself partly responsible for the continued dominance of nationalist parties within Bosnia. As the International Crisis Group argued in the aftermath of the elections, the results were entirely predictable, since the electoral system reinforced the same nationalist dynamic that had dominated Bosnia political life for almost a decade:

> The results are simply the latest manifestation of a political system which panders to extremists and does not afford Bosnians the luxury of forsaking nationalism. Electors fear living under the ethnic rule of another community and therefore vote for the most robust defence of their own interests, this sustaining a vicious cycle of fear and insecurity.[62]

The ICG's own recipe for overcoming this cycle was to adopt a new electoral system which would first and foremost guarantee the 'ethnic security' of all three of Bosnia's ethnic groups, thereby freeing up voters to concern themselves with other political issues. By early 1999, in fact, electoral reform was very much on the agenda in Bosnia, as a draft Permanent Election Law for the country began to take shape. By late in the year, however, it became clear that the new law – drafted by a team of international and national experts under the guidance of the OSCE and the Office of the High Representative – would not radically change the Bosnian electoral system. In large measure, this was due to the fact that the working group was mandated to produce a law that was consistent with Bosnia's Dayton constitution, which entrenched many of the principles which critics of the system found so objectionable. In any case, a major overhaul of the electoral system would require a constitutional amendment, an unlikely prospect given the ongoing control over Parliament by the major nationalist parties and the fact that any amendment would almost by definition undermine the interests of the nationalists. If the hold of the nationalists over the Bosnian political system is to be broken any time in the near future, therefore, is seems improbable that electoral reform will be the vehicle through which this feat will be accomplished.

In the aftermath of the 1998 elections, Bosnian democracy seemed increasingly under siege on another front, this time from the international community itself. In late 1997 the Bonn Peace Implementation Council granted the High Representative increased powers to impose decisions on key political issues and to remove obstructionist local officials. Given the failure of Bosnia's elected representatives to come to agreement on any key issues – from the look of the national flag to the design of the Bosnian currency – the High Representative increasingly used his new powers to impose solutions. This trend has led many observers to suggest that Bosnia is becoming a 'creeping protectorate', with executive authority increasingly concentrated within the person of the High Representative as Bosnia's own elected authorities continue to be either unwilling or unable to carry out normal decision-making functions. As the ICG has noted, "In order to get the peace process moving in Bosnia, the international community has had to run roughshod over the country's democratic institutions," thus allowing Bosnian politicians "to wash their hands of responsibility for the reconstruction of their own country."[63]

The most dramatic manifestation of this dilemma was the dismissal of Republika Srpska President Nikola Poplašen in March 1999 for his continued obstruction of Dayton implementation. Not only did the dismissal raise serious questions about the legitimacy of an internationally-appointed High Representative firing a democratically-elected President, it also touched off a constitutional and governance crisis in Republika Srpska that continued to fester throughout the remainder of 1999. While the new High Representative, Wolfgang Petritsch, has made 'local ownership' of the Dayton process the cornerstone of his mandate, the dilemma between allowing ongoing political paralysis and imposing decisions by international fiat remains. The real challenge, as a recent report by the European Stability Initiative suggests, is to ensure that the High Representative's powers are used to promote democracy over the long run rather than undermine it.[64]

Beyond Ballot-Box Democracy

While war-torn countries, as James Schear has rightly pointed out, are never fertile ground for elections,[65] the conditions under which Bosnia's post-Dayton elections have taken place could scarcely have been less hospitable. In addition to the difficulties inherent in holding democratic elections in the aftermath of any violent civil conflict, the inability of Dayton's architects to settle the fundamental conflict over which the war was fought ensured that post-Dayton Bosnia would continue to be dominated by the same logic of ethnic nationalism that produced the war in the first place. Held within the context of an unresolved conflict over territory, sovereignty, and self-determination, and subject to the manipulation of dominant nationalist forces more intent on pursuing their wartime goals than on creating conditions for genuinely free and fair elections, Bosnia's elections were almost inevitably hijacked by the ongoing struggle over the shape of post-Yugoslavia Bosnia.

Dayton's uncomfortable compromise between partition and unity also generated ongoing tensions between national and individual rights within post-war Bosnia. By carving out ethnically-defined territories on Bosnian soil, the Dayton deal implicitly legitimized collective Bosnian Serb claims to nearly half of Bosnia. At the same time, however, the peace accord also recognized the rights of ethnically-cleansed individuals to return to their original homes. The continual collision between the rights of refugees and displaced persons to 'uncleanse' large areas of the country by going home, on the one hand, and the collective rights of particular ethnic groups over specific territories, on the other, further entrenched ethnicity and territory as the central issues of Bosnia's recent electoral campaigns.

In the absence of even the barest outlines of a post-war Bosnian 'political community', the central political institutions created at Dayton have not been sufficient in themselves to ensure that all parties have a stake in an effectively functioning Bosnian state. Intransigence and confrontation have characterized Bosnia's central institutions from the start, and much of the progress that has been made has required intense international pressure. Bosnia's Serbs, meanwhile, have spent much of the post-Dayton period fiercely resisting any reintegration of their entity with the rest of Bosnia, while much international energy has been expended shoring up the chronically fragile Muslim-Croat Federation.

At the same time, Bosnia's recent electoral experiences have also underlined the reality that building democracies from the ground up in war-torn societies means more than simply organizing elections. Equally fundamental are efforts to open up political space in which the nascent institutions of a democratic civil society – including non-governmental organizations, independent media, and a viable political opposition – can develop and prosper. Indeed, given the virtual gridlock at the top of Bosnia's political system as a result of the mutually-hostile and the largely uncooperative behaviour of the ruling nationalist parties, progress towards a unified and democratic Bosnia may ultimately require less dependence on the ability of Bosnia's nationalists to work together and more emphasis on activities further down the political spectrum. As Michael Sells noted in the aftermath of the 1996 elections, "if peace is still possible, it will rely on the strength of non-state democratic institutions: unions, independent media, political coalitions and other groups."[66]

In fact, the key to creating a united, democratic Bosnia may lie not in holding nominally democratic elections but in doing more to build up the institutions of civil society that are essential both to the effective functioning of a democratic society and to the creation of a cohesive political community. Bogdan Denitch, for example, has argued that rather than continuing to expend political capital legitimizing Bosnia's nationalists, the international community should instead be pouring its support into Bosnia's emerging institutions of civil society. Without massive support for such core elements of democracy, Denitch suggests that it is 'rank hypocrisy' for the West to expect Bosnia to remain united and to develop stable, democratic institutions.[67] At the same time, other observers have suggested that the international community's obsession with elections has in fact come at the expense of civil society development. As a recent U.S. Institute of Peace report noted: "Managing numerous complex elections has diverted resources and human capital from other equally important areas of democratic development, such as building civil institutions and encouraging civic education programmes."[68]

While attempts to foster democratic, non-nationalist alternatives through the development of a strong Bosnian civil society hardly represent a 'quick fix' to the problems of Bosnian unity, they are more promising than the hope that Bosnia's nationalists will suddenly discover the virtues of inter-ethnic tolerance. While civil society development has been a component of the international intervention in Bosnia since Dayton – with some notable successes – there is little doubt that much more could be done in this area. A re-focussed peace process which prioritized support for women's groups, students' and workers' organizations, civic fora, multi-ethnic opposition parties and independent media outlets would offer some hope of gradually re-building links between Bosnia's divided communities. Such a strategy would also help propel new voices of tolerance and political moderation onto the Bosnian political stage, giving voters the possibility of real choice in future elections. Such a result, as Jonathan Landay has suggested, "would be welcomed by ordinary Bosnians of all ethnic stripes, who are exhausted by the politics of division and hate, including moderate Serbs forced to remain silent by Karadzić's police and thugs."[69]

The minimalist strategy with which NATO has approached its peacekeeping tasks has also hampered the implementation of the Dayton Accords and limited the peacebuilding potential of Bosnia's recent elections. As one former U.S. diplomat commented in the aftermath of the 1996 elections: "The defining moment of the post-Dayton process was the flat refusal of NATO to do anything other than defend itself and enforce the military separation line."[70] Indeed, a more proactive approach by NATO forces could have done much to improve the environment within which elections were held. For example, a more aggressive NATO stand against war criminals, while not without its risks, could have made a valuable contribution to the peacebuilding process. In addition to removing some of Bosnia's worst nationalists from the political scene, the dispatch of prominent war crime suspects to The Hague to stand trial offers a potentially vital tool of Bosnian reconciliation, by underlining the fact that "individuals – not nationalities – are guilty of war crimes, and that the momentary political leaders do not necessarily speak for the national communities they claim to represent."[71]

NATO's peacekeeping troops could have made an equally valuable contribution to establishing a neutral political environment for elections by facilitating freedom of movement and protecting vulnerable minority populations, including returning refugees and displaced persons. Yet despite the fact that the Dayton Accords granted NATO considerable authority to "observe and prevent interference with the movement of civilian populations, refugees and displaced persons, and to respond appropriately to deliberate violence to life and person,"[72] NATO commanders have consistently refused to take on responsibilities for facilitating freedom of movement, creating a secure human rights environment, or for protecting returnees. Ironically, therefore, while the electoral regulations encouraged Bosnians of all ethnicities to vote in their pre-war municipalities, little has been done to guarantee the safety of those who wish not only to vote in their pre-war communities, but to actually return to live in them.

If, as Michael Williams has argued, "the implementation of the Dayton Accords has not done enough to create the space in which 'good guys' could emerge as meaningful political actors,"[73] at least part of the responsibility for this failure must lie with the design of Bosnia's post-war electoral system. Rather than making political moderation pay, Bosnia's electoral rules have in fact contributed to further ethnic polarization by allowing candidates of all three ethnic groups to achieve electoral success through narrow appeals to ethnic solidarity. This structure has benefitted nationalists on all sides of the ethnic divide, who have been able to neutralize their opponents by portraying themselves as the true defenders of the 'nation', while charging their non-nationalist challengers with ethnic betrayal.

The International Crisis Group, for one, has called for Bosnia's electoral rules to be revised in order to make successful political candidates reliant on support across the ethnic spectrum.[74] In a 1997 report, the ICG suggested that in future elections, the proportion of seats to be held by each ethnic group should be allocated in advance on the basis of population distribution, and that every voter be allowed to cast ballots for candidates of each ethnic group. Rather than the current system in which candidates can be elected solely on the support of their own ethnic group, this system would require successful candidates to appeal to voters of all ethnicities, thereby rewarding moderation and discouraging ethnic extremism.

Logistically and theoretically, it would be relatively simple to revise Bosnia's electoral laws to allow the country's tripartite presidency to be elected along the lines proposed by the ICG. Since the joint presidency is made up of one member from each of Bosnia's main ethnic groups, the ICG proposal could be implemented simply by allowing all Bosnians a vote for each of the three presidency positions. Bosniaks, for example, would be able to choose not only among the Bosniak presidential candidates, but among the Serb and Croat candidates as well. Such a change would fundamentally alter the dynamics of the electoral contest, and provide political moderates with a distinct advantage over their less tolerant rivals. As noted above, however, constitutional and political obstacles have so far thwarted radical revisions to Bosnia's electoral system.

CONCLUSION

While elections at the local, entity, and national levels were to be the vehicles through which Bosnians could freely and collectively decide their political future, the inability or unwillingness of Bosnia's international peace-brokers to offset the political advantages enjoyed by the country's ruling nationalists has compromised both the freedom and the fairness of post-war elections. As Bogdan Denitch has suggested, by accepting the continuing dominance of the nationalist leaderships of all three of Bosnia's ethnic communities, the Dayton agreement made ordinary Bosnians hostages to their own political leaderships.[75]

Despite some encouraging signs of progress – such as the splits and fractures within Bosnia's main Serb and Croat nationalist parties – Bosnia remains a divided and volatile country in the midst of a turbulent and unstable region, where the forces of ethnic nationalism remain powerful. The election of the Serb hardliner Nikola Poplasen as President of Republika Srpska in September 1998 sent shivers through the international community in Bosnia, and showed clearly that there is nothing linear or inevitable about Bosnia's path to peace. The crisis in Kosovo in the spring of 1999 similarly revealed the ease with which regional instability can cross national borders in Southeastern Europe. Surprisingly, however, while the Kosovo crisis significantly chilled the political climate in Bosnia for a time, there has so far been relatively little long-term political fall-out.

Any assessment of Bosnia's post-Dayton elections must also take into consideration the harsh realities of pre-Dayton Bosnia. While its inherent flaws and subsequent implementation problems must be acknowledged, Dayton did silence the guns and stop the killing in Bosnia, succeeding where four years of diplomatic efforts before it had failed. And while the agreement failed to determine the ultimate political configuration of post-war Bosnia, it did provide valuable time in which to accomplish that task.[76]

Nearly four years after the signing of the Dayton Accords, the balance-sheet on Bosnia's peacebuilding experience revealed clearly that appropriate institutional mechanisms are necessary but not sufficient to reconcile democracy and nationalism within the boundaries of a deeply-divided state. While Bosnia's current political architecture includes, in one form or another, all of the elements of Lijphart's power-sharing approach, such mechanisms have arguably been used as much to pull Bosnia apart as to keep it together. For example, group autonomy, the most visible manifestation of which is the very existence of Republika Srpska, has reinforced Serb claims to sovereignty over the 49 per cent of Bosnia under their control. The fact that Dayton legitimized a Serb entity within Bosnia has also undermined subsequent efforts to promote return of non-Serbs to Republika

Srpska, and has prompted Bosnia's Croats to push claims for an exclusively Croat entity in Western Bosnia. At the same time, the failure of executive power-sharing, exemplified by the discordant and acrimonious three-member Bosnian presidency, has provided ammunition for those who would argue that Serbs, Croats, and Bosniaks cannot live together within a single state. And while Bosnia's post-Dayton constitution includes minority vetoes aimed at protecting the vital interests of each of the country's constituent ethnic groups, such mechanisms have been insufficient to prevent the ongoing process of 'majoritization' within Bosnia.

Ultimately, therefore, while Bosnia's post-war experience does not invalidate Lijphart's power-sharing approach as a means of reconciling national interests within a single state, it does suggest that such an approach cannot succeed in the absence of key supporting conditions. These conditions include a social and political climate more conducive to cooperation than confrontation, the existence of mechanisms that promote a shared sense of political community and that give all relevant groups a strong interest in the continued existence of the state, and the presence of flexible, capable, and tolerant political leaders willing to work together in the interests of all citizens.

These conditions cannot, however, be created merely through the act of voting, and Bosnia's democratic experience underlines the fact that elections themselves do not produce democracy. Premature elections, in fact, can seriously undermine both the process of democratization and peacebuilding in divided societies. At the same time, electoral systems that encourage moderation and tolerance can play a key role in placing post-conflict societies on a sound and stable democratic footing, although this role should probably not be overestimated. In post-conflict situations such as Bosnia, ultimately, democratization must be viewed in the widest possible terms: the organization of elections must be accompanied by vigorous efforts to strengthen and foster vibrant civil society organizations such as a free and independent media, active non-government organizations, and viable political opposition parties. Creating a secure environment in which elections can be held must also go beyond military security to encompass human security, since elections in which fear and uncertainty are the main factors motivating voters are unlikely to produce positive political change. In the absence of substantial progress towards the creation of the fundamental institutions and conditions which underpin effectively-functioning democratic societies, the goal of consolidating peace and democracy in places like Bosnia is likely to remain elusive. Creating this kind of democratic space, however, will not happen overnight. One of post-Dayton Bosnia's most important lessons, therefore, may be that post-conflict peacebuilding is a process better measured in years or even decades rather than in days and months, meaning that the task of constructing a peaceful, united, and democratic Bosnia remains in its early stages.

NOTES

1. Ian Traynor, 'Bosnia Muslims win revenge', in *The Guardian Weekly*, October 19, 1997, p. 3.

2. cited in *This Week in Bosnia*, September 24, 1997. (Internet version: http://world.std.com/~slm/twib0924.html).

3. The term 'Bosniac' has also been used to described Bosnia's Muslim population. The two terms will be used interchangeably in this paper.

4. This process, in fact, had begun somewhat earlier in Yugoslavia. By 1990, a decade after the death of Tito, the combination of nationalism and electoral politics was already placing significant strains on the Yugoslav federation.

5. cited in John F. Stack, Jr., 'The Ethnic Challenge to International Relations Theory', in David Carment and Patrick James, (eds.), *Wars in the Midst of Peace: The International Politics of Ethnic Conflict*. (Pittsburgh: University of Pittsburgh Press, 1997), p. 11.

6. Ghia Nodia, 'Nationalism and Democracy', in Larry Diamond and Marc F. Plattner (eds.), *Nationalism, Ethnic Conflict, and Democracy*. (Baltimore: The Johns Hopkins University Press, 1994), p. 4.

7. ibid., pp. 7–8.

8. Robert Kaplan, 'Was Democracy Just a Moment?', in *The Atlantic Monthly*, vol. 280, no. 6, December 1997, pp. 60–61.

9. David Welch, 'Domestic Politics and Ethnic Conflict', in Michael E. Brown (ed.), *Ethnic Conflict and International Security*. (Princeton: Princeton University Press, 1993), p. 53.

10. Nodia, op.cit., p. 9; Francis Fukuyama, 'Comments on Nationalism and Democracy', in Diamond and Plattner, op.cit., p. 24.

11. Fukuyama, op.cit., p. 24.

12. cited in Zoran Pajić, 'Bosnia-Hercegovina: From Multiethnic Coexistence to 'Apartheid' . . . and Back', in Payam Akhavan and Robert Howse (eds.), Yugoslavia: *The Former and Future: Reflections by Scholars from the Region*. (Geneva: UN Research Institute for Social Development, 1995), p. 160.

13. ibid., p. 162.

14. Donald Horowitz, 'Democracy in Divided Societies', in *Journal of Democracy*, vol. 4, no. 4, October 1993, p. 18.

15. Vicki Hesli, 'Political Institutions and Democratic Governance', in Robert D. Grey (ed.), *Democratic Theory and Post-Communist Change*. (Upper Saddle River, NJ: Prentice Hall, 1997), p. 202.

16. Horowitz, op.cit., p. 32.

17. James Schear, 'Bosnia's Post-Dayton Traumas', in *Foreign Policy*, No. 104, Fall 1996, p. 88.

18. Arend Lipjhart, 'The Power-Sharing Approach', in Joseph V. Montville (ed.), *Conflict and Peacemaking in Multiethnic Societies*. (Lexington: Lexington Books, 1990) p. 493.

19. The cartographic gymnastics of the Bosnian peace negotiations, in fact, often verged on the absurd. As the American journalist Peter Maass described part of one doomed plan: 'The 'Muslim-majority republic,' as it was called, would cover 30 per cent of the country, and be divided into four parcels of land, connected to one another by tunnels, elevated roads and bridges running under, or over, huge stretches

of territory belonging to the Serbs and Croats. Please do not laugh. I possess a series of official maps outlining the path of these projects.' Peter Maass, *Love Thy Neighbor: A Story of War.* (New York: Alfred A. Knopf, 1996), p. 256.

20. Lijphart, op.cit., p. 494. For a more detailed examination of the history of partition, see Radha Kumar, 'The Troubled History of Partition', in *Foreign Affairs*, vol. 76, no. 1, January/February 1997, pp. 22–34.

21. Lijphart, ibid., p. 495.

22. ibid. p. 494.

23. ibid., p. 495.

24. Horowitz, op.cit., p. 36.

25. ibid., p. 36.

26. ibid., p. 35.

27. Donald Horowitz, 'Making Moderation Pay', in Montville, op.cit., p. 471.

28. Milton J. Esman, *Ethnic Politics*. (Ithaca: Cornell University Press, 1994), p. 44.

29. Horowitz, 'Making Moderation Pay', op.cit., p. 457.

30. Susan Woodward, 'Foreign Policy Challenges: Bosnia', in *The Brookings Review*, vol. 15, no. 2, Spring 1997, p. 29.

31. Chester Crocker and Fen Osler Hampson, 'Making Peace Settlements Work', in *Foreign Policy*, #104, Fall 1996, p. 60.

32. Janusz Bugujski, 'Policy Forum: Bosnia – After the Troops Leave', in *Washington Quarterly*, vol. 19, no. 3, Summer 1996, p. 61.

33. General Framework Agreement for Peace in Bosnia and Hercegovina, Annex 4 (Constitution of Bosnia and Hercegovina). (Internet version – http://www.nato.int/ifor/gfa/gfa-an4.htm)

34. Bugajski, op.cit., p. 64.

35. Crocker and Hampson, op.cit., p. 65.

36. Bogdan Denitch, Ethnic Nationalism: *The Tragic Death of Yugoslavia.* (Minneapolis: University of Minnesota Press, 1996), p. 213.

37. Jonathan Landay, 'Policy Forum: Bosnia – After the Troops Leave', in *Washington Quarterly*, vol. 19, no. 3, Summer 1996, p. 66.

38. see, for example, David Rieff, 'Abandoning Bosnia – Again', in Newsweek, September 16, 1996, p. 19; and Human Rights Watch, 'A Failure in the Making – Human Rights and the Dayton Agreement', in *RefWorld*, (UNHCR Centre for Documentation and Research), vol. 8, no. 8, June 1996 (CD-ROM Version), p. 2.

39. Michael Williams, 'Polling for Partition', in *Index on Censorship*, vol. 25, no. 4, July/August 1996, p. 15.

40. Rieff, op.cit., p. 63.

41. cited in International Crisis Group (ICG), 'Elections in Bosnia and Herzegovina – Part 2 of 3: The Lead-Up to Elections', (Internet edition: http://www.intl-crisis-group.org/projects/bosnia/-report/bh16rep2.htm), September 22, 1996, p. 15 of 25.

42. cited in ICG, 'The Lead-Up to Elections', op.cit., p. 17 of 25.

43. Michael Dobbs, 'Karadzic is Still Thorn in NATO's Side', in *The Guardian Weekly*, May 11, 1997, p. 19.

44. ICG, 'The Lead-Up to Elections', op.cit., p. 9 of 25.

45. ibid., p. 7 of 25.

46. Jane M.O. Sharp, 'Bosnia: Begin Again', in *Bulletin of the Atomic Scientists*, vol. 53, no. 2, March/April 1997, p. 18.

47. ICG, 'The Lead-Up to Elections', op.cit., p. 7 of 25.

48. ibid., p. 21 of 25.

49. ibid., p. 20 of 25.

50. cited in Chris Hedges, 'Municipal Elections in Bosnia Postponed, Raising Doubts about Troop Pullout', in *The New York Times*, August 28, 1996, p. A6.

51. Human Rights Watch, op.cit., p. 2.

52. ICG, 'Elections in Bosnia and Herzegovina – Part 3: Results and Conclusions', (Internet Edition: http://www.intl-crisis-group.org/projects/bosnia/report/bh16rep3.htm), September 22, 1996, p. 5 of 14.

53. Personal Interview, Rijeka, Croatia, August 30, 1996.

54. ICG, 'Results and Conclusions', op.cit., p. 8 of 14.

55. Susan Woodward, 'Bosnia After Dayton: Year Two', in *Current History*, vol. 96, no. 608, March 1997, p. 97.

56. Lee Hockstader, 'Scattered Signs of Progress in Bosnia', in *The Washington Post*, September 16, 1997, p. A12.

57. ibid., p. A12.

58. ICG, 'Beyond Ballot Boxes: Municipal Elections in Bosnia and Herzegovina', (Internet edition: http://www.intl-crisis-group.org/projects/bosnia/report/bh26main.htm), September 10, 1997, p. 10 of 25.

59. ICG, 'ICG Analysis of 1997 Municipal Election Results', (Internet edition: http://www.intl-crisis-group.org/projects/bosnia/report/bhxxpr10.htm), October 14, 1997, p. 1 of 4.

60. 'Progress in Bosnia', *The Washington Post*, January 22, 1998, page A20.

61. 'Seat Allocation – Cantons and Municipalities' OSCE Mission to BiH website: (http://www.oscebih.org/98results/seat-all-cantonsMun.htm).

62. ICG, 'State of the Balkans', (Internet edition: http://www.crisisweb.org/projects/sbalkans/reports/ba01rep.htm), November 4, 1998, p. 12 of 15.

63. ICG, 'Doing Democracy a Disservice: 1998 Elections in Bosnia and Herzegovina' (Internet edition: http://www.crisisweb.org/projects/bosnia/reports/bh40rep.htm), September 9, 1998, p. 21 of 21; and ICG, 'State of the Balkans', op.cit., p. 11 of 15.

64. European Stability Initiative, 'Reshaping International Priorities in Bosnia and Herzegovina', 14 October 1999, p. 2.

65. Schear, op.cit., p. 97.

66. Michael Sells, 'Backing the Balkans', in *The Nation*, December 23, 1996, p. 3.

67. Bogdan Denitch, 'What can we learn from the terrible peace in Bosnia-Hercegovina?', speaking notes for the conference *The Lessons of Yugoslavia*, Innis College, University of Toronto, March 20–23, 1997.

68. US Institute of Peace, 'Bosnia Report Card: Pass, Fail, or Incomplete' (Internet edition: http://www.usip.org) November 12, 1998, p. 6 of 20.

69. Denitch, 'What can we learn?', op.cit.

70. cited in Massimo Calabresi, 'Tallying the Hate', in *Time*, September 16, 1996, p. 53.

71. Denitch, *Ethnic Nationalism*, op.cit., p. 224.

72. General Framework Agreement for Peace in Bosnia and Hercegovina, Annex A–1 (Military Aspects). (Internet version – http://www.nato.int/ifor/gfa/gfa-an1a.htm).

73. Williams, op.cit., p. 17.

74. ICG, 'Beyond Ballot Boxes', op.cit., pp. 24–25 of 25.

75. Bogdan Denitch, 'What can we learn?', op.cit., pp. 5–6.

76. Maynard Glitman, 'US Policy in Bosnia: Rethinking a Flawed Approach', in *Survival,* vol. 38, no. 4, Winter 1996–97, p. 76.

14. WHO GAINS, WHO LOSES? THE EFFECT OF YUGOSLAVIA'S BREAKUP ON WOMEN'S PARTICIPATION IN THE ECONOMY

Milica Z. Bookman

The 1990s were very difficult for the population of Yugoslavia (currently composed of Serbia and Montenegro). Economically, the people suffered from the breakup of the former Yugoslavia, the wars in Croatia and Bosnia, international sanctions, hyperinflation, the inflow of refugees and the out-migration of skilled labor. Politically, they suffered as a result of a leadership crisis, a wave of nationalist and often xenophobic sentiment, and the curtailment of some political and human rights. Moreover, they suffered from international ostracism emanating from the world perception of overall Serbian blame in both the breakup and its ensuing crisis. Socially, the population suffered as increases in crime, divorce rates, and suicides reflected the mass mood of desperation and apathy.

This chapter focuses on one segment of that population, namely women. It contains an analysis of how Yugoslav women have been affected by the breakup of the former Yugoslavia, war, international sanctions and the transition from socialism to capitalism. It also assesses the changes since 1990 in women's roles in the economy. Since women's economic participation takes place in the workplace (as producers) and in the home (as producers and consumers), their roles have been altered as a result of fundamental changes in work and home

Research on Russia and Eastern Europe, Volume 3, pages 259–276.
2000 by Elsevier Science Inc.
ISBN: 0-7623-0280-1

conditions. The emphasis of this study is on the early period, namely 1990 until the mid-1990s, since it is during those years that the impact of the economic crisis on women was the strongest.[1]

A few words are in order about women in the economy before 1990. As in other East European countries, women in Yugoslavia enjoyed numerous legal rights that conveyed *de jure* equality with men. They were guaranteed equality by the constitution, they had laws that protected them against discrimination, allowed them the right to voluntarily choose a marriage partner, granted them equal rights in divorce and child custody, dictated that they receive equal pay for equal work, granted them equal rights to property, both with respect to ownership and inheritance, etc. *De facto* however, this equality was not as pervasive as it might seem since the traditional division of labor, according to which women were entirely responsible for the home activities, prevailed (especially in the rural areas and among the less educated population). For this reason, gender equality in Yugoslavia has often been called 'false emancipation'.[2] However, in comparison with their counterparts in most industrialized or less developed countries, women in the former Yugoslavia enjoyed greater legal protection and had higher levels of participation in the economy (this was part of socialist ideology. With the demise of communism, there was a backlash against the idea of female equality associated with the communist period). Also, in comparison with the deterioration of women's conditions over the past few years, even their 'false emancipation' of the pre-1990 period seemed preferable to their condition in 1996.

THE ORIGINS OF THE ECONOMIC CRISIS

In the early 1990s, several shocks to the Yugoslav economy produced an overall state of crisis: the real income per capita in 1993 dropped to 44% of its level in 1989 (when it was $3300);[3] inflation was among the highest in world economic history, culminating with over five hundred quadrillion (5.5×10 to the 17th) in January 1994;[4] the national product decreased to one half of its 1989 value and the capital goods industry fell to 13% of its pre-breakup value.[5] Moreover, industrial output in 1995 was 36.6% of its value in 1989.[6] By 1995, 35.6% of the population had fallen below the poverty line.[7]

What were the shocks that caused a crisis of such magnitude? First, the *breakup* of the former Yugoslavia decreased the size of markets in all the newly formed countries.[8] It caused a disruption in inter-republic trade and capital flows, both of which were extensive despite decentralization of the economy. Second, the *war* in Croatia and Bosnia caused loss of lives and the destruction of property. It also entailed a diversion of resources from civil to war activities, despite the fact that

Yugoslavia was not formally involved in those wars. Moreover, the massive displacement of human populations necessitated enormous quantities of humanitarian aid to alleviate the plight of the refugees, of which some one million settled in Yugoslavia. Third, the imposition of international *sanctions*[9] in 1992 caused the sudden closing of an otherwise internationally oriented economy. Sanctions resulted in the loss of foreign markets and rapidly outmoded the existing technology. That translated into decreased production, unemployment and poverty. Sanctions also decreased personal income and thus decreased the demand for goods and services. At the same time, sanctions decreased the supply of goods and services, especially basics such as food, heating, housing, transportation. Fourth, *macroeconomic policies* were implemented in an effort to stabilize the economy. However, those often entailed desperate measures aimed simply at enabling the government to survive the short term. The state incurred a deficit which was monetized, leading to hyperinflation. This inappropriate economic policy was in part the result of the lag in recognizing the effects of disintegration and sanctions. Fifth, while some steps had been taken in 1992 in the direction of a *transition* to a capitalist economy, including privatization and freeing of prices, all these efforts were slowed down or reversed because of the priority placed on stabilizing the economy. Thus, instead of opening up, the government increased regulation of selected economic activity. For example, international trade activities become strictly monitored by the state, resulting in increased monopolistic activity (while at the same time, privatization occurred in the service sector).

   Not one of these events is alone responsible for the crisis that characterized the Yugoslav economy. Rather, the combination of all these factors caused the worst economic conditions the country ever experienced.

   All Yugoslavs were affected by the economic and humanitarian crisis, yet men and women were not affected in the same way. There are at least three reasons underlying these differences between genders. First, the demise of the communist party brought to power political parties with more traditional (read: patriarchal and patrilocal) orientations which, in reference to women, implied a change in women's role to one that accentuated biological reproduction and the raising of children. Second, the drop in production due to sanctions and war (and to a lesser extent, the economic reforms that are part of the transition process) entailed large scale layoffs. Many of those laid off were women, not only because of discrimination or because of the view that it is more important for men (as head of families) to hold jobs than women, but primarily because the jobs that were cut have been from the sectors in which women are highly represented. Third, under conditions of economic and financial crises, the social policies that women have come to depend on were negatively affected (policies such as maternity leave, child supplements, free education, social help, etc.).

WOMEN IN THE LABOR FORCE

Evidence of changes in women's participation in the economy will be assessed first in the labor force and then in the household.

Employment and Unemployment.

In 1991, women accounted for 50.5% of the population in Yugoslavia yet only 39% of the labor force was female. Women's participation rate was approximately 53%, roughly the rate in OECD countries but significantly lower than the rate in East European countries (by contrast, in Albania it was 86.3% in 1990).[10] Both open unemployment and underemployment increased during the years of the crisis in Yugoslavia. The unemployment rate in 1993 was 15% of the labor force (of which 11.4% for males and 20.7% for females). This represents an increase of 0.4% over what it was in 1990 for males, while for females there was a 2.5% increase.[11] Of the people seeking employment in 1992, 53.7% were women.[12] A disaggregation according to skill levels indicates that the highest proportion of women seeking work is in the highly skilled category (66.3%) and the lowest in the unskilled category (40.6%).[13] During the early 1990s, women at all skill levels waited for work approximately one to three years.[14] On average, women waited for employment longer than men.[15]

This unemployment did not go away after the Dayton Peace Accords, the stabilization of the dinar, and the suspension of sanctions. There is evidence that total unemployment in fact rose during the year of mid-May 1995 to mid-1996 by 1.5%.[16]

While there are no data for underemployment, there is plenty of anecdotal evidence indicating that in many work locations, during the peak crisis period (1993 and early 1994) workers were asked by their employers to stay home. This was called unpaid vacation time, and it occurred under a variety of arrangements between employer and employee. The bottom line was always the same: because firms stopped producing (since they could not get inputs for production (due to sanctions) and since they could not sell their output in the global markets (due to sanctions and domestically, due to poverty) they had no need for their labor. Moreover, they also lacked the funds with which to pay their workers. So workers stayed home and were paid sporadically, when some money 'happened'. What happened to the women who were laid off, to the ones who were waiting for their first employment and to the ones who were on forced vacations without pay? Two new phenomena developed in the Yugoslav context to relieve the pressure on the labor market. The first is that women (as well as men) massively out-migrated during the crisis years. The evidence is that in

1992 alone, 100,000 to 150,000 professionals, roughly half of them female, left Serbia.[17] The profile of the outmigrating woman was as follows: she was in her twenties and thirties, childless and highly skilled (this coincides with evidence that it is the highly skilled women who are most frequently unemployed). As a result of the outmigration, Yugoslav unemployment statistics are actually an underestimate of the true severity of the labor situation.

The second phenomenon that relieved pressure in the Yugoslav labor markets was the proliferation of an underground economy over the past five years. This sector, operating at the edge of legality, arose in part because the legal economy was incapable of producing the consumer goods that the population demanded. It also gave vent to creativity, initiative, and private endeavors. Bogosavljevic estimated that the size of this underground economy was in the order of 34.7% of total economic activity (or 53.2% of social product).[18] The underground economy absorbed labor during the crisis years when the regular economy could not – in fact, while there is an increase in unemployment in the regular economy, there is a rise in employment in the underground economy (for example, during mid-1995 to mid-1996, unemployment increased by 2.7% while employment in the underground economy increased 3.4%).[19] For women, the underground economy offers the possibility of retail trade in sidewalk kiosks (possibly of imported goods brought into the country by a variety of sanction-busting schemes), domestic service work, prostitution, and other petty activity. It offers flexibility in work time and work location, it offers work with low barriers to entry. A term has come to be associated with people who, as a result of the crisis, have begun doing all sorts of jobs that they never would have done before: svestariti. Women trying to survive the crisis have been especially motivated to svestare. While the numbers of women in this parallel economy has not been estimated, simple visual observation indicates that many women during 1990–1996 have found work or supplemented their work in the underground economy.

Income.

Without a doubt, women's income plummeted in the aftermath of the Yugoslav breakup. During the ensuing scarcity and drop in production, business and enterprise earnings fell so the funds for payment to workers, pensioners and welfare recipients were depleted. Very often, especially during 1993, workers were not paid for six months. Alternatively, they were paid in goods.

Before the economic crisis of the 1990s, there is no doubt that a wage differential existed between male and female workers. Occupational segregation explained this phenomenon, as women were concentrated in sectors in which productivity (and therefore income) was low (sectors such as retail, some manufacturing, services, education and culture).[20] Women's lower income was also due to their exclusion from most positions of power in the economy. What happened to the wage differential between men and women as a result of the crisis? In the absence of disaggregated data to support a deterioration in the wage ratios, indirect evidence points to such an outcome. First, the sectors women were concentrated in are ones whose derived demand dropped most significantly during the crisis years (indeed, in the knitwear and garment sectors, highly represented by women, production in 1992 dropped to 20 and 40% of its pre-1991 levels).[21] As a result, women suffered from layoffs as well as tardy and/or partial wages. Second, the 1990s brought in a new openness in discrimination against women, especially in hiring. While this is discussed below, suffice it to say here that as a result of such discrimination, fewer women were hired and thus it is likely that the total wage bill paid out to women was lower than that paid to men.

Economic Reforms.

The economic reforms that were introduced after 1991 did not follow a clear path towards capitalism as they did in other formerly communist countries. In Yugoslavia, the greatest shock to the economy came from the demise of the system of self-management. Privatization occurred, but it was selective; domestic prices were freed, but controls were reimposed when it was deemed necessary; structural and institutional reforms occurred, but incompletely and haltingly.

How did these changes affect women? On the whole, they had negative effects. To the extent that reforms entailed the end of state subsidies to unproductive and inefficient firms, they resulted in layoffs that affected women. To the extent that austerity measures associated with macroeconomic stabilization policies restricted employment requirements, women waited for work longer than men. Displaced women tended not be requalified, and because of family obligations, tended not to accept relocation. Reforms also entailed changes in work conditions that did not improve the quality of life of women. Primarily, reforms introduced a new kind of competition, not only between firms but also between workers. For example, women were expected to improve their performance in their workplace. In the past, many women used some of their work time to run errands and take care of domestic activities. This was accepted by their co-workers as a necessary component of survival given the particular social conditions. However, in private enterprise, it is not acceptable, it cannot be tolerated in an atmosphere of efficiency and profit maximization. Hence, women found themselves having to work more and harder in the workplace than they did before. Moreover, the private sector had come to be associated with open discrimination.

Private employers were no longer bound by constraints in hiring, as they were during the communist period. In the 1990s, it was possible to see employment ads that specify age and personality characteristics of its female applicants.

What about women employers in private enterprise? With privatization, some women have emerged as employers, owners and entrepreneurs. While the data on private initiative are not disaggregated by sex, there is some indication that women are leaders of new business, although not to the same extent as men.[22] Women tend to be concentrated in retail, clothing and food provision (ranging from street kiosks to cafes and restaurants). In order to provide support to each other and to share information, the Women's Club of Private Entrepreneurs has been established in Belgrade.

WOMEN IN THE HOUSEHOLD

Despite the high degree of equality that Yugoslav women enjoy on paper, they remain the primary, if not the sole, careproviders in the household. Even when married and living with a spouse, they are almost entirely responsible for two economic activities in the household: consumption and rearing of children. Many Yugoslav women are not married (indeed, in 1991, 21.8% of all families were supported entirely by women)[23] and are thus also solely responsible for all household expenses. Whether married or single, women's role with respect to consumption and children has been affected by the crisis.

Consumption.

Women are responsible for the satisfaction of basic needs of a group of individuals. In this role, women are the primary consumers of foodstuffs, household items, and utilities. Their task has been dramatically altered by the crisis. Households have less income with which to buy goods and services due to layoffs and the inability of enterprises to pay their workers. As a result, their demand for luxury and unnecessary items has been significantly reduced if not altogether eliminated. Their demand for food, however, has increased. This paradoxical effect occurred largely because of uncertainty pertaining to the future and expectations of worse times ahead. The rational response among women was to make stocks of goods such as sugar, flour, oil, coffee, etc.

Another source of the upward pressure on demand is the often increased size of the household, which often supported refugee relatives or displaced kin from other parts of the country. At the same time, the supply of goods has decreased due to the lack of imports (due to sanctions) and the decrease in domestic production. Indeed, there was less food produced and therefore less available to the consumers (for example, the production of wheat dropped in 1993 by 22.7% over its value in 1991, corn dropped by 48.8% and sugar by 79.7% during the same period).[24] This meant that necessities were harder to come by. In practice, these shortages translated into standing in long lines for detergent only to find it is sold out. It also means expending much time and energy canvassing numerous stores before finding milk for sale. As a result of these market conditions, prices of food and necessities skyrocketed during the crisis. They did so despite governments supports and the sporadic supply injections from national reserves. It became more difficult for women to perform the duties associated with running a household. Since over 40% of the population couldn't satisfy the minimal expenditures for food (and that does not include housing utilities and transportation costs),[25] women began to economize in a way that they hadn't at any time since World War II. They began spending an inordinate amount of time thinking about, researching, and planning where to buy goods and how to pay for them. Moreover, transportation services were reduced so getting around to buy necessities posed enormous hurdles (buses were fewer, less frequent and more crowded; the lack of gasoline prevented the use of private automobiles). All in all, it seemed like women in 1993 had returned to an existence akin to what it was during times of hunting and gathering, when subsistence economic activity filled up the entire day.

The travails of everyday life are described in the following anecdote. In 1993, I was on a bus in Belgrade when a conductor came around to inspect tickets. While the majority of the population rode without valid passes, he began harassing the woman seated next to me because she refused to pay for the ticket he offered her. She said to him, "I am retired and have not received my miserable pension for over a month. I am trying to get to the other side of town because I heard there was some detergent for sale. I have no money to pay for the bus ticket. In fact, I would be grateful to you if you would take me to the station and put me in jail because at least I will be warm and there will be one less mouth to feed at my home tonight." The conductor walked away.

What effect does this decreased ability to perform their consumptive duties have on women? Women bear a disproportionate burden with respect to the emotional strains of uncertainly because they are the principal care providers and deal with the basic needs of the household. Women also bear the primary responsibility for the provision of goods. Paradoxically, at a time when women need income more than ever, some women have voluntarily withdrawn from the labor force where they were on unpaid leave in order to devote more time to the purchase and preparation of food. There is even evidence that women took early retirement in order to perform this task.

Children.

There are two issues pertaining to children that have changed as a result of the economic crisis of the past few years – population growth rates and childcare services. Both have ramifications for the economy.

While population growth rates decreased in Yugoslavia during the late 1980s, they plummeted during 1990–1996. Poverty, insecurity, and apathy have caused the Yugoslav population to lose interest in procreation (this is a very different result from the induced by poverty in many less developed countries, where people often have many children because they are poor). In 1990, the natality of Serbs dropped to 11.4% (while that of the Croats is 12.5%, of Muslims is 18.4% and Albanians is 28.8%).[26] It is estimated that in ten years (mid-2010s), the Serbian population over 65 will be larger than that under 15. During the 1990s, the marriage rate dropped, young people outmigrated, and pregnancies were terminated at unprecedented rates (Serbia now has the highest abortion rate in Europe). As a result, the next population count will indicate even lower natality rates than in 1990.

The realization that population growth is decreasing in all the new countries of former Yugoslavia has sufficiently worried their leaders that most have introduced some form of pronatalist policy. In their mildest form, these include adding several years of pensionable service for each child born.[27] They also include altering the tax structure to discourage no-child families, as in Croatia.[28] Among all former Yugoslav peoples, there is concern about the inter-ethnic differences in population growth. For example, there is alarm among Serbs because estimates are that by 2050, Serbs will be a minority in Serbia. While no clear legislation has been passed with demographic concerns in mind, nationalist leaders are aware of the problem and are trying to turn resolutions into action.[29] In this line of thinking, Zeljko Raznjatovic (better known as Arkan) said he would like to see all Serb women to have four children if possible, to offset the Albanians "who produce like rabbits."[30]

In addition to the nationalist view that greater population numbers imply strength of a given ethnic group, the economic considerations of the decrease in natality are great. For example, the future labor supply will be affected as there will be less people entering the labor force. Moreover, large numbers of workers are needed to provide the revenue for social programs to support those outside the labor force. Also, given the decreasing ability of the state to adequately provide for social security to its population, children perform that function for their parents. Without children, who will care for the old?

The availability and accessibility of childcare facilities has also undergone a transformation as a result of the crisis in Yugoslavia. Childcare is at once less

available and less accessible. With respect to availability, the general state of
economic crisis, hyperinflation, drop in production and the virtual shutting down
of workplaces has resulted in a drop in pre-school care. There is evidence that
at the republic levels, the country is making every effort to finance preschool
institutions from budgets for extraordinary use and to make foods available in
these centers from its reserves.[31] Childcare is also less accessible because, as
all forms of schooling, it is no longer free. Some payment, albeit minimal, is
required in order for children to attend day care. That payment is sufficient to
make daycare inaccessible to those hovering at the poverty line, namely some
one half of the population. So, given that it is less available, women had to
rely on kin for care of their children. But kin were then, in the early 1990s,
least able to provide help given how difficult and time consuming mere food
marketing and preparation had become.

What are the implications of the childcare crisis for the economy? Women
who cannot have childcare for their children, either because it is unavailable
or because they cannot afford it, will respond in several ways. They will either
leave the labor market (with negative effects on their household income), or
they will stay in with high levels of absenteeism and low productivity (with
negative effects on the economy). Moreover, they are likely to have fewer children in the future, as the personal cost to them of more children rises (with
negative demographic and thus economic effects in the long run). Under non-
crisis conditions, childcare facilities answer to the needs of employers as much
as parents and children. To the extent that the economy benefits from mothers
working, then childcare also becomes a collective responsibility. The state has
failed in its commitment to this responsibility during this time of crisis.

FACTORS AFFECTING WOMEN'S PERFORMANCE IN THE LABOR MARKET AND IN THE HOUSEHOLD

There are three factors that have emerged as major issues in women's well
being and therefore in their ability to perform their functions in the labor force
and in the household: education, health and violence. All of these have deteriorated since the crisis began and thus have had negative effects on women's
participation in the economy.

Education.

The crisis of the early 1990s resulted in a decrease in the education of women
in Yugoslavia. The only available evidence pertains to higher education.[32] While
53.3% of all students at the university level in 1992–93 were females, [33] there

was a drop in enrollments from before the crisis period.[34] There was also a decrease from 1990 to 1992 in the number of female students who graduated from college.[35]

Several factors associated with the crisis can be brought to bear to explain this trend. There is no more free schooling in Yugoslavia, there are no more generous scholarships for room and board, there is no more expansion of student housing, and the opportunity cost of students' time has increased because the student-aged population must engage in economic activity in order to make household ends meet. Under these circumstances, parents will be unlikely to educate all their children, so it is likely they will educate their sons rather than their daughters.

The decrease in human capital that will result from this decrease in education at the higher level will supplement the trend of outmigration among the highly skilled workers and will have devastating implications for the future of reconstruction efforts. But, specifically for women, how will the decrease in education prevent women from performing their economic duties? They will be less skilled, they will be less likely to have jobs in sectors that pay adequately and they will generally have fewer employment options.[36]

It is expected that the next census will clearly show how detrimental the early 1990s have been to the education of women. Women at all levels of education.

Health.

While data on female employment is hard to come by, data on female health is not. The evidence is clear: according to a variety of indicators, there has been a marked decrease in the health of Yugoslav women. There has been a deterioration in nutrition of mothers as the average calorie intake decreased by 10.1% during just the first year of the crisis.[37] Women are also receiving less health care since 1990: for example, women's visits to doctors had decreased by 32% in 1992 from its 1986 level. Moreover, there has been a decrease (from 1988 to 1992) in the number of health organizations that provide health care to women (461 and 427 respectively) and a decrease in the medical staff that treats them (1183 to 1171 respectively).[38] Women are using less birth control,[39] largely because there is less available since it used to be imported from other republics or from abroad (indeed, some 60% of sanitary materials and medicines have been lost because they were previously manufactured in the other Yugoslav republics). They are also using less birth control because that which was available was no longer covered by health insurance, hence had a positive cost (money that women do not have).

As a result of the decrease in birth control use, there has been an increase in the number of abortions, especially those without anesthesia (anesthesia is usually unavailable, and if it is available through private channels, it costs money). For the first time since World War II, there is an increase in the number of botched abortions and women are dying as a result of having them performed in non-medical establishments.[40] Mothers dying in labor increased from 9 per 100,000 in 1992 to 16 per 100,000, indicating a 183.3% increase in one year.[41] Women are giving birth at home, not because it is fashionable but rather because they do not have the money to pay for the care in the hospital. All this coincides with overall health expenditure: according to Posarac, there was a drop of 31.5% from 1990 to 1994 of public expenditure on health care.[42]

Violence.

There seems to have been an increase in violence against women since 1991. According to a recent study, there has been a consistent annual increase every year since 1990.[43] This violence is committed by husbands or live-in partners (65%), by ex-husbands (13.3%), by sons (6.4%) and by strangers.[44] A very small portion of this violence is sexual and an even smaller portion results in death. However, it is unclear how much of this rise in violence is due to the particular conditions of apathy and frustration due to poverty, unemployment, and the sudden decrease in standards of income and how much is due to the increased reporting of such crimes.

Clearly, under conditions in which women have less access to education, in which they are unhealthier and in which violence against them is on the rise, their ability to perform effectively in the workplace and in the household comes into question. They will be less productive workers in the labor force and less likely to have the energy, the health and the motivation to provide their families with not only survival necessities but also an environment conducive for emotional and intellectual growth.

WOMEN AND REFUGEES; WOMEN REFUGEES

A special discussion of refugees is warranted in the assessment of women's participation in the economy for two reasons. Both have to do with increased competition during times of increased scarcity. The first is because women in Serbia and Montenegro have been affected by the influx of refugees in both the workplace and in their household activities. They are affected in the workplace insofar as there is additional competition for work. Often younger women who are waiting for their first employment come into competition with women who are more qualified and more experienced who have been displaced from their jobs in Croatia or Bosnia. The competition has been quite acute in petty services, especially in the areas of domestic service and sidewalk retail. In the household, women are affected by refugees primarily because many have accepted relatives and friends to live with them (indeed, 94% of refugees from Croatia and Bosnia live with kin or friends). That translates into an increased demand for food and services in many homes exactly at a time when there is a decreased ability to supply those goods and services.[45]

The second reason to study refugees in this assessment of women is because most refugees are women. Indeed, there are some one million refugees in Yugoslavia in 1996[46] of which the majority are women: in Serbia they account for 58% of all refugees, while in Montenegro they are 66.6%.[47] Women were 85.2% of all adult refugees in 1993.[48] Children under sixteen are 42% of the refugee population in Serbia and some 50% in Montenegro. All these women refugees have left their jobs and their homes and are at the mercy of the state or relatives and friends. They have no source of income, no pensions and no unemployment insurance. Their health insurance is pending the resolution of inter-republic negotiations (some of these have only been resolved in 1995–1996). As a result of their insecurity and loss, these women are under permanent stress, they fear for their future, and they deal with problems of survival and separation from family and home. All the changes in women's roles and conditions since the crisis began are magnified in the case of refugee women.

LESSONS FROM THE YUGOSLAV EXPERIENCE

Three lessons can be drawn from the experience Yugoslav women had during the 1990s. First, change does not necessarily imply improvement. The 1990s were not a mixed blessing for Yugoslav women, as they were for women in many other socialist countries. Rather, there is no doubt that Yugoslav women were better off before 1991 than they were in 1996. Women in the labor force faced greater risk of unemployment, new competition from men in the work-place, open discrimination and increased demands for better performance. Moreover, the increased poverty and the decrease in the standard of living complicated the task of providing for their families' daily needs. Caring for the household has become a full time job. At the same time, women became less healthy (and in the future, as a group they will also be less educated). It is clear that women were the principal losers in the post-breakup period.

However, not all women lost equally in the 1990s and not all women lost in the same way. For example, while rural women were more hurt than urban women with respect to work in the organized sector and with respect to access to education, urban women were more hurt with respect to procuring consumption goods for the household, especially food. Rural women, given their proximity to the source of the supply as well as the existence of family plots for domestic consumption, had greater capacity to satisfy the basic food needs of their households. Moreover, rural women relied less on modern conveniences (such as electricity and central heating) for their daily survival, hence were more adaptable when they lost them due to sanctions and poverty. However, irrespective of where a woman found herself on the rural/urban divide, older women, women with fewer skills, and women heads of households were always worse off.

The second lesson has to do with economics. When the economic pie shrinks and scarcity prevails, two consequences follow. First, there is increased competition for the decreasing slices. This competition occurs along all social cleavages: there is increased competition between members of different ethnic groups, between residents of different regions, and between males and females. In this competition, women very often get squeezed. As in numerous less developed countries and as in periods of recession in industrialized economies, Yugoslav women became marginalized as a result of the crisis. Second, when the economic pie shrinks, adjustment to the new scarcity results in a natural tendency to revert to what is, for many, the 'natural state'. In Yugoslavia, this 'natural state' entails traditional values and traditional economic roles for women. That means more women outside of the labor force, more time in child-care, in home activities and in childbirth.

When the economic pie begins to grow again, as it will one day, women's lives will not necessarily be as good as they were before 1990. While some women will undoubtedly have greater opportunities than they did in the past, the majority will face gender inequality in employment, unemployment, income, education, etc. There is no reason to believe that the attention to women and women's issues that prevailed under communism will return. Indeed, while communist ideology was committed to equality, capitalist ideology has no such commitment and nationalist ideology firmly places women in their traditional role. Moreover, there is ample evidence across the globe of economic prosperity without gender equality in the economy. While economic growth will undoubtedly provide more employment opportunities, it will increase the standard of living and generally make life easier, it is unclear that it will address issues such as discrimination and women's health care. While economic recovery, the transition to capitalism, and renewed foreign investment may all succeed in bringing about a rehiring of women, will they in fact provide women with the quality of life that they had before? Not necessarily.

The final lesson has to do with the political involvement of women in the 1990s. Communism was not repressive in Yugoslavia the way it was in other East European countries, so there has been no significant increase since 1990 in the ability of women to voice their opinions in the political arena. Instead, there has been a change in the focus of their concerns as well as the context within which it occurs.

With respect to focus, Yugoslav women have become vociferous in their attempt to underscore discrimination against women in general and in hiring and firing in particular. They have also focused on issues such as war, rape, and refugees. They are pressing for the restoration of the gender equality that existed before and that they perceive to have been lost.

With respect to the context in which women are speaking out, there has been an increase in participation in both governmental and non-governmental organizations. In the former, there are more women in elected positions in politics (1990, 1.6% of parliament in Serbia, while in 1993 the number rose to 2.8%).[49] Also, there has been increased women's participation in positions of power in non-elected office, both within the country as well as in the diplomatic corps. While it is possible that this trend is due to the increased awareness and recognition of women as equal partners in the political arena, it is also possible that with war, displacement and outmigration of populations, there is a personnel vacuum created by the departed men that is of necessity filled by women. Non-governmental organizations (that previously did not exist) have arisen to address women's issues. Before 1991 (but after the Nairobi conference on women of 1987), a working group was set up in Yugoslavia to deal with women's issues, to review concerns and to make recommendations. This was the only body that dealt explicitly with women. Since the breakup of Yugoslavia, there has been a proliferation of new non-governmental groups. These include *Savez Zena Jugoslavije, Aliansa Zena Jugoslavije, Kolo Srpskih Sestara, Humanitarno Drustvo Zena Dobre Volje*, as well as many others that are war, refugee and rape related. To date, they remain localized in Belgrade and several other big cities. Women, especially the urban and educated ones, are increasingly speaking out in these organizations in order to rectify some of the changes that have occurred in the last few years.

Finally, given the political awakening of women (especially the urban and educated women) due to the deterioration of their overall condition during the 1990s, it is likely that Yugoslavia will witness some gender differentiation in its voting patterns. It is likely that Yugoslav women, like their counterparts in other East European countries, will vote for parties that address their concerns.[50] These concerns include childcare, equality in hiring, and education – in short, all those concerns that are most often addressed by the old communist party. The party of

Yugo-nostalgia, 'JUL', headed by Mira Markovic, has been particularly popular among urban, young and educated women. It remains to be seen how women voters will exercise their vote in the upcoming decade.

CONCLUSION

By way of conclusion, it is noted that the lessons drawn from the Yugoslav experience have broad applicability across all societies in the throes of economic crises, not only in those experiencing a breakup. Indeed, the manifestation of galloping inflation, the breakdown of trade relations, the imposition of economic sanctions, and the inflow of war refugees are not exculsively associated with breakups. Whatever their origin, such economic phenomena tend universally to have adverse effects on women. Indeed, they increase social instability and exacerbate the inter-gender competition for resources, both of which tend to hurt women. Moreover, economic crises tend to alter the political focus of women, entailing a step back as they strive to redress the regressive effects of economic hardship. If it were possible to acknowledge, a priori, the increased risk to women of economic crises, then action could be taken before the nega-tive effects percolate to women. In other words, policy makers who are on alert might have the opportunity to address some of those adverse effects before they occur. Such preventive measures are likely to have positive long run social, economic and political ramifications.

NOTES

1. Very few data on women's economic participation are available – the census is not due until the year 2000 and the surveys conducted in the course of the past few years rarely disaggregated by sex.

2. Women were not a uniform category before 1991 in Yugoslavia. There were large differences according to location of residence (namely rural/urban) and also often according to ethnic orientation. Rural women tended to be less educated, to be in more traditional roles and to have fewer opportunities than urban women. Moreover, rural Muslim women tended to be more traditional in their economic roles than either rural Serb or Croat women. A difference also existed according to level of development, namely the less developed regions were more traditional than the more developed regions.

3. Aleksandra Posarac et al., 'Socio-Ekonomski Polozaj Zena i Dece u Saveznoj Republici Jugoslaviji: Analiza Situacije'; unpublished paper prepared for UNICEF, Belgrade, 1994, p. 5.

4. James Lyon, 'Hyperinflation in Serbia, Casues and Consequences,' paper presented to the meetings of the American Association for the Advancement of Slavic Studies, Washington, 1995.

5. Posarac, op.cit., p. 7.

6. Ibid. p. 7.

7. Ibid. p. 11.
8. Moreover, the demand in the Yugoslav market has further decreased because the population became poorer. See Svetlana Adamovic, 'Efforts Towards Economic Recovery and Monetary Stabilization in FR Yugoslavia', *Communist Economies and Economic Transformation*, vol 7 #4, 1995.
9. Sanctions were introduced under United Nations resolutions 757, 787, and 820.
10. Susan Wolchik, 'Gender Issues During Transition', *East-Central European Economies in Transition*. Joint Economic Committee of the United States Congress, p. 152.
11. Savezna Vlada, Nacionalni Izvestaj o Sprovodjenju Najrobijskih Skategija za Unapredjivanje Polozaja Zena do 2000 te Godine, Belgrade, 1995, p. 17.
12. Ibid., p. 25.
13. Ibid. Table 6.
14. Ibid. p. 16.
15. All these findings on unemployment coincide with the evidence for all Eastern European countries during transition: Wolchik found that during the early 1990s, women's unemployment increased eleven times while that of men increased only six. Moreover, women tend to stay unemployed longer than men. Wolchik, *op. cit.*, pp. 156,158.
16. Savezni Zavod Za Statistiku, *EkonomskiTrend*, June 1996, p. 36.
17. *RFE / RL News Brief*, December 28-January 8, 1993, p. 14.
18. Srdjan Bogosavljevic and Gorana Bozovic, 'Uticaj Sive Ekonomije na Projekcije Drustvenog Proizvoda', *Ekonomski Trend*, June 1996, p. 13.
19. Savezni Zavod ZaStatistiku, *EkonomskiTrend*, June l996, p. 36.
20. Savezna Vlada, *op. cit.*, p. 3.
21. *Vreme*, May 3, 1993
22. At the same time, there has been no clear change in women in positions of power in companies that have not been privatized. Savezna Vlada, *op. cit.*, p. 7.
23. 1bid p. 16.
24. 1bid. p. 33.
25. Posarac, *op. cit.*, p. 17.
26. *Duga*, May 28, 1994, p. 4.
27. Susan Dill, quoted in Pembroke Associates' News, vol 11 #1, 1994, p. 3. She adds that such pronatalist policies have been perceived in Croatia as asking women to 'raise soldiers for Croatia'.
28. Lawmakers passed a package of laws in the early 1990s under the program of 'Renewal of the Republic of Croatia' aimed clearly at producing more births by inducing young women to procreate. These include: the increase in child benefits; the replacement of working mothers with children from jobs 'unsuitable' for them; the alteration of the tax structure so as to benefit couples with children and discourage one or no child families; an effort to make divorces more difficult; and lastly, the elimination of childcare programs to induce mothers to care for their children. These have been described in detail by Cynthia Enloe, *The Morning After: Sexual Politics at the End of the Cold War*, Berkeley: University of California Press, 1993, pp. 241 243.
29. While president, Dobrica Cosic submitted to Parliament a request for emergency deliberations on the population crisis; the Serbian Academy of Sciences and Arts had a meeting in May 1994 to discuss population issues.
30. *Vanity Fair,* June 1–4, p.170.
31. Savezna Vlada, *op. cit.*, p. 30.

32. The trend is the same for high school and vocational schools. All females (and males) have eight years of compulsory schooling. It should be noted that illiteracy rates are four times higher for women than they are for men.

33. Savezna Vlada, *op. cit.*, p. 21.

34. Ibid, p. 21.

35. Ibid., Table 9.

36. What these female students study will have implications for what work they will perform in the future. Women tend to be in female professions as they continued to study and train for jobs in those professions. Indeed, very few females are in the fields of technology, engineering, and computer science. A division of university majors indicates that women make up 68.8% of social science majors and only 33.4% of majors in 'technological' fields, including engineering. Savezna Vlada, *op. cit.*, p. 21.

37. Posarac, *op. cit.*, p 32.

38.. In 1991, only 12% of the women used birth control (Ibid, p. 24).

40. There was also an increase in death due to complications with childbirth. Infant mortality has increased (in 1992, in the capital it was 16% and in 1991 it was 14%, while throughout the entire country the numbers were 20.9% and 21.7%, respectively. Ibid p. 25).

41. Posarac, *op. cit.*, p. 31.

42. Ibid, p. 21.

43. Savema Vlada, *op. cit.*, pp. 26–29.

44. Ibid., p. 28. The percent of violence perpetrated by sons increased from 6.4% in 1992 to 11.4% in 1994.

45. There was resentment among resident women when refugees began receiving rationed goods while permanent residents did not have those same benefits. Anecdotal evidence exists of violent confrontations between women refugees and women residents in food stores, especially during the harsh winter of 1993.

46. It is very difficult to know the exact number of refugees since only a small portion of them are actually registered. Many prefer to be unregistered so they do not run the risk of deportation. However, if they are not registered then they do not receive any state attention. Savezna Vlada, *op. cit.*, p. 31.

47. Ibid, p. 30.

48. Posarac, *op. cit.*, p. 32.

49. Savezna Vlada, *op. cit.*, p. 7. There has been a rise in the number of women diplomats: in 1983, 12 out of 419 were female, while in 1994 they are 16% of total.

50. For other countries in Eastern Europe, see Wolchik, *op. cit.*

PART IV

KOSOVO: THE POST-WAR WAR

15. KOSOVO: BETWEEN CO-EXISTENCE AND PERMANENT CONFLICT*

Srećko Mihailović

The conflict in the 1990s between Serbs and Albanian Kosovars over the control of Kosovo not only precipitated the break-up of the former Yugoslavia and fighting in Slovenia, Croatia, and Bosnia, but at the end of the decade still threatens to spread violence throughout the Balkans. That is why anyone who wants to learn from these events must study closely the painful lessons of Kosovo.

Kosovo is the southern province of the Republic of Serbia, which, together with Montenegro, constitutes the Federal Republic of Yugoslavia (FRY). Kosovo's outer borders are Serbia's borders with Macedonia, Albania, and Montenegro. In area, Kosovo is about 11,000 square kilometers, which amounts to 12% of Serbia. Kosovo encompasses the plains of Kosovo and Metohija, which are situated between Mt. Šara, the Prokletija Mountains, and the Kopaonik Massif. A mild Mediterranean climate penetrates Metohija.

Kosovo is at the crossroads of major north-south and east-west communication routes, with rail links to Thessaloniki and Belgrade. Although it has long been the poorest and least developed region of Yugoslavia, it has coal (lignite), deposits of lead, zinc, nickel and magnetite to be exploited. In the Middle Ages, mining in Kosovo was already the most important branch of the economy.

Kosovo was the center of the medieval Serbian state, with Prizren serving for some time as the capital city and Peć as the seat of the patriarchate of the Serbian Orthodox Church. Cultural monuments can still can be found in Kosovo

* This article was written early in 1997.

Research on Russia and Eastern Europe, Volume 3, pages 279–307.
2000 by Elsevier Science Inc.
ISBN: 0–7623-0280–1

that bear witness to the greatness of that state, as well as to the centrality of the Kosovo and Metohija region in it. The biggest towns in Kosovo and in Metohija date back to the Middle Ages or even earlier.

On June 28, 1389 Serbian and Turkish armies fought the battle of Kosovo Polje (Gazimestan) on a plain near the present-day city of Pristina. In accordance with medieval custom (and this should be noted nowadays) groups of Serbs from the regions already conquered by the Turks fought beside the Sultan Murad, while Bosnians, Croats, and even Albanians fought on the side of the Serbian Prince Lazar. The defeat of the Serbs on this occasion was to determine the future of Serbia and thus Kosovo for the next four to five centuries. The national interpretation of that battle, its role and significance in the creation of Serbian national identity, are still relevant political factors.

The Turks finally established their rule in Kosovo in 1455. From the Middle Ages until the end of the seventeenth century the Serbian population was numerically predominant in Kosovo and Metohija. However, the situation dramatically changed in the following three centuries, primarily due to migrations, Islamization, and Albanization (conversion from Orthodox Christianity to Islam and inclusion into the Albanian ethnic group). This is a matter of contention between Serbian and Albanian historians and the point will be taken up again below.

During the First Balkan War Kosovo was 'liberated' in 1912 by the Serbs and Montenegrins. (According to Albanian historians Kosovo was 're-conquered.') The London Peace Treaty (July 29, 1913) recognized an independent Albania which excluded Kosovo. From then on, except during World Wars I and II, the borders remained unchanged; Kosovo and Metohija remained a part of Serbia and, for that matter, of Yugoslavia.

ALBANIANS, MONUMENTS AND SERBIAN POLICEMEN

Cynically, though not incorrectly speaking, in Kosovo and Metohija there exist Albanians, Serbian monuments, and Serbian policemen. The province of Kosovo has a population of over two million. It is impossible to be more precise since the Albanians refused to participate in the population census in 1991. They justified their boycott by expressing the fear that due to its organization the census would under-report the number of Albanians in the province.[1]

The overwhelming majority of Europe's ethnic Albanian population is found in three countries: Yugoslavia (Kosovo and southeastern Montenegro), Albania proper, and Macedonia (the western region). According to the data provided by Shkelzen Maliqui, Kosovo Albanians account for 38% of all Albanians.

Even before the war of the 1990s Kosovo had become an unusually homogeneous territory. The Albanian population is the dominant group, accounting for 81.6% of the total inhabitants. Serbs account for 9.9%, Slav Muslims for 3.4%, Gypsies 2.3%, Montenegrins for 1.0% and Turks for 0.5%. In place of the former multiplicity of cultures, the number of communities shared by Albanians, Serbs, Montenegrins, and others is constantly decreasing. This ethnic homogenization is illustrated by additional data: inasmuch as 93.4% of all settlements a single ethnic group is the dominant one.[2]

In contrast to Central Serbia, where the average age of men is 36.2 and of women 38.3 years, in Kosovo it is 24.9 and 25.8 years respectively. The population is so young that 23% of Serbia's conscripts have been from Kosovo (17% of the population of Serbia) despite the fact that Albanian youths have not been called up. The remarkable youth of the Albanian population, together with the stability of marriage and family, indicates an unusually rapid demographic growth.[3]

Despite the evident differences between ethnic groups, it is important to emphasize their common features: the principal economic activity, the typical relations of production, the authoritarian intra-group relations, their shared law, the maintenance of tradition, certain moral values, Turkish (Asiatic-Mediterranean) influences, the desire to get rid of Turkish domination, the peripheral place in Europe, the status of a small people, the status of a victim of those bigger and stronger.

Nevertheless, the rivalry between groups is long-standing. According to the Register of Cultural Monuments, in 1986 there were 372 cultural monuments in Kosovo and Metohija, of which 48% are Serbian, 20% Turkish, and 11% Albanian. Such a distribution of historic sites and monuments forms the basis for the legitimacy of Serbian rule over these territories, which are usually described as 'the historic Serbian territories'.

Branislav Krstić compares this legitimation to that of "the similar right of the Jewish and Armenian people, who also preserved such rights through their cultural monuments and, as the Jews did, built them into the sources of their religion. And as is well known," he adds, "none of the aforementioned peoples agreed to cede its historic land to some other people, at least not without waging a war. That cannot be expected from the Serbian people either."[4]

This historical justification has been the prevailing 'national atmosphere' among the Serbian intellectuals, among the political and even ecclesiastic elites, who portray Kosovo as necessary for the existence of Serbs and Serbia itself. "Kosovo is the cradle of Serbdom," "Kosovo is the heart of Serbia," "Kosovo is the Holy Serbian Land," "Kosovo is the Serbian Jerusalem."[5] Conceived in this manner, Kosovo has been the legitimation for everything and for everybody.

From 1989 on, Serbia started gradually abolishing the autonomy of Kosovo
and the Albanian response was the declaration of the Republic of Kosovo and
the creation of a parallel system of government. From the beginning of 1998
on, Albanians engaged in armed attacks (terrorist according to Serbian reports,
defensive and liberating, according to the Albanian view). Serbs and Albanians
were caught up in a territorial dispute which could have led either to a zero
sum game or to cohabitation.

The conflict in Kosovo expanded with increasing rapidity and intensity from
1989 to 1998. This acceleration began with the abolition of Kosovo's provincial autonomy. In place of the self-government provisions of the 1974
constitution, Kosovo came to be governed by a Serbian administration supported
by a huge police force.

The Albanian response to this form of direct rule was the total rejection of
all things Serbian or Yugoslav and the establishment of a parallel state, government, police constabulary, health service, and educational system, including a
university. In short, "as far as the Albanians are concerned, Kosovo is no more
a part of Serbia."[6]

Kosovo is torn on the one hand by Albanian secessionism, and on the other
hand by Serbian unitarism. The fact is that two fanatical nationalisms have
acted upon one another, have polarized their respective populations, and now
are involved in a fierce conflict. There are two camps only, with nothing in
the middle. All identities are reduced to national identity, and all interests to
what is perceived as the 'national interest'.

Albanians and Serbs in Kosovo have been waging a war over territories, or
rather over the Albanian claim for an independent state on the Kosovo territory: the Serbs are not giving in. Ibrahim Rugova's non-violent approach to the
conflict dominated seven or eight years, but with the emergence of the Kosova
Liberation Army (UCK or KLA) in late 1997 and early 1998, it became obvious
that the Albanian response to the conflict was no longer under Rugova's control;
it escalated into armed confrontations. "What happened in Kosovo was the escalation of violence since the section of the Albanian majority subjected to violence
turned its back on Serbs, Serbian police officers, and Albanians loyal to the
Serbian state."[7]

The winter and spring of 1998 were marked by terrorist actions by Albanian
armed groups, and by random violence and reprisals against civilians by the
Serbian police. The new phase was labeled 'liberation' by Albanians but
'terrorism' by Serbians, who played down the conflict, while the Albanians
exaggerated it. In any case, the facts speak for themselves concerning the number
of victims. According to the Albanian sources, 1,934 ethnic-Albanians were
killed in 1998. According to the Serbian sources, 169 civilians were killed and

157 wounded, and 115 policemen killed and 399 wounded. In 95 serious incidents at the Albanian border 14 soldiers (border guards) were killed and 29 wounded. Obviously, each side is counting only its own casualties. Put together, the Serbian and Albanian figures suggest that in 1998 around 2,300 people were killed, and some 1,600 abducted or gone missing. However, these are approximate figures only. Accurate lists with each person's details – like the one with the names of around 240 abducted Serbs and some 60 ethnic-Albanians, submitted to Ibrahim Rugova, the leader of Kosovo Albanians by European Union envoy Wolfgang Petrich on 26th November last year – are the only reliable source of this type of information.

The civilians (both Serbs and ethnic-Albanians) fleeing their homes due to war activities and violence are an aspect of the Kosovo tragedy – dubbed 'humanitarian disaster' by the international public. Thus, for instance, on 13th November 1998 UNHCR spokesperson Chris Ianowsky said that in Kosovo there were 100,000 displaced persons, while another 125,000 had left the province. Some 65,000 returned to their homes following the agreement between Yugoslav President Milošević and US envoy Richard Holbrooke (by mid-October 1998) putting an end to the Serb offensive against the illegal Kosovo Liberation Army (KLA).

What brought about this escalation? Could it have been avoided? The essence of the conflict between the Serbs and the Albanians has been spelled out in *Serbia: The Milošević Factor*, a report issued by the International Crisis Group.[8] It notes that the conflict has been there for centuries and then states: "Serbs justify their possession of Kosovo by international agreements and historical right because this had been a part of medieval Serbian kingdom and had once been inhabited by a Serbian majority. Everywhere in Kosovo are centuries-old Serbian Orthodox monasteries that Serbs consider as their national treasure. Albanian scholars, on the other hand, challenge Serbian historical claims by arguing that the Albanians are descendants of ancient Illyrians and now constitute over 90% of Kosovo's population."[9]

The Serbian response is that secession cannot be based on the principle of ethnic majority. This would be unacceptable to the international community since to endorse it would open the doors to numerous other separatist movements that would prove fatal to European integration. Besides, argue the Serbs, state borders are inviolable. The border between Serbia/Montenegro and Albania was agreed on May 30, 1913 by the London Peace Treaty after the First Balkan War and no changes were made after World Wars I or II.

The recent events in Kosovo and Metohija called for firm action of the international community on two occasions: in October 1998 and by the end of January and the beginning of February 1999. The goals of the international activity were outlined with precision: to stop violence, prevent humanitarian catastrophe, and stop the conflict from spreading to the other countries in the region.

In October 1998 an agreement was reached between U.S. envoy Richard Holbrooke and President Milošević, putting an end to the Serb military and police offensive against KLA. The Serbian part agreed to fully abide by the Resolution 1199 of the UN Security Council; create the conditions for a cease-fire; withdraw (as of March) special police units and army troops sent to the province and have the rest return to the barracks; allow access to Kosovo to relief workers and co-operate with the war crimes investigators; accept a verifying mission, including OSCE monitors in the field and unrestricted NATO monitoring flights over Kosovo; and engage in political dialogue about the province's autonomy. Faced with NATO's firmness, Milošević accepted the above conditions (warned that "NATO remains ready to act if Milošević doesn't stick to the agreement"). On the other side, it was made clear to the Kosovo Albanians that they were expected to fulfill their obligations by refraining from violence, taking part in the negotiations, and supporting the verifying mission.

The agreement resulted in, among other things, the end of the Serbian military offensive, a large number of refugees returning to their homes, which prevented an impending humanitarian disaster in the winter conditions, and the establishment of the verifying mission. Nevertheless, the fighting resumed shortly. The information published in the Serbian and Albanian media suggest that from October 13th, when Yugoslav President Milošević and U.S. envoy Richard Holbrooke reached the cease-fire agreement, until October 19th, 78 people were killed and 27 wounded. This is further corroborated by the following figures from the Serbian sources: during the first eleven days of 1999, 10 people were killed and 5 wounded in 80 Albanian guerrilla raids on the police, army, and civilians in Kosovo and Metohija. These raids left 4 policemen and 6 civilians – 4 ethnic-Albanians and 2 Serbs – dead. In January, ethnic-Albanian guerrillas abducted one civilian and eight Yugoslav soldiers.

The escalation of violence led to the Contact Group ministerial meeting in London on 29th January, 1999. The Group concluded that the violence in Kosovo has become a daily occurrence: "Despite all-out efforts by the international community, violence in Kosovo is happening every day. The escalation of violence – for which both the Belgrade security forces and the KLA are responsible – must stop. Repression against the civilian population from the part of the security forces must stop and these forces must withdraw. The ministers call upon the two sides to break away from the circle of violence and negotiate a political solution. 'This requirement has a military backing." In its statement, the Council of NATO put special emphasis on this: "NATO pledges its full support for an urgent political solution through the mediation of the Contact Group, which will provide a better status for Kosovo, preserve the territorial integrity of FR Yugoslavia and protect the rights of all ethnic groups. The Yugoslav authorities must at once conform the level, position and activities of the Yugoslav Army and police to the obligations they took in relation to the NATO on October 25th, 1998, and they must, in compliance with these obligations, stop the use of excessive and disproportional force. All the armed elements among the Kosovo Albanians must at once suspend hostilities and all provocation, including hostage-taking. All sides must stop the violence and concentrate on a peaceful solution to their goals." The negotiations between the Serb and ethnic-Albanian delegations, with the direct participation of the Contact Group, started in Rambouillet near Paris on 6th February, 1999.

KOSOVO HAS A SURPLUS OF HISTORY

Kosovo has rarely been an object of scholarly research,[10] though for centuries history has been made in Kosovo on a daily basis through conquests and liberations; revolts and suppression of revolts; alliances of different peoples in struggles against third peoples; emigration and immigration; persecution of aliens and settling of one's own folk; grand agreements concerning a common struggle against the occupying force and breaches of such agreements; wars against yesterday's friends and allies; changes of alliance during one and the same battle; setting villages ablaze; erection and destruction of sacred and other cultural monuments; pillage and murder; blood feuds and genocides. Only after World War II, within the institutional framework of socialist Yugoslavia, can one speak of modernization processes in Kosovo.[11]

For the peoples of Kosovo and the Balkans in general, history is not just an empty tale. Such a concept of the past may be popular in America but not in the Balkans, where history is being lived and one cannot easily distinguish the past from the present. From the point of view of the future one can reproach individuals for living in the past, but they do so all the same and reproaches change nothing. In rational conceptions of life the future is one of the most important orientation points and determinants of behavior. All planning necessarily includes the future as the measure of things.

But in Kosovo and generally in the Balkans (and certainly not only there) rationality does not rule. The prevailing type of identity is ethnic or national, which indicates the dominance of the logic of emotions instead of the logic of interests. In the traditional view of life in Kosovo and other parts of the Balkans, the past is the measure of all things and it largely shapes behavior. This traditionalist pattern is confirmed by the frequent instances of danger to property and even to the life of Kosovars. Traditional behavior requires that one remain on one's soil and to fight for it, not emigrate and 'betray' the fraternity, the *fis*, the tribe.[12]

In countries of recent origin, history can be less relevant for understanding social phenomena than in those countries whose peoples are the oldest and "in immediate contact with Adam and Eve." Dimitrije Bogdanović has written of such countries that "political phenomena have their history [and...] without understanding their genesis and conditions under which certain relations were formed, it is impossible to solve the main open questions. That holds true in the case of Kosovo, too."[13]

In the Balkans there are two conflicting views of history. In June of 1998 Italy's defense minister Benjamino Andreata declared that "the world community is getting sick of their [Serbian] insistence on the historical reasons for intolerance and waging wars with peoples surrounding them."[14]

On the other side, the Member of Serbian Academy of Science, Miodrag Jovičić, referred (May 13, 1998 in the daily Demokratija) to the international community, saying: "When some of them state that they are sick and tired of our talking about history and our invocation of historical rights, such upstarts only demonstrate that they do not comprehend the significance every nation attaches to its origins and all the genetic and emotional bonds that connect generations across centuries."

Until 1981 postwar Serbian historiography showed little interest in the history of Kosovo, leaving it in the charge of historians in Kosovo. For some Serbian historians that is one reason for the emergence of the 'distorted historical consciousness' concerning the Albanians.[15] Venceslav Glišić, himself a historian, partly explains this as resulting from the absence of contact since 1981 between historians who are Serbian and those who deal with Kosovo Albanians.[16] Serbians also say that for more than two decades, Albanian historians in Kosovo have created a conflict with Yugoslav historians concerning the history of Albanians and Albanian-Yugoslav relations.[17] Momčilo Zečević claims that this trend began on the day in 1971 when there was a suppression of the writings of a distinguished revolutionary who had witnessed Kosovo events during the People's Liberation War of 1941-1945, Svetozar Vukmanović Tempo. "From that time on, the Great-Albanian indoctrination was legalized in Kosovo's historiography. The dialogue with Yugoslav historians had stopped and the deafness of Kosovo historiography to any other, different opinion and argument, made it more and more convincing that ideological and political goals and interests had become more important than science."[18]

In Serbia there exist of course also serious historians who distinguish history as a scholarly discipline from the use of history for propaganda and nationalist purposes. As Boško Kovačević has observed, "Digging up the distant past in order to establish priorities and settlements is unfortunately usually only serving the struggle for power."[19]

Albanian historians of course blame the Serbian side. Thus the Albanian historian Mark Krasniqi writes that since 1981 Serbian scientists have begun denying the Illyrian origin of Albanians, so as to deny their being indigenous to these their ethnic lands, where they live today, particularly in Kosova and Macedonia, although these same scholars had previously written that the Albanians are of Illyrian origin and, with the Greeks, the oldest nations of the Balkans, while the Slavs are immigrants in these lands.[20]

Mark Krasniqi also argues: "The present Serbian pretensions about Kosova based on 'historical rights, because it is well known that the Serbian state and culture had its 'cradle' far from Kosova, somewhere north of old Ras and not in Kosova. Only at the end of twelfth century did the Serbs penetrate into Kosova after a battle in 1190 against Byzantium, which ruled Kosova at the time, inhabited always by Albanian people."[21]

Serbia uses another argument to justify its expansionist pretensions over Kosova. It points to monasteries and medieval orthodox churches, which Serbian propaganda calls its cultural religious monuments, to prove that "Kosova is a cradle of Serbian state and culture." The Albanian nationalists, however, insist that almost all the Orthodox churches of today had been built long before the Serbs came to inhabit these territories. Thus Krasniqi writes, "They were built during the Byzantine rule of these territories, which were then inhabited by Albanians. Consequently, these are monuments that belong to the culture of other people, which were usurped by Serbia and presented today to the world as its own historical monuments. . . The Serbian rulers . . . only repaired or enlarged some of these churches, but did not build them from the foundations."[22]

When it comes to ethnic mobilization, it is not important what the truth is nor which historians are more objective. The proliferation of nationalist history has effects that distort the significance of particular ethnic groups. This is in accordance with the famous 'Thomas theorem', of self-fulfilling prophecy: "If men define situations as real, they are real in their consequences. . . Public definitions of a situation (prophecies or predictions) become an integral part of the situation and thus affect subsequent developments."[23]

History was practically the only ground for closing down schools and Kosovo University in the year 1970–71. The Serbs demanded that history be taught in accordance with the program valid in Serbia, and the Albanians demanded their own curriculum.[24]

Nationalistic histories produce ethnic narcissism, perhaps even a sort of ethnic paranoia, that could be discerned even some years back, when research revealed that 70% of Albanians and 87% of Serbs agreed with the statement "My people are not perfect. Our tradition, however, is superior to other traditions."

THE PENDULUM OF DOMINATION

The conflict between Serbs and Albanians has persisted for centuries in Kosovo, each side sometimes dominating the other and at other times finding itself subordinate. Indeed, the phases of dominance and subordination have oscillated and can be described almost as a pendulum in motion.[25] A historiographic and sociological analysis yields the following division into periods, taking as the point of departure the change of agents of domination.

- Ninth to Fourteenth centuries: Domination of Serbs
- Fourteenth to Seventeenth centuries: Domination of Turks (and Serbs?)
- End of Seventeenth century to 1912: Domination of Turks and Albanians
- 1912–1918: Domination of various ethnic groups and/or states (first Serbs until after the withdrawal of Serbian troops from Kosovo and Metohija, then Bulgarians, Austrians and Albanians
- 1918–1941: Domination of Serbs
- 1941–1945: Domination of Albanians, Germans, Italians, Bulgarians.
- 1945 to 1966: Domination of Serbs
- 1966 to 1988: Domination of Albanians
- 1988 to 1997: Domination of Serbs.

This periodization shows that ethnic agents of domination succeed one another with striking regularity, and that the periods of domination get ever shorter. The change of the agents of domination is ever more frequent.[26]

According to the Serbian interpretation, Kosovo and Metohija entered the twentieth century in the turmoil of the disintegrating Turkish Empire, with Albanians terrorizing Serbs, just as was the case at the end of the nineteenth century. According to the Albanian interpretation, on the other hand, those were the times of uprising against 500 years of Turkish oppression.[27]

In 1912 an alliance including Serbia, Montenegro, Bulgaria and Greece – all Christian (Orthodox) nations – was formed to oppose the Turkish army and Ottoman Empire. On the eve of the war Serbia was unsuccessfully trying to win over the Albanians so that they at least would not participate at all in the approaching war preparations.[28]

In October 1912 the Serbs liberated (or, according to the Albanians, conquered) Kosovo and Metohija. At the time of the first battle 12,000 Albanian volunteers from Kosovo fought on the Turkish side. "Well-known and numerous are the data on massive participation of Albanians from Kosovo and Metohija in the battles against Serbian and Turkish armies in this war. In historiography and in school history textbooks in the Albanian language, one cannot find a single word about it, nor can one intimate from what is written there on the First Balkan War where the Albanians were and on whose side they fought."[29] The war was terminated by the London Peace Treaty, which recognized the new state of Albania and the new international borders: the mutual borders of Serbia and Montenegro and their respective borders with Albania. The International Border Commission seems to have been led by the historical instead of the ethnic principle of partition: Peć and Prizren, claimed by Albania with the support of Austria and Italy, became a part of Serbia, as did Priština, Skopje, and Bitolj.

During the inter-war period the Albanian question was not seriously posed – striking indeed when one considers its great significance in the second half of the century. The balance of power was favorable to Yugoslavia and this precluded the 'internationalization' of the Albanian question in Yugoslavia.[30]

Albanians who favored the Axis side – the occupying power – dominated during the period 1941-45.[31] Svetozar Vukmanović-Tempo, Tito's envoy in Kosovo during the Second World War testifies: "The main problem was the attitude of Albanian masses. Albanian population was en bloc against the partisans. . . . all there was were some twenty to thirty partisans under the command of Fadil Hoxha. . . . every village was against us like a fortress. . . . Albanian masses were on the side of the occupying powers" (1989: 12–13). During the war Kosovo and Metohija were held by the Germans, Bulgarians and Italians except for one part that was annexed to 'Great Albania'. Historians estimate that 30,000 Serbs and Montenegrins were driven out of this area – most of them colonists brought in there after the First World War and who would later be forbidden by the new Communist authorities to return to Kosovo.

For Serbs and Montenegrins the worst war criminals were "balisti" (from Balli Kombetar) and the Albanian SS-division Skender-bey. In December 1944 (Kosmet had just been liberated) Albanians revolted against partisan authorities. Martial law was introduced and the revolt was suppressed. On the other side, Albanian historians try to diminish the collaboration of the Albanians with the fascist occupying powers and to blow up Albanians' participation in the partisan troops.

The Yugoslav Constitution in 1946 established the Autonomous Region of Kosovo-Metohija within Serbia. This period is marked by two contradictory trends: considerable progress in education, culture, industrial development, and even living standards. On the other side the discrimination against Albanians is evident[32] in the forced confiscation of their weapons, the lack of proportional representation in power, the ban on hoisting the Albanian flag (one could end up in prison for that) and other abuses of police power. This period ends at the

Fourth Plenary Session of the Central Committee of the League of Communists of Yugoslavia held on July 1, 1966 in Brioni. Aleksandar Ranković, Vice-President of the Republic, and member of the central committee charge of internal security, was deposed and a reorganization of the security department was launched.[33]

Domination of Albanians – from 1966 to 1988

A series of 1968 amendments to article 2 of the 1963 Constitution introduced a novelty: the Socialist Republic of Kosovo was to be a constituent part of Serbia (Metohija disappeared from the name). Legislative and judicial powers were transferred to the autonomous regions, which were directly represented in the Chamber of Nationalities of the Federal Assembly. Rights and duties of autonomous regions were to be established by the constitution instead of by statute, as had been the case until then.

Further constitutional amendments in 1971 put the autonomous regions on nearly the same footing as the republics. "An autonomous region has executive, legislative and judicial powers, in other words, 'state power.'" Since the two autonomous regions now had nearly the same authority as that enjoyed by the republics, one started talking of autonomous regions as "constitutive elements of the Yugoslav federation." Yet the provision from the 1963 Constitution which stipulated that the territory of SFRY was "indivisible and made up of territories of socialist republics" (Article 2) remained unchanged. A republic is a state (Article 3) and has its own territory (Article 5), while the autonomous region is only a 'socio-political community' (Article 4) and thus a part of a Republic (Article 2) – in other words, an integral part of that Republic's sovereign territory. Furthermore, according the Constitution, the citizens of an autonomous Province remained citizens of the Province and exercised their 'sovereign rights' there, while the Province itself did not have a sovereign territory. According to legal experts, this Constitution turned Serbia into a federation of three republics (two Provinces and 'the rest' of Serbia, in which one of the three ('the rest') had less authority than the other two (the Provinces).

The position of Kosovo according to the Constitution from 1974 had three traits:

(1) constitutional-political independence;
(2) status as a constituent part of the federation and
(3) autonomy as a separate body within the constitutional-legal structure of the Republic of Serbia.

This third trait actually submerged the political subjectivity of the Albanian people in the multi-nationality of Yugoslavia and provoked Serbian possessiveness towards Kosovo. Nevertheless, the power in Kosovo fell into the hands of the Albanians: the pendulum changed direction.

Three forms of discrimination can be seen against Serbian and Montenegrin populations in Kosovo: informal, institutional and ideological.[34] Informal discrimination took place at the level of primary social groups, in everyday life and everywhere. Its aims were to make each individual insecure in every respect (property, life itself) to frighten him or her so that it would be safer to leave than to stay. Physical violence, danger to children, destruction of crops, threats – these are the most effective push factors behind Serb emigration from Kosovo.[35]

Institutional discrimination took place in the workplace (the relations among employees ranging from verbal conflicts to fighting over the allocation of positions), in schools, in health service institutions, in institutions of state power, in the League of Communists. Discrimination, though formally illegitimate, was an established practice.[36]

There is no consensus on how many Serbs and Montenegrins left Kosovo. Serbs, for example, state that between 1941 and 1981 over 100,000 emigrated from Kosovo and Metohija; some other estimates state even higher figures. Albanian sources tend to cite lower figures. For example, Hivzi Islami claims 52,000 Serbian emigrants for the period 1966-1971 and from 1981 on some 20,000.[37] According to the data of the Federal Bureau for Statistics over 6,000 Serbs, almost 1,200 Montenegrins and some 9,700 Albanians left Kosovo from 1989 to 1993.

Albanians explain Serbian emigration by economic factors and by the desire to return to their places of origin – according to this account, nearly all emigrants are found among those who had immigrated to Kosovo some decades earlier. However, demographic research, carried out by Ruža Petrović and Marina Blagojević on a sample of emigrated Serbs, demonstrates that only some 15–25% of migrations were motivated by economic factors, while the rest were non-economic in nature, being primarily due to discrimination.

Organized groups of Serbs and Montenegrins started arriving in Belgrade at the end of the eighties, demanding their rights and security in Kosovo. Propaganda created an atmosphere in Serbia which made plain that only those individual politicians and groups that carried out pro-Serbian changes in the southern province could count on legitimacy. On this wave of national sentiment, Slobodan Milošević – at the end of the eighties and the beginning of the nineties the leader of the Socialist Party of Serbia and virtually uncontested as the leader of Serbia as a whole – swam ashore. His actions would lead to the disintegration of Yugoslavia and the wars waged in the first half of the nineties.

Domination of Serbs – from 1988 to 1997

The pendulum of domination swung in this period to the Serbian side. Serbia took steps to control the province. In 1989 they started by amending the Constitution, eliminating Kosovo's legislative power as a sovereign republic and limiting the prerogatives of the provincial parliament. The provincial constitution should no longer contradict the constitution of the republic; the republic's influence over international relations was to be regulated. These amendments were introduced with the approval of the provincial parliaments of Kosovo and Vojvodina on March 24, 1989. Out of the total number of deputies (180) only ten voted against and two abstained.

The new Serbian Constitution of 1990 defined the provinces as 'a form of territorial autonomy'. The province of Kosovo got back its old name: Kosovo and Metohija. Legislative power came to reside with the Parliament of Serbia, executive power with the Government of Serbia and the supreme judicial power with the Supreme Court of Serbia. The provinces got their own statutes – passed with the prior agreement of the Serbian parliament (Article 10) – and exercised power through their own parliaments and executive councils.

According to Paskal Milo, the Albanians were purged from all institutions by a forcible act that had a basis in neither the Serbian constitution nor the federal constitution. Within a few months the majority of Albanians were sacked from their jobs in all sectors of public and economic life: health service, education, media, culture, economy and some 500,000 pupils and university students of Albanian nationality were thrown out of the educational system. The sacking from jobs was followed by intensified police repression, and extensive emigration – as with the earlier Serbian and Montenegrin exoduses, it is extremely difficult to establish how many emigrated.[38]

The same author says that further measures abolished the judicial and administrative systems and local self-government. Ethnic Albanians were deprived of the right to develop their national culture or use the Albanian language officially. Even more discriminatory were the measures in education and health care. Serbian authorities closed down all schools, including the Albanian University. Six thousand teachers, 850 university professors, and 15,000 health service workers lost their jobs for political reasons. The total number of dismissed Albanians was 100,000.[39] The Serbs usurped all public goods and citizens' facilities. Hotels, restaurants, sport facilities, stadiums, all school institutions for secondary education, all university buildings, dormitories for secondary school pupils and university students, all public halls, trade union rest homes, swimming pools, became 'Serbian'.[40]

Social unrest increased sharply at the end of February 1989, when demonstrations in front of the federal parliament building in Belgrade were attended by almost one million Serbian participants demanding the arrest of Azem Vllasi, the Albanian leader. Milošević complied with their demands. Albanian protesters in Kosovo were met by the army, with clashes resulting in 24 deaths. A state of emergency was proclaimed there and a curfew was imposed.

Nevertheless, Albanians demonstrated in March 1989 after Kosovo's parliament approved amendments to the Serbian constitution curtailing Kosovo's autonomy and proclaiming Serbia a unitary state.

In January 1990 a new wave of demonstrations took place in Kosovo, when some 40,000 university students demanded that state of emergency be lifted and Vllasi be released from prison. At the beginning of February, Kosovo was on the verge of civil war and the Yugoslav People's Army intervened, killing around 30 people.

In July 1990 the Albanian part of Kosovo's parliament proclaimed Kosovo an independent unit equal with the other constituent parts of Yugoslavia at that time. In September the Albanians organized a referendum; 87% of those voting favored Kosovo as an independent republic. Nevertheless, Serbia soon adopted a new constitution which declared that Serbia was not a nation-state, but a state of all her citizens, irrespective of their ethnicity. Autonomy was reduced to 'territorial autonomy' with a narrow circle of jurisdictions. In October the Kosovo parliament and the Albanian parties formed the provisional government of Kosovo and, seven months later, held presidential and parliamentary elections.

The period from 1992 until the beginning of 1998 can be considered as the period of passive Albanian protests. The situation radically changed in early 1998, when terrorist attacks began against police patrols. Within six months the conflict escalated into a war.

ETHNIC ISOLATION AND INTER-ETHNIC HATRED AS RESULTS AND AS PROJECTED GOALS

My research has involved conducting and analyzing opinion polls of about 1997 in Kosovo and Metohija with respect to four factors:

(1) the distance between two principal ethnic groups, Albanians and Serbs;
(2) perceptions of the legitimacy of Yugoslavia and/or Serbia as political units, as well as the legitimacy of the existing authorities;
(3) ethnic nationalism in Kosovo, and
(4) attitudes toward possible solutions to the problem of the status of Kosovo.[41]

There exist two parallel worlds which come in contact only in case of conflict. Our research only confirmed what was already known about this; it was designed to measure distance, and in our final report it was possible to present some concrete suggestions at least as to how to improve communications.

Except for neighborhood quarrels, continuous contacts between Serbs and Albanians are extremely rare. We deal here with two completely separated ethnic communities. Not only are Serbs and Albanians socially separated, they no longer even have the possibility to quarrel since they do not know each other's languages. Only 5% of Serbs declare that they have sufficient knowledge of Albanian to take part in a simple conversation. Many more Albanians – as much as 54% – declare that they know Serbian.

For Kosovo Albanians there is no such thing as Serbia or Serbian authorities. A complete absence of even a slight support for the institutional political system is a sufficient indicator of Albanians' denial of the legitimacy of both the political system and the very states of Serbia and Yugoslavia. They also massively refuse to participate in elections organized by Serbia and/or Yugoslavia.

In our 1997 study as many as 99% of Albanians declared that they had no confidence whatsoever in the President of Serbia; 98% declared a total lack of confidence in the Serbian parliament, 98% in the Yugoslav Army and police; 96% had no confidence in the president of Yugoslavia and 96% in the judiciary. In short, from the perspective of Albanian public opinion, Yugoslavia was not legitimate.

The reverse of this de-legitimation of Yugoslavia is the legitimation the Albanians bestow on their own political institutions. Rugova, the President of Kosovo enjoyed the confidence of 66% of Albanians; 29% had confidence in the Government of Kosovo; and 27% had strong confidence in the prime minister.

Confidence in three of the Albanian political parties was also measured. The most confidence was attached to DSK (The Democratic Alliance of Kosovo) (47% strong confidence). Next follows the Christian-Democratic Party of Kosovo (16% strong confidence), and, finally, the Parliamentary Party of Kosovo (12% strong confidence). Evidently, the confidence of the Albanian public opinion in Kosovo was directed toward their own independent political life.

The other side of this denial of legitimacy was the support the system in Yugoslavia and Serbia received from the Kosovo Serbs, which is rather different from what we found with respect to the citizens of Serbia proper.

Kosovo nationalism

The existence of ethnic nationalisms in Kosovo is not a matter of contention. Of course, one can find many Serbs who claim they are not nationalists and that those on the other side who are. Many Albanians say that they are not but that the Serbs are nationalists. However, very few Serbs and Albanians deny that mutual relations are bad; only 1% of Albanians and 10% of Serbs did that, in contrast to 86% of Serbs and 98% of Albanians who find mutual relations bad. Ironically, this is one thing about which Serbs and Albanians did agree.

The picture changes when one asked, for example, whether there was repression against Albanians: 76% Serbs declared that nothing of the kind took place (and 20% said they did not know anything about it, while 88% of the Albanians asserted that there was a heavy repression (9% claim that the repression did exist – but that it is not as bad as usually portrayed). As many as 53% cited discrimination in education as an example of repression. Besides, the Albanians saw the solution to this problem in the implementation of the Milošević/Rugova agreement concerning the return to the existing education facilities, regardless of the manner in which it should occur.

Most Albanians (75%) complain that the right to the use of Albanian in public institutions has not been realized. They blame bad inter-ethnic relations most frequently on Serbian politicians (49%) and intellectuals (23%), while the Serbs put the blame on the Albanian politicians (28%), both Serbian and Albanian politicians (24%) and foreign intelligence agencies (11%).

Nationalism is induced by both positive and negative stereotypes. Respondents were offered ten positive and ten negative traits and asked to choose three of each which give the best description of their own and other peoples. Naturally, both Serbs and Albanians have the most favorable opinion of their own group, even with respect to the very same categories, such as hospitality, courage, love of peace, cleanliness. When thinking of Serbs, Albanians assume that they hate other peoples, that they are pushers, egoistic, rough. When asked to portray the Serbs, Albanians use 7% positive traits and 93% negative ones. Serbs think that Albanians stick together (Serbs perceive this quality as a danger to their own ethnicity and therefore accord it a negative valence), that they hate other peoples, that they are backward, rough, and industrious. Overall, the traits ascribed to Albanians are 32% positive and 68% negative.

Acceptable and Unacceptable Solutions to the Kosovo question

Respondents could choose among eleven models that had been debated as possible solutions (usually within one of the two communities, rarely between them). We found that the Serbs saw only one acceptable solution – the abolition of any form of autonomy. All other options were rejected by more than half of the Serbs. The Albanians saw several acceptable options: first of all they overwhelmingly accepted Kosovo's independence (98%); 68% favored an international protectorate until a more permanent solution is found; 60% favored secession and union with Albania. Finally, 51% saw the solution in the granting of special status to Kosovo under international guarantees.

None of these options enjoyed the support of both parties. Unanimity could be found, however, in the rejection of single options, such as the re-activation of the Constitution of 1974; division of the territory, coupled with an exchange of populations; establishment of a Balkan federation; regional organization of Yugoslavia with Kosovo as one of the regions; Kosovo as a federal republic within FRY and Adem Demaçi's proposed 'Balkania', an equal federation of Serbia, Montenegro, and Kosovo. (In the current situation, however, Demaçi argues only for independence as a political solution.)

Our opinion poll yielded many interesting findings, only two of which will be mentioned here. One concerns the perception of the future and the other concerns ethnic narcissism.

Both ethnic groups displayed a need for a leader. The statement "We should be grateful to the leaders who tell us exactly what to do and how to do it" was supported by 61% of Albanians and 52% of Serbs. The Albanian leader with the highest confidence rating is Ibrahim Rugova – medium or strong confidence from 86% of respondents; followed by Fehmi Agani (close assistant of Rugova) – 75%; Marq Krasniqi (leader of the Christian-Democratic Party of Kosovo) – 67%; Redzep Qosja (leader of United Democratic Movement) – 65%; and Mahmut Bakali (ex communist leader) – 61%.

The situation is somewhat different in the case of Serbs, who had confidence in the 'external' political elite (Milošević, with a 65% approval rating, and the radical nationalist leader Vojislav Šešelj, with 62%) and 'domestic' ecclesiastic elite (Bishop Artemije, with 58%).

According to the research conducted for the U.S. Information Agency, Serbs unanimously reject the secession of Kosovo from Yugoslavia (98%). A similar unanimity, but this time in favor of secession, is shown by the Albanians (93%). Both populations claim to be highly motivated to participate in political actions: Serbs are willing to engage in signing petitions (59%), attending rallies (60%) and demonstrating (77%), while even more Albanians say they will attend rallies (83%), sign petitions (84%), and take part in demonstrations (85%).

One more finding is worth mentioning in this context: 83% of Albanians agreed with the following statement: "Independence of Kosmet (Kosovo and Metohija) from Serbia is something worth dying for" (11% partly agreed and 6% disagreed). The events of 1998 substantiated these research findings.

Inter-ethnic relations as they now stand in Kosovo and Metohija certainly result from numerous factors, but neither from God's will nor from merciless historical laws. What has been happening for centuries in this region had the potential for a different outcome. I should like to mention in the end two actors who could build a different outcome on the narrow (and yet existing) basis that was promising a different result. What I have in mind is one internal factor – the role of national elites of Serbs and Albanians – and one external factor, the role of Europe in the solving of these problems.

During this century, Serbian and Albanian national elites have created the idea of preserving the 'cradle of the nation' within the state borders, and of Kosovo as the basis of national identity. They aspired to a territorial political realization of the numerical superiority of their own ethnic group. Whenever they started realizing such goals "at any cost," war broke out. The main determinants of the crisis as well as of its solution are the political, intellectual and ecclesiastic elites of both ethnic groups. It is unrealistic to expect these elites (who by and large are ethnically exclusive, non-rational, burdened with traditionalism, authoritarian, un-cooperative, intolerant, and incompetent) to find a solution. Small autonomous groups of those who see the problems in a different light have practically no power and minimal influence.

The influence of Europe on the solution to the Kosovo problem cannot be underrated. Three times in this century Europe was directly involved in the fate of Kosovo and Metohija: after the First Balkan war and after World Wars I and II. Yet the attitude of Europe toward this part of Europe has been one of distance. Even those countries that do not belong to the European elite are at pains to prove their Eurocentrism and modernity by their distance from the Balkans. The Balkans became the measure for volatility, militarism, ethnic conflicts, tensions, and wars resulting from them – a measure of sub-civilization. However, these 'metric qualities' of the Balkans have another side as well. The Balkans are also a black spot on the modern face of Europe; this peninsula is a measure of Europe's failure. Whether Europe likes it or nor, the Balkans are a part of Europe – maybe her ugly face, but all the same hers! Europe must determine its role in this and similar cases, for it is an illusion that 'Kosovo' cannot happen to anybody else. Europe must choose between the role of a mentor or the role of a policeman. There remains, however, a justifiable fear that the problem of multi-ethnic Kosovo will in the long run be solved only by means of 'naked' – either internal or external – force.

OPEN QUESTIONS AND LESSONS

Kosovo is just a tip of an iceberg. The nature of the problem is universal. There are only three nationally homogenous countries in Europe – Portugal, Iceland, and Slovenia. Out of the total number of the states in the world, the ratio between those which are nationally homogenous and those which are not is one to ten. It is estimated that there are some 5,000 ethnic groups on this planet.

With the collapse of the Soviet bloc and the disappearance of the militant discipline ensuing from the division into two blocs, Communist federations (the Soviet Union, Yugoslavia and Czechoslovakia) disintegrated and 22 new states were formed. However, this trend still persists in the post-Communist countries: one-third of the population in Serbia are non-Serbs and one fifth in the Russian Federation are non-Russians. Twelve million Russians are living in Ukraine.

In that context it is the practical rather than theoretical questions that are thrown into relief: what to do to prevent ethnic conflicts in multiethnic states – i.e. how to prevent ethnic conflicts from escalating into wars? In other words, what lesson may be drawn from the ethnic conflicts in multiethnic states so far?

It is noteworthy that ethnic groups never learn from their previous conflicts. If they could learn the lessons, there would be no conflicts in the first place! Let it also be noted that totalitarian and authoritarian systems cannot learn from the conflicts in which they participate. This also holds true for ethnic political and military (including terrorist) organizations of that provenance. The main principles upon which totalitarian and authoritarian systems and organizations are based are ideological by nature and not inclined towards experiential learning. Nothing they may learn can alter their nature; such systems collapse due to either internal frictions or external intervention.

On the other hand, democratic systems and democratically constituted ethnic organizations can learn from their own experience and the experience of others. Consequently, they engage in ethnic conflicts only sporadically. Moreover, they are more amenable to peaceful settlements.

The lesson issue is also central for a number of supranational integration processes oriented toward preventing inter-state conflicts and ethnic conflicts within multiethnic countries. Learning one's lesson is important for those organizations and social groups in totalitarian and autocratic systems that try to solve ethnic conflicts in a democratic way.

Internal and external actors in ethnic conflicts have different lessons to learn. Internal actors need to understand the nature of conflict, its escalation patterns – the contributing factors and other elements relevant to the break-out and deterioration of ethnic conflicts. It is particularly important to know which options are available for preventing conflict escalation and for solving it.

The external actors (various supranational organizations) need to agree upon the principles for managing events in this sphere. The principles we have in mind are above all the inviolability of state borders and the observance of the rights of all ethnic groups. The methods for implementing these principles include weighing the effectiveness of political (diplomatic) means as compared to military force aimed at resolving a conflict. When appraising a conflict, do the international public, Security Council, Council of Europe and other international institutions use the same standards for all? Do the United Nations and NATO apply universal standards when deciding about specific military actions? In other words, is the use of military force against the IRA in Northern Ireland, the PLO in Israel, the KLA in Yugoslavia, the Chechen rebels in Chechnya and the Kurds (be it by the Turks, Iranians or Iraqis) equally acceptable for the world?

A lesson drawn from the management of inter-ethnic conflicts is that military measures provide a partial solution, with limited effects (they do not solve everything) and limited duration (they do not solve the problem once and for all). In contrast, democratic management of conflict and negotiated settlement are, as experience shows, the best options. In the resolution of inter-ethnic conflicts, only democracy – with full respect for ethnic and all other minorities – is effective.

NOTES

1. This justification of the census boycott is indirectly tested by the census in Macedonia 1994. The Macedonian census, financed and monitored by the Council of Europe, showed that there were there 22.9% Albanians (in 1991: 21.7%), and not, as was claimed by the Albanian ethno-politicians, 40%. This test proved the validity of the 1991 census data. The census in 1994 gave "legitimacy to the basic statistical data from 1991" – Friedman, 106.

2. Marina Blagojević, 'Iseljavanje Srba sa Kosova: trauma i/ili katarza' [Serbian emigration from Kosovo: trauma and/or catharsis], in *Republika* Ogledi (1-15.11.1995), 127.

3. Hivzi Islami, 'Zakasnjela demografska tranzicija na Kosovu u okviru demograf-skog prelaznog razdoblja u Jugoslaviji i Evropi' [Belated demographic transition in Kosovo seen in the framework of demographic transitional phase in Yugoslavia and Europe] (Beograd: *Sociologija*. Vol.XXVII, No.3 [1985]), 32.

4. Krstić, Branislav. *Kosovo izmedju istorijskog i etnickog prava* [Kosovo between historical and ethnic rights] (Beograd: Kuca Vid, 1994), 43-44.

5. There is sufficient ground to claim that there is something like the 'Cult of the Scene of the Holy Defeat' and that a considerable portion of the Serbian elite is obsessed by it.

6. Janjić, Dušan. 'Nacionalni identitet, pokret i nacionalizam Srba i Albanaca.' [National identity, national movement and nationalism of Serbs and Albanians]. In *Sukob ili dijalog* [Conflict or dialogue] (Subotica: Open University and European Civic Centre for the resolution of conflicts, 1994), 154.

7. Report of International Crisis Group (ICG), *Danas*, April 6, 1998.

8. Internet document. URI http://www.crisisweb.org/projects/sbalkans/reports/yu01main, viewed November 23, 1998.

9. *Danas*, April 6, 1998.

10. Branko Horvat wrote in 1987 in the preface to his book *The Kosovo Question* (Zagreb: Globus, 1987) that all that had been published on Kosovo was 'extremely unsatisfactory,' and the authors of the book 'Kosovski cvor' [*The Kosovo Knot*] complain about 'little reliable knowledge of actual goings-on' in Kosovo and stress "the lack of any solid basis of information. Namely, both empirical research and official evidence are very much lacking." Report of independent commission, 'Kosovski cvor: dresiti ili seci' [The Kosovo knot: untie or cut] (Beograd: Chronos, 1990), 1, 3.

11. Many indicators support this. For example, data on education show that the percentage of illiteracy amounted to 95% prior to World War II (74% in 1948) but decreased to 35% in 1971 (Islami, 1978). More impressive still is the time series showing the pace of growth in literacy by years 1921 (5% men, 2% women), 1931 (25% men, 6% women), 1948 (53% men, 22% women) 1961 (70% men, 42% women) and 1971 (79% men, 56% women) by calculations based on Islami's data. Hivzi Islami, 'Osvrt na razvitak stanovnistva Kosova' [A review of the population development in Kosovo] in *Sociologija*. Vol.XIX, No.1 [1977], 169.

The number of elementary schools increased from 278 in 1945 to 824 in 1971, while the number of intermediate schools grew from 11 to 69 (Kosova, Kosova, 1973:552). A university too was established in Pristina and its language of instruction for Albanians was Albanian.

12. Thus on June 20, 1998 mass media carried a press release of the Headquarters of the 'Kosovo Liberation Army' calling upon "all sound Albanian political forces to put themselves at the disposal of the fatherland and of the war of liberation." The Albanian people should mobilize and show readiness "to finally confront the enemy' and not to leave their ethnic hearths."

13. Dimitrije Bogdanović, *Knjiga o Kosovu* [A book on Kosovo] (Beograd: SANU, 1985), 3.

Venceslav Glišić writes the same: "Without knowledge of Kosovo's history, one cannot comprehend some contemporary events and processes either" in his 'Neki aspekti ideologizirane istoriografije o Kosovu' [Some aspects of the ideologized historiography on Kosovo], in *Kosovska kriza. Uzroci i putevi izlaska* [The Kosovo crisis: its causes and the roads to its resolution] (Pristina: Jedinstvo, 1989), 122.

That is so because "when the Serbs still claim that neither there is, nor there can be Serbs and Serbia without Kosovo and Metohija, it is not a matter of simple relapse into pathetic nationalism and romanticism of the nineteenth century type. Kosovo is not just the territory of the promised land, nor is it the leading idea in the struggle for national and state independence. In the Kosovo oath, the crucial determinant of Serbian identity, rests a powerful metaphysical charge." Dusan Bataković, "Anarhija i genocid nad Srbima 1897–1912." [Anarchy and the genocide on the Serbs 1897-1912], in *Kosovo i Metohija u srpskoj istoriji* [Kosovo and Metohija in the history of Serbia] (Beograd: Srpska knjizevna zadruga, 1989), 7.

On the Albanian side a Member of the Kosovo Academy of Science correctly observes: "The past history belongs to past centuries and that cannot be a valid argument to solve big contemporary problems." Mark Krasniqi, 'Kosova today', report presented to Belgian Senate (Prishtinë, 1992), 9. However, on the following pages he corrects this thought: "The Albanians do not turn to history in order to defend today their ethnic rights, because the past centuries could not be brought back again. But, if we base our demands of nowadays upon historical arguments, these anyhow are in favor of the Albanians. It is well known that the Albanians are direct descendants of ancient Illyrians, who had lived where the Albanians live now but in much broader lands then these of nowadays." Ibid., 10.

14. In the context of NATO's threatening intervention in case Kosovo Albanians would not be granted autonomy, Italy's defense minister Benjamino Andreata declared: "Serbs have to realize that the world community is getting sick of their insistence on the historical reasons for intolerance and waging wars with peoples surrounding them. We live in this century and we want solutions to be sought and life with peoples living next to us to develop in a democratic manner."

15. Glišić in *Kosovska kriza*, 122.

16. Ibid., 121.

17. Momčilo Zečević, 'Kosovska istoriografija kao ideologija i politika' [Historiography in Kosovo as ideology and politics], in *Kosovska kriza*, 135.

18. Ibid., 136. "In search of the origins of their ethnic group, these historians made use of all disposable means, not only historical, but archeological as well, to prove its existence before Slavs migrated to these lands, although Albanians appear in written sources only in the eleventh century" (ibid., 122).

19. Boško Kovačević. '(Ne)mogucnosti srpsko-albanskog dijaloga' [The (im)possibilities of Serbo-Albanian dialogue]. In *Sukob ili dijalog*, 316.

20. Krasniqi, 'Kosovo Today', 11.

21. Ibid., 10. A view different from the just quoted one, as well as from the views of some other Serbian historians, is represented by Sima Ćirković: "In any case, meticulous and unbiased research demonstrates the untenability of the thesis defended in the older historiography, namely that there had been no Albanians at all in the Kosovo territory prior to the Turkish conquest. The thesis represented by some Albanian historians and publicists, of a permanent and massive Albanian presence in Kosovo since the classical period cannot be successfully defended either. The territories of present-day Kosovo were far from the regions where the Albanians originated and were formed in early Middle Ages. By the end of the twelfth century, however, the Albanians spread gradually to reach the region of Gornji and Donji Pilat in the plains and the mountainous hinterlands of Lake Scutari and thus came into direct vicinity of the region which would later constitute Metohija . . . The picture one gets from the Turkish censuses from 1455, immediately after the Turkish conquest, shows that typical Albanian names in heads of households were found in 80 out of over 600 villages and that they were not territorially grouped.... Mixing of populations and migrations of the Albanians into the area of present-day Kosovo were facilitated by the circumstance that both these territories and the regions of northern Albania were under Serbian authority." Sima Cirkovic, 'Kosovo i Metohija u srednjem veku' [Kosovo and Metohija in the Middle Ages], in *Kosovo i Metohija u srpskoj* istoriji, 36-37.

22. Krasniqi, 'Kosovo Today', 13-14. When one deals with historians and subjects like these, one always runs into a sort of a 'virtual dialogue': instead of a real, genuine, communication of two scholars, one is faced with an unreal contact of 'ours' and 'theirs' on an almost metaphysical level, an exchange of abstract accusations and, as response, counter accusations: "At the time when the Serbian rulers and archpriests were building Gracanica and all other monasteries, churches, castles, Siptars [note: pejorative for Albanians!] practically did not exist there. Historical documents are unambiguous about that. The Šiptars were immigrants to Kosovo, who, making use of the friendliness of their fellow-Moslems – the Turkish authorities – came from their fatherland Albania and settled in Kosovo and Metohija" (Jovicic, 1998).

23. Robert Merton has explained the Thomas theorem in *Social Theory and Social Structure* (New York: The Free Press, 1967), 421, 423.

24. ". . . history was to be taught according to Serbian, not Albanian, canons because the latter are considered biased by a nationalistic view (which may be partially true, but the Albanians say the same thing of the Serbian curricula)." Alberto L´ Abate, 'Kosovo: a war not fought,' Dipartamento di studii sociali, Universita degli Studii di Firenze, 4. The Albanian position in these matters was that every people has the right to give its youth historical education based on the history of its own people, while Serbian authors involved in this polemics claimed that the problem was in its anti-Yugoslav spirit. "Education is directed at a creation of illusions among Albanian children and youth about a fatherland outside of Serbia and Yugoslavia. Glorification of Albanian national values, romanticism, mythomania and similar phenomena are well calculated to bring about albanization of Kosovo and to strengthen the integrative ties with the mother nation and eventually merge with it." Milos Sekulović, 'Prilog raspravi o Kosovu' [A contribution to the Kosovo debate], in *Kosovska kriza*, 233-234. "In the history textbooks in Albanian for all ages and generations one reads that Serbia and Montenegro had occupied Kosovo and Metohija during the First Balkan War in 1912. Literally so! More precise interpretation of historical events in this area one can find in the history textbooks in Albanian, written, published and authorized here, is that Serbia and Montenegro had then re-occupied Kosovo and Metohija." Milenko Jevtović, 'Naopaka istorija' [Inverted history], in Kosovska kriza., 390.

25. A periodization of Serb-Albanian common life in the same area has not been made so far for the simple reason that hardly anybody dealt with the complex as a whole. Instead, only different temporal section were taken up and then subdivided in periods. For example Dušan Janjić (in *Sukob ili dijalog*, 124 and passim) studied Serb-Albanian relations in the Yugoslav state and provided the following periodization: 1918-1941/5; 1945-1966; 1966-1981, and, finally, from 1981 till the present-day developments. Blagojević, in 'Iseljavanje Srba sa Kosova', divides the same period in a slightly different manner: from 1945 till 1966; from 1966 till the end of the eighties, from the end of the eighties till the present day.

26. I borrowed the pendulum metaphor from Marina Blagojevic who, in her analysis of migrations in this area, writes: "Serbian migrations from Kosovo are a component part of the general social and ethic context in the Province. They themselves are an aspect of the migration 'pendulum' characterized by the 'pendulum' of the relations between Serbs and Albanians in Kosovo.' Blagojević, 'Iseljavanje Srba sa Kosova," vi.

27. A French contemporary, V. Berard, wrote in *La Macedonia* (Paris, 1900), 138-9: 'it is quite obvious that neither the Sultan, nor the Porte, will interfere against the Albanians and will not establish order in the Kosovo Vilayet. In this Slav country Albanians play and will continue to play the same role as the Kurds in Armenian land. Advocates of Islam and servants of the Lord (Sultan) will on this very grounds exact impunity whatever their crimes might be'. Similar observations were made by G. Gaulis in *La ruine d'un Empire, Abd'ul Hamid ses amis et ses peuples* (Paris, 1913),.325-326: "Those from Debar kill in order to steal, those from Djakovica kill because of mad fanaticism, those from Peć kill for pleasure, those from Prizren kill due to their vicious instincts, and those from Tetovo kill in order to test their carbines." Quoted by Bataković in Kosovo i Metohija u srpskoj istoriji , 251.

28. Ibid., 251. Serbian Prime Minister Nikola Pašić offered Albanian notables "a contract concerning community of Serbs and Albanians in the Kosovo Vilayet," which

"guaranteed freedom of confession, the use of Albanian as instruction language in schools, society, administration of Albanian municipalities and districts, preservation of common law and, finally, a separate Albanian parliament which would pass laws on religious, legal and educational issues. This proposal was rejected. The Albanians decided to defend their Ottoman fatherland by force of arms and to turn the weapons received from Serbia against the Serbian army."

29. Milenko Jevtović, 'Naopaka istorija' [Inverted history], in *Kosovska kriza. Uzroci i putevi izlaska*, 391.

30. Ranko Petković, "Odnosi Jugoslavije i Albanije od zavrsetka prvog svetskog rata do tzv. jugoslovenske krize" [Relations of Yugoslavia and Albania from the end of the World War I until the so-called Yugoslav crisis], in *Sukob ili dijalog*, 237

31. Svetozar Vukmanović-Tempo, "Odnosi Jugoslavije i Albanije od zavrsetka prvog svetskog rata do tzv. jugoslovenske krize" [Relations of Yugoslavia and Albania from the end of the World War I until the so-called Yugoslav crisis], in *Sukob ili dijalog*, 12-13.

32. In 1956 60.8% of police personnel was Serbian while they accounted for 23.5% in the total population of Kosovo. The figures for Albanians are 31.3% and 64.9% respectively. Data from the weekly *Intervju*, Sept. 4, 1987, p. 64.

33. The Report of the Federal Public Prosecutor, directed at the federal government in December 1966 illustrates the background of the reorganizations: "A special system of intelligence regarding the situation in Kosovo and Metohija was spreading Serbian-nationalist and chauvinist views aimed essentially at smashing the unity of the peoples of Yugoslavia. 'Data' for such intelligence was procured by various inadmissible means – extraction, shadowing, eavesdropping, etc. These measures were primarily applied to intelligentsia and school teachers, in particular to the teachers of national language and history. SDB [Internal Security Service] monitored and controlled ideological contents of curricula and a considerable number of journalists and publicists was characterized as disloyal." However, one should stress that the same document shows such goings-on outside of Kosovo too.

34. Blagojević, "Iseljavanje Srba sa Kosova."

35. Ibid., viii.

36. Ibid., viii.

37. Islami, "Osvrt na razvitak stanovnistva Kosova," 47.

38. Muhamedin Kulashi, 'Kosovo i raspad Jugoslavije' [Kosovo and the disintegra-tion of Yugoslavia], in *Sukob ili dijalog*, 171.

39. Paskal Milo, 'Albanski faktor u balkanskoj i evropskoj politici.' [Albanian factor in the Balkan and European politics], in *Sukob ili dijalog*, 272.

40. Shkelzen Maliqi, "Samorazumevanje Albanaca u nenasilju-izgradnja nacionalnog identiteta naspram Srba" [Non-violent self-conception of Albanians: the construction of national identity as against the Serbs]. In *Sukob ili dijalog*, 220.

41. The Forum for Ethnic Relations in Belgrade in co-operation with the Institute for Philosophy and Sociology in Pristina, conducted in June 1997 opinion polls in Kosovo and Metohija in order to contribute to the answering of the questions regarding the anatomy of the Kosovo crisis. The polls were carried out using subsamples of Albanians (816) and Serbs (405). A two-stage stratified sample was applied. For Kosovo and Metohija as a whole the level of significance of the obtained apprisals was below 5% for traits that account for 3% and more as a proportion of the basic set. For subsam-ples (the stratum level) the apprisals are reliable with a level of significance amounting to 7% for all traits which occur more frequently than 3%. Indeed, the contents of the

polls and the character of the Kosovo population do not allow for a sensible merger of data taken from these two samples. A comparative interpretation, though, makes sense and in fact only such comparative interpretation should be trusted.

REFERENCES

L´Abate, A. *Kosovo: a war not fought*. Dipartamento di studii sociali. Universita degli Studii di Firenze.

Ahmeti, S. (1994). Oblici aparthejda na Kosovu [Forms of Apartheid in Kosovo]. In: *Sukob ili dijalog* [Conflict or dialogue]. Subotica: Otvoreni univerzitet i Evropski gradjanski centar za resavannja konflikta [Open University and European civic centre for the resolution of conflicts].

Baćević, L. (1989). Medjunacionalni odnosi [Inter-ethnic relations]. In: V. Goati, et al.: *Jugosloveni o drustvenoj krizi* [Yugoslavs on the crisis of their society]. Beograd: IC Komunist.

Baćević, L. (1990). Nacionalna svest omladine [National consciousness of youth]. In: S. Mihailovic, et al. *Deca krize* [The children of the crisis]. Beograd: Insitut drustvenih nauka.

Bataković, D. (1989). Ulazak u sferu evropskog interesovanja [Entering the sphere of Europe's interest]. In: *Kosovo i Metohija u srpskoj istoriji* [Kosovo and Metohija in the history of Serbia]. Beograd: Srpska knjizevna zadruga.

Bataković, D. (1989). Anarhija i genocid nad Srbima 1897-1912. [Anarchy and the genocide on the Serbs 1897–1912]. In *Kosovo i Metohija u srpskoj istoriji*.

Blagojević, M. (1995). 'Iseljavanje Srba sa Kosova: trauma i/ili katarza' [Serbian emigration from Kosovo: trauma and/or catharsis]. *Republika*. Ogledi, 1-15.11, 127.

Bogdanović, D (1985). *Knjiga o Kosovu* [A book on Kosovo]. Beograd: SANU.

Bogosavljević, S. (1994). Statisticka slika srpsko-albanskih odnosa' [Statistical picture of the Serbo-Albanian relations]. In *Sukob ili dijalog*.

Ćimić, E. (1989). Socijalno-duhovni doseg islamizacije [Socio-spiritual scope of Islamization]. Zagreb: *Nase teme*. Vol. XXXIII. No 9.

Ćirković, S. (1989). Kosovo i Metohija u srednjem veku [Kosovo and Metohija in the Middle Ages]. In *Kosovo i Metohija u srpskoj istoriji*.

Dinko T., (1948). Personality and Culture in Eastern European Politics. New York: George W. Stewart, Publisher, Inc.

Djurić, V. (1993). Srbija na zapadu' [Serbia in the west]. *Gledista*, 1-6.

Flere, S. (1986). Odnos mladih prema etnosu [The attitudes of the youth to the ethnos] In: S. Vrcan, et al. *Polozaj, svest i ponasanje mlade generacije Jugoslavije* [The position, consiousness and behavior of the young generations of Yugoslavia]. Zagreb-Beograd: IDIS-CIDID.

Glišić, V. (1989). Neki aspekti ideologizirane istoriografije o Kosovu [Some aspects of the ideologized historiography on Kosovo]. In *Kosovska kriza. Uzroci i putevi izlaska* [The Kosovo crisis: its causes and the roads to its resolution]. Pristina: Jedinstvo.

Golubović, Z., Kuzmanović, B., & Vasović, M. (1995). *Drustveni karakter i drustvene promene u svetlu nacionalnih sukoba* [Social character and social change in the light of national conflicts]. Beograd: Institut za filozofiju i drustvenu teoriju [Insitute for philosophy and social theory] and 'Filip Visnjic'.

Hadri, A. (1968). Nacionalni pokret albanskog naroda od tridesetih godina XIX veka do kraja 1912' [Albanian national movement from the 1830s until the end of 1912]' in *Iz istorije Albanaca* [From the history of Albanians]. Beograd: Zavod za izdavanje udzbenika SRS.

Hivzi, I. (1978). Kretanje nepismenosti u Albanaca u Jugoslaviji [Changes in illiteracy among Albanians in Yugoslavia]. Beograd: Sociologija. Vol.XX, No.2-3.

Horvat, B. (1988). *Kosovsko pitanje* [The Kosovo Question]. Zagreb: Globus.

Hrabak, B. (1978). Prvi izvestaji diplomata velikih sila o Prizrenskoj ligi [Early reports of the diplomats of great powers on the Prizren League].' *Balcanica* IX.

Igić, Z. (1989). Kontarevolucija albanskih nacionalista i separatista na Kosovu 1981. godine kao kontinuitet i sastavni deo velikoalbanskih pretenzija na srpske i jugoslovenske zemlje [Counterrevolution of Albanian nationalists and separatists in Kosovo 1981 seen as continuity and component part of Great-Albanian pretensions on Serbian and Yugoslav lands]. In: *Kosovska kriza. Uzroci i putevi izlaska.*

Islami, H. (1977). Osvrt na razvitak stanovnistva Kosova. [A review of the population development in Kosovo] Beograd: *Sociologija*. Vol.XIX, No.1.

Islami, H. (1985). Zakasnjela demografska tranzicija na Kosovu u okviru demografskog prelaznog razdoblja u Jugoslaviji i Evropi' [Belated demographic transition in Kosovo seen in the framework of demographic transitional phase in Yugoslavia and Europe]. Beograd: *Sociologija*. Vol.XXVII, No.3.

Islami, H. (1994). Demografska stvarnost Kosova [Demographic reality in Kosovo]. In: *Sukob ili dijalo.*

Ismajli, R. (1994). Albanci i Jugoistocna Evropa (Aspekti identiteta) [Albanians and south-eastern Europe: Aspects of Identity]. In: *Sukob ili dijalog.*

Janjić, D. (1994). Nacionalni identitet, pokret i nacionalizam Srba i Albanaca.' [National identity, national movement and nationalism of Serbs and Albanians]. In *Sukob ili dijalog.*

Jastrebov, I. S. (1875). Podatci za istoriju crkve u Staroj Srbiji. [Data for the history of the church in Old Serbia]. *Glasnik* SUD.

Jevtović, M. (1989). Naopaka istorija [Inverted history]. In: *Kosovska kriza. Uzroci i putevi izlaska.*

Kosovski cvor: dresiti ili seci [The Kosovo knot: untie or cut]. *Izvestaj nezavisne komisije* [Report of independent commission]. Beograd: Chronos, 1990.

Kovačević, B. (1994). (Ne)mogucnosti srpsko-albanskog dijaloga [The (im)possibilities of Serbo-Albanian dialogue]. In: *Sukob ili dijalog.*

Krasniqi, M. (1992). Kosova today. Report presented to Belgian Senate. Prishtinë.

Krstić, B. (1994). *Kosovo izmedju istorijskog i etnickog prava* [Kosovo between historical and ethnic rights]. Beograd: Kuca Vid.

Kullashi, M. (1994). Kosovo i raspad Jugoslavije' [Kosovo and the disintegration of Yugoslavia]. In: *Sukob ili dijalog.*

Kuzmanović, B. (1994). Socijalna distanca prema pojedinim nacijama' [Social distance towards various nations]. In: M. Lazic et al. *Razaranje drustva* [The distruction of society]. Beograd: Filip Visnjic.

Maliqi, S. (1994). Samorazumevanje Albanaca u nenasilju -izgradnja nacionalnog identiteta naspram Srba [Non-violent self-conception of Albanians: the construction of national identity as against the Serbs]. In: *Sukob ili dijalog.*

Marković, M. (1989). Nacionalno-oslobodilacki pokret Albanaca i kosovska kriza [National liberation movement of the Albanians and the Kosovo crisis]. In: *Kosovska kriza. Uzroci i putevi izlaska.*

Merton, R. (1967). *Social Theory and Social Structure.* New York: The Free Press.

Milo, P. (1994). Albanski faktor u balkanskoj i evropskoj politici. [Albanian factor in the Balkan and European politics]. In: *Sukob ili dijalog*, 1994.

Milosavljević, O. (1996). Jugoslavija kao zabluda [Yugoslavia as delusion]. *Republika*. Ogledi. 1-15.3., 135-136.

Mladenović, M. (1989). Kontrarevolucija na Kosovu, demografska politika i planiranje porodice' [Conterrevolution in Kosovo: demographic policy and family planning]. In: Kosovska kriza. *Uzroci i putevi izlaska.*

Nušić, B. (1986). *Kosovo.* Beograd: Prosveta.

Pantić, D. (1967). *Etnicka distanca u SFRJ* [Ehnic distance in the SFRY]. Beograd: IDN-CPIJM.

Pantić, D. (1974). Neke vrednosne orijentacije omladine [Some value orientattins among the youth]. In: S. Joksimovic et al., *Stavovi i opredeljenja jugoslovenske omladine* [Attitudes and outlook of the Yugoslav youth] . Beograd: Mladost.

Pantić, D. (1986). Odnos mladih prema inter/nacionalnom: zatvorenost -otvorenost prema svetu [The attitude of the youth concerning the inter/national sentiment -closedness: openness to the world]. In: S. Mihailovic et al., *Omladina '86 – sondaza javnog mnenja* [Youth '86: opinion poll]. Beograd: IIC SSO Srbije.

Pantić, D. (1987). *Nacionalna svest mladih u SR Srbiji bez SAP* [National consciousness of the youth in Serbia minus autonomous provinces] . Beograd: IIC SSO Srbije and IDN-CPIJM.

Pantić, D. (1989). Drustveno distanciranje'[Social distance]. In: V. Goati et al., *Jugosloveni o drustvenoj krizi.* [Yugoslavs on the crisis of their society]. Beograd: IC Komunist.

Pantić, D. (1991). Nacionalna distanca gradjana Jugoslavije' [National distance of the citizens of Yugoslavia]. In: L. Bacevic et al., *Jugoslavija na kriznoj prekretnici* [Yugoslavia at a crisis crossroads] . Beograd: IDN-CPIJM.

Petković, R. (1994). Odnosi Jugoslavije i Albanije od zavrsetka prvog svetskog rata do tzv. jugoslovenske krize' [Relations of Yugoslavia and Albania from the end of the World War I until the so-called Yugoslav crisis]. In *Sukob ili dijalog.*

Petranović, B. (1989). *Kosovska kriza i istoriografija'* [Kosovo crisis and historiography]. In: Kosovska kriza. Uzroci i putevi izlaska.

Petrović, R., & Blagojevic, M. (1989). *Seobe Srba i Crnogoraca sa Kosova i iz Metohije.* [Migrations of Serbs and Montenegrins from Kosovo and Metohija] Beograd: SANU.

Puto, A. (1994). Albansko pitanje na Balkanu -proslost i sadasnjost [Albanian question in the Balkans past and present]. In *Sukob ili dijalog.*

Radovanović, M. Kosovo i Metohija kao geografska i etnokulturna celina Republike Srbije, Savezne republike Jugoslavije i jugoistocne Evrope [Kosovo and Metohija as a geographic and ethno-cultural totality in the Republic of Serbia, Federal Republic of Yugoslavia and south-eastern Europe].

Raduški, N. (1995). Etnicki sastav stanovnistva opstina Kosova i Metohije i Vojvodine 1981-1991' [Ethnic composition of the population of Kosovo and Metohija and Vojvodina 1981-1991] Beograd: *Stanovnistvo.* Vol. XXXIII. No 1-4.

Samardžić, R. (1989). Kosovo i Metohija: uspon i propadanje srpskog naroda' [Kosovo and Metohija: rise and fall of the Serbian people]. In *Kosovo i Metohija u srpskoj istoriji.*

Sekulović, M. (1989). Prilog raspravi o Kosovu [A contribution to the Kosovo debate]. In: *Kosovska kriza. Uzroci i putevi izlaska.*

Tričković, R. (1989). Velika seoba Srba [Great Serbian Migration]. In Kosovo i Metohija u srpskoj istoriji.

Vukanović, T. (1986). *Srbi na Kosovu* [The Serbs in Kosovo]. I. Vranje: 'Nova Jugoslavija'.

Vukanović, T. (1986). *Srbi na Kosovu.* II. Vranje: 'Nova Jugoslavija'.

Vukanović, T. (1998). *Drenica. Druga srpska Sveta Gora* [Drenica: the second Serbian Mount Athos]. Pristina: Muzej u Pristini i Narodna univerzitetska biblioteka u Pristini.

Vukmanović, S, T. (1989). O situaciji na Kosovu i Metohiji 1943. godine [On the situation in Kosovo and Metohija in 1943]. In: *Kosovska kriza. Uzroci i putevi izlaska.*

Zajmi, G. (1994). Potiskivanje albanskog pitanja kao faktor degradiranja i raspadanja Jugoslavije [The suppression of Albanian question as a factor of degradation and disintegration of Kosovo]. In: *Sukob ili dijalog*, 1994.

Zečević, M. (1989). Kosovska istoriografija kao ideologija i politika [Historiography in Kosovo as ideology and politics]. In *Kosovska kriza. Uzroci i putevi izlaska*.

Zirojević, O. (1989). Prvi vekovi tudjinske vlasti [The first centuries of foreign power]. In: *Kosovo i Metohija u srpskoj istoriji*.

Žarković, D. (1989). Kosovska kriza' [The Kosovo crisis]. In *Kosovska kriza. Uzroci i putevi izlaska*.

16. CIVIL SOCIETY AND THE KOSOVO CRISIS, 1981–1999

Ken Simons

When the Yugoslav wars of secession began in 1991, the formerly autonomous province of Kosovo had already gone through more than a decade of social and political upheaval. Since 1987, this upheaval had been characterized by open repression of the majority Albanian population by the officials and security forces of the Republic of Serbia.[1]

Nevertheless, it took another seven years before the Kosovo conflict was transformed into a fully militarized war, involving first the rebel Kosova Liberation Army (KLA – UÇK in Albanian) and eventually, after the failure of the Rambouillet talks, the armed forces of the NATO powers.

What happened in the meantime, and how did the majority-Albanian population manage to avoid going down the road of open warfare for so long? The answer is a complex and at times deceptive one – that the primary response to increased repression by the Serbian state was one of non-violent resistance and non-cooperation, but that this strategy was itself inflexible and arbitrarily administered. Moreover, there was no place within the strategy for dialogue with minorities within Kosovo, an omission which continues to have damaging long-term results.

At the same time, however, the parallel government did build a sense of social unity and gave Kosovar society the strength to break with destructive and divisive traditions, most notably in the successful 1990–92 campaign to end blood feuds.

This part of the Kosovo story – the success of non-violence in holding a community together in resistance, but its failure to adapt to changing internal

Research on Russia and Eastern Europe, Volume 3, pages 309–319.
2000 by Elsevier Science Inc.
ISBN: 0-7623-0280-1

and external factors – is one which has not been fully explored by conventional strategic analysts. The evolution of Kosovar politics up to and through the current conflict has been the subject of several recent studies; there are important lessons to be learned both about the particularity of the Kosovo conflict and about the wider field of strategic non-violence.[2]

THE LEAGUE OF COMMUNISTS, THE RADICAL STUDENTS, AND THE SERBS

The period leading up to Marshal Tito's death in 1980 was one of optimism for politically active and educated Albanians in Kosovo. The province's League of Communists, formerly a stronghold of the Serbian minority, had gone through a rapid Albanianization since the 1971 constitutional reforms. The proportion of Albanians in the public sector increased, notably in the police force – formerly almost wholly Serbian – but also in the education, health, and judicial sectors. A quota system limited Montenegrins and Serbs – 20.9% of the province's population in 1971 – to 20% of the jobs.[3]

The quota system led to widespread dismissals of Kosovo Serbs (and Montenegrins), who were also handicapped by being unable to speak or read Albanian. The Serbian government in Belgrade made its unease public as early as 1977, with the commissioning of a document known as the 'Blue Book'.[4] This catalogue of arguments against provincial autonomy received no formal sanction or public discussion, but provided the basis for the far more systematic and ultimately successful campaign against autonomy in the 1980s.

The local party leadership was clearly committed to strengthening Albanian culture, including the renewal of cultural links with Enver Hoxha's Albania, still isolationist during this period but moving away from its unlikely alliance with the People's Republic of China. Nonetheless, it failed to strengthen the local economy – one of the failures in Yugoslavia's 'worker self-management' system was that there was no mechanism for ensuring that less-developed regions diversified their economies, and that regional planning tended to be haphazard and short-term.

In March 1981, students frustrated with overcrowding and bad food at the university had begun a protest which quickly turned into massive street protests, which spread to other cities in the province and occasionally degenerated into riots. The unifying slogan at most demonstrations was 'Kosovo republic' – that is, a Kosovar republic within the Socialist Federal Republic of Yugoslavia.

After two months, the protests stopped after tear gas was used against a university sit-in, the university itself was closed, and students were ordered home.

Similar protests in Croatia, Slovenia, and Serbia began to be infused with the identity politics of those republics' dominant groups. But Kosovo differed from the northern republics in an important sense. A non-Slav people, the Kosovar Albanians were considered ineligible for republican status within Yugoslavia because an Albanian state already existed outside the country; therefore the Albanian 'nationality' was not a 'nation' in the Yugoslav sense of the word. This condition also fed fears of irredentism; near the end of the 1981 riots there were indeed isolated incidents of pro-irredentist (that is, pro-Tirana) slogans being shouted or displayed.

Had the Enver Hoxha government been involved in fomenting the unrest in Kosovo? For a complex of reasons – including the fear that a united Albania would be dominated by the politically more sophisticated Kosovars – successive governments in Tirana have had little interest in unification with Kosovo,[5] but the willingness of Albanian governments to use Kosovo for their own ends, and of Kosovar, Serb, and Yugoslav politicians to play the Greater Albania card, is a continuing problem.

CANCELLATION OF AUTONOMY:
THE CREEPING COUP

In July 1990, the doors to the provincial legislative building were locked and the assembly, which had a two-thirds Albanian majority, dissolved. This was the culmination of a large-scale campaign over the preceding two years to remove control of key institutions from local elected officials, at the same time firing ethnic Albanians en masse. The education and health systems were particularly badly hit. Health care in Yugoslavia was dependent on social insurance contributions; dismissed workers were deemed to be outside the system, so 750,000 Albanians were now unable to seek treatment. Those who still qualified for treatment were often reluctant to do so for fear that Serbian doctors would maltreat them; pregnant women were particularly unwilling to trust the government clinics.[6] Parents refused to have their children immunized after rumors spread that the vaccinations had been doctored to cause sterility.

Albanian health professionals set up private clinics as a stopgap measure. The Mother Teresa clinics – named for the ethnic-Albanian nun from Macedonia but unconnected with her religious order – covered most of the country but facilities were often inadequate. Moreover, treatment was not free.

The Serbian state authorities still provided primary school classes in Albanian, but the secondary school curriculum was now taught only in Serbian. Initially, some schools tolerated the use of their classrooms after hours by the excluded Albanian teachers and their Albanian-speaking students, but by 1993 this was

impossible. Classes moved into private homes, vacant buildings, or other locations, where they were regularly broken up by police. Albanian-language
textbooks were systematically confiscated, making it more difficult for teachers
to deliver anything more than the most rudimentary lessons.

Albanian-language press and radio outlets were closed. Albanians once again
experienced random police searches of their homes, ostensibly for weapons. If
no weapons were found, they were beaten up. This technique had been used
before in Kosovo, in particular in the 1950s; some Albanians at that time even
bought guns and left them out in plain view so that the police would find them
and not have an excuse to inflict a beating.

The political and cultural disenfranchisement of the Albanians of Kosovo
was more or less complete by late 1990. The patterns of institutionalized repression, guaranteed by an all-Serb police force which answered to Belgrade,
remained in place throughout the next decade. Disappearances, random attacks,
and assassinations were common, and together with the lack of job opportunities and a contracting economy, led many Albanians to emigrate to Western
Europe or North America.

The direct response to political disenfranchisement was the formation of a
parallel administration – the government of the 'Republic of Kosovo', based
initially on members of the provincial assembly who had been locked out in
July 1990 – and the simultaneous non-recognition of the institutions of Serbian
or Yugoslav power in the province. The parallel government administered education, health, and other public services, collecting a modest tax from every
Kosovar Albanian family (who were, however, still liable to pay taxes to Serbia
as well). It also gave an additional measure of legitimacy to the Kosovar cause
– it was not a 'government in exile' but a functioning, if illegal, administration within the province itself.

The 1992 London Conference on former-Yugoslavia was attended by a delegation of Kosovars led by Ibrahim Rugova.[7] The limitations of the parallel
government strategy were immediately apparent in the reception given the
Kosovars at this meeting. Not only did they fail to be recognized as a national
delegation on a par with the former republic delegations (in fact, as an observer
group they had to follow the plenaries on closed-circuit television in a side-
room),[8] but the potential for serious conflict in Kosovo was scarcely mentioned
in the conference.

The LDK continued to try to raise awareness of Kosovo's situation at international fora, in the media, and in representations to governments. Some progress
was made around the need for NGO and media monitoring of the repression,
and small NGO efforts began within the province. It was not until 1996 that
the Organization for Security and Cooperation in Europe, the most credible

para-governmental human rights monitoring body, began work in Kosovo, so the presence of non-governmental bodies, even those which were unable to remain for extended periods, was all the more important.

Some outside NGOs expressed frustration over the LDK's reluctance to establish dialogue with moderate Kosovo Serbs. It is, however, true that any such dialogue would have been premature without extensive preparation on both sides. Even by 1996–97, it was less dangerous for Kosovo Albanians to seek contact with opposition groups in Serbia proper than with members of the Kosovo Serb community. Belgrade-based NGOs which became involved in human rights monitoring and mediation during this later period did so on the basis of established trust with groups (most, but not all, of them independent of the LDK) in Kosovo.

CIVIL SOCIETY ENDS THE BLOOD FEUDS

In the late 1980s, blood feuds were still endemic to the Albanian-speaking regions, including Kosovo. Up to 17,000 Kosovar Albanian men were at risk of death or injury because of recent or long-standing grievances against their families by members of another clan or family.[9]

The Serbian crackdown in Kosovo gave a new impetus to attempts to end the feuds and the divisive attitudes which they represented. Students from the city of Pec, horrified when colleagues were killed by feuding clans in the late 1980s, called on the Council for the Defence of Human Rights and Freedoms, the main human-rights group in the province.

The Council launched a concerted campaign – the 'Year of Reconciliation' – in 1990, but continued with follow-up work through to 1992. Teams of volunteers visited village families, returning as many times as was necessary, to persuade them to absolve the blood debts on their men by issuing 'pardoning the blood'. While the volunteers were often young people, there was always at least one visit by elders, and often by Anton Çetta, the legendary ethnologist who headed the campaign.

The *besa*, or word of honour which ended the feud, was given publicly at a ceremonial meeting. The largest of these ceremonies, on the plain of Decani in May 1990, attracted between 100,000 and 500,000 people from all parts of the Albanian-speaking world.

Police frequently harassed the participants and spectators at the ceremonies. In August 1990, Serbian authorities banned all large gatherings in Kosovo, breaking up one reconciliation ceremony at Pec in the west.

After the ban on public demonstrations, activities continued in private houses and so forth. Not all feuds were settled, but the tradition of retribution had been largely laid to rest, and positive traditions (for instance, "it is the Strong who pardons the Weak") were emphasized in its place.[10]

THE STUDENTS, *KOHA DITORE*, AND THE KLA

After the 1995 Dayton agreement, a UN preventive mission was deployed to Macedonia, in part because of the widely held expectation that Kosovo would erupt into a one-, two- or multi-sided war. Kosovo was itself not viewed as a containable struggle, despite the best efforts of Rugova and the LDK, which had lobbied foreign ministries intensively during the three-year period between London and Dayton.

By removing the Serbian militaries and paramilitaries from combat but not disarming them or otherwise attenuating their fighting capacities, the Dayton Accord increased the odds that Kosovo would be the scene of the next war in the Balkans.

Education remained a contentious issue, and one which often took on a highly symbolic role; education had, after all, been central to the 1981 and 1987–88 protests in the province, and the parallel schools were a visible gesture of defiance against Serbian exclusionism and repression. But the parallel schools were increasingly a liability for the LDK; while most Albanians were reconciled to the necessity of the parallel school system in the short term,[11] once the first year had passed there was concern for the long-term effects of a substandard and under-supplied education system. Talks on reopening the state system to Albanian-speakers began in 1993, but there was little progress until 1997–98.

In the summer and autumn of 1997, students at the University of Pristina attempted to take the initiative in refocusing discontent with continued Serbian repression and intransigence by holding public protests. The LDK advised student leaders against taking a more activist stance – 'active non-violence' in contrast to the institutionalized passive resistance of the parallel government – but the protests went ahead anyway. Elaborate preparations were made to ensure that non-violent principles were respected and only agreed slogans and demands were used; while the student demonstrations were focused on the education issue, they also attempted to redirect the wider civil struggle for rights and against police repression.

Prior to the 29 October demonstration, student leaders had meetings with Belgrade student groups, themselves engaged in a sustained protest against the Milosevic regime. There were hopes that the Pristina students would change the rhythm of political struggle in Kosovo, but these hopes were dealt a blow when the KLA made its first public appearance, at the funeral of a murdered teacher, in December of that year.

There was, however, another important change in Kosovar civil society politics, and one which clearly had more of an effect on post-war political alignments in the province than the student protests. Journalists have occupied prominent positions both in the leadership and as dissenters within Kosovar political institutions; but from 1991 to 1996 there was no daily press in which their positions could be set out.

Somewhat ironically, a privatization drive in Serbia was responsible for reviving independent media in Kosovo. Private publishing companies were permitted in Serbia and its provinces, although the state still had control over the licensing of new titles. In 1996, a license was granted to *Koha Ditore*, a tabloid-style daily newspaper, defiantly opposed to Belgrade but also independent of Rugova and the LDK. The core of its staff were young journalists and political activists who had been involved in the formerly dominant Rilindja publishing house and with the LDK. An entertaining and politically engaged paper was thus able to provide an alternative locus of power to the LDK, at the same time becoming the dominant media voice in the province.

Koha Ditore gave extensive and supportive coverage to the student opposition movements of 1996–97 and – somewhat more controversially – to the post–1996 armed resistance.

There was no organized paramilitary force operating on the Kosovar Albanian side between the mid-1980s and 1995. This is remarkable given the events in Bosnia and Croatia during this period, the continued repression by the Serbian police and the federal army after 1990–91, and the presence of Serbian paramilitaries (such as Arkan's Tigers) in Kosovo itself. In essence, there was a high degree of discipline both inside and outside the LDK to avoid any organized violence which would give Belgrade a rationale for violently suppressing all dissent in the province.

Isolated guerrilla attacks took place in 1995–96, though there appeared to be no pattern or recognizable command structure. In February 1996, a group called the 'National Movement for the Liberation of Kosova' claimed responsibility for attacks on Bosnian and Croatian Serb refugees in Kosovo. The first attacks credited to the 'Kosova Liberation Army' followed soon after in late April 1996. Eight Serbian police and civilians were killed in a series of attacks over a few days, apparently in reprisal for the killing of an Albanian student.

The LDK denied knowing anything about what was clearly an emergent guerrilla force, and suggested that the killings (of civilian Serbs and police officers) were a provocation organized from Belgrade. The groups were small, but well-armed with light arms and mortars, and were commanded by veterans of the Yugoslav People's Army (JNA).

The KLA soon emerged as the most important of these forces. It was largely organized and financed by the Kosovar Albanian community in Western Europe, particularly Germany. Donations from diaspora sources, which included both legitimate business and criminal activity, combined with the flood of cheap light weapons from Albania (where armed groups threatened civil war through most of 1997) allowed for rapid growth in the army's strength. There is no indication that the KLA received funding from Western government sources or was otherwise backed by the West; indeed, the U.S. special envoy to Kosovo in 1998 had labeled the KLA a 'terrorist organization'.

ENDURING THE WAR

Throughout 1998 a medium-level war was fought in the rural areas of Kosovo. A February-March offensive by Serbian forces against what were believed to be KLA strongholds in the Drenica region sparked massive protest in the cities and strengthened the political position of the KLA (notwithstanding the lack of a political party identified with the insurgents). The LDK under Rugova was forced into an even stronger pro-independence stance by this growing radicalism, and came under sustained criticism for their long-term strategy as well as for their inflexibility and stratified organizational structures.

By July, the KLA claimed to control one-third of Kosovo, and a pattern of skirmishes, KLA raids and Serbian reprisals had been established. One thing which was missing, however, was an ability to retain control of territory. KLA units knew how to harass but not how to fight a war; moreover, a failure to engage with the Serbian forces led them to strike out at people they identified as 'collaborators.'

Serbian attacks on villages, under the pretext of flushing out KLA fighters, continued. The massacre at Raçak in February 1999 took on crucial importance in that it immediately preceded the Rambouillet talks and the March ultimatum by NATO – which this time was not a bluff.

The forced deportation of nearly half the population, which was concentrated in the first two weeks after the NATO bombings began on 23 March, appeared through its speed and thoroughness to be based on an existing contingency plan. Under cover of war, a far more sweeping clearance of Albanian-majority villages and cities was possible than in the months or years before. Not only were media and human rights observers absent,[12] but alternative 'explanations' for the refugee flow could be advanced. Official Serbian state media persisted with the explanation that hundreds of thousands of refugees were leaving Kosovo voluntarily, to escape the NATO bombing and KLA violence.[13]

The non-violent opposition suffered doubly from the NATO intervention – not only were they largely incapable of directing resistance to the Serbian deportations and killings, but those who stayed in Kosovo ran the risk of assassination and capture. Within hours of the NATO bombing, the human rights lawyer Bajram Kelmendi was found murdered. Fehmi Agani, elder statesman of the LDK, was killed in early May. Ibrahim Rugova was detained, in circumstances which continue to be unclear, and taken to Belgrade where he was quoted as calling for an end to the NATO bombings and was photographed, clearly uncomfortable, with Milosevic. As clear as it may have been that he was acting under duress, his wartime meetings with the Belgrade government further damaged his already weakened authority.

The *Koha Ditore* journalists appeared to handle the nightmare of the war rather better than the organized politicians, despite being clearly targeted for arrest or assassination. The day the NATO bombings began, Serbian police shot and killed the paper's security guard and burnt down the office. Gjeraqina Tahina continued to file reports for the London-based Institute for War and Peace Reporting until her expulsion in early April. Editor Baton Haxhiu escaped in disguise from Pristina later that month, helped in part by reports that he had been killed shortly after the air war began. Working out of Tetovo (an Albanian-majority town in Macedonia) he and his staff published a limited edition of the newspaper for distribution to the refugee community. Publisher Veton Surroi, meanwhile, spent the entire 80-day war underground in Pristina. Surroi had been part of the Kosovar delegation at Rambouillet: "As someone who had signed the Rambouillet accords, I felt I had a responsibility to stay," he explained in a postwar interview.[14]

THE POSTWAR POLITICAL ORDER

At the time of writing, Kosovo was a UN protectorate, still under formal Yugoslav sovereignty but governed by three international bodies: the United Nations Mission in Kosovo (UNMIK), responsible for interim administration; the KFOR military mission, responsible for maintaining security; and the Organization for Security and Cooperation in Europe (OSCE), responsible for organizing elections, supporting an independent media, and drawing up laws. The bodies are interdependent, but have frequently been in conflict with one another as well as with Albanian and Serbian political forces in the province.[15]

Hashim Thaci, the 29-year-old KLA leader who positioned himself as 'head of the provisional government' shortly after UN administration of the province began in June 1999, failed initially to be recognized as the primary voice of the Kosovar resistance, despite the KLA's somewhat unconvincing disarmament and transformation into a civil police force. Rugova supporters continued to claim equality, if not supremacy, of authority for the LDK and the structures of the pre–1999 parallel administration; opinion polls in October 1999 showed that a large majority of Kosovars would prefer Rugova to Thaci as leader of their community.[16]

Meanwhile, the growing lawlessness within the province, and the seeming impunity with which former KLA fighters terrorized and stole from non-Albanians still resident in the province, seemed to indicate that the pendulum had merely swung the other way and the Kosovar Albanians were now treating the Serbs as the Serbs had once treated them. But this apparent brutalization is balanced by an openness about the problems of the postwar period. In a March 2000 interview, Veton Surroi notes that "debate is becoming more vibrant: criminality, political monopolies, the unpreparedness of Kosovo political organizations – we have been dealing with all of these things in public. That shows a strength actually in civil society."[17]

How much remains of the historic strength of civil society politics in Kosovo, and of the political will which delayed the province's descent into open war for more than a decade? As brutalized as Kosovar society has become through war and the loss of political infrastructure, the more imaginative aspects of Ibrahim Rugova's strategic nonviolence – in particular, the ending of the blood feuds – have probably had a lasting effect on the political culture of the province.

NOTES

1. Hugh Poulton, *The Balkans*, London: Minority Rights Group 1991 examines the pre-1991 status of the province in some detail, as do Miranda Vickers, *Between Serb and Albanian: A History of Kosovo*, New York: Columbia University Press 1998; and Julie Mertus, *Kosovo: How Myths and Truths Started a War*, Berkeley: University of California Press 1999.

2. In particular, see Howard Clark, *Civil Resistance in Kosovo*, London: Zed Press 2000.

3. Vickers, p. 182.

4. Vickers, p.183.

5. Or with western Macedonia. In 1981, the last year for which census figures can be considered reliable, there were 1.227 million Albanians in Kosovo; 377,000 in Macedonia; 40,000 in Montenegro; and 50,000 in other parts of Serbia, for a total of 1.7 million in Yugoslavia as a whole. The population of Albania was 2.75 million in the same census year. The modern Albanian republic and the contiguous Albanian-inhabited areas of Yugoslavia thus had a total Albanian population of 4.45 million, of which approximately 28% lived in Kosovo and 10% in other parts of the Socialist Federal Republic of Yugoslavia. Figures derived from Poulton, pp. 57, 75, 76.

6. An estimated 85% of births to Kosovar Albanian women during this period took place without any professional medical care. Vickers, p. 274.

7. Rugova led the umbrella national movement, the LDK, from 1989. He was elected president of Kosovo in unofficial elections in 1992.

8. Vickers, pp. 265–266.

9. Howard Clark, 'The Campaign to Reconcile Blood Feuds in Kosov@', in *Peace News*, No. 2438 (Feb-April 2000), pp. 28–29.

10. The magnificently named 'Gjilan Community Council to Avoid Negative Phenomena' reported in 1998 that it had settled 541 of the 778 disputes brought before it in the previous six years. Clark, p. 29.

11. "Albanians are prepared to lose a year or two of school," Zenun Celaj of the Council for the Defence of Human Rights and Freedom stated in 1991. Vickers, p. 301.

12. Most foreign NGOs withdrew in the week before the expiration of the NATO ultimatum. Some Serbian NGOs, most notably Belgrade's Humanitarian Law Centre, continued operating quietly in the region, but the most outspoken of the Belgrade news media -- the student radio station B-92 – was shut down on 24 March, thus restricting most Serbs from non-foreign news coverage of the explusions and killings in Kosovo. A further warning to journalists was delivered on 11 April, with the assassination of *Dnevni Telegraf* publisher Slavko Curuvija in Belgrade.

13. Goran Matic, Minister of Information in the Yugoslav government, put forward an even more baroque explanation. There were, he said, "3,000 to 4,000 ethnic Albanians [who] were paid $5.50 each to act the parts of Kosovo refugees during the first 10 days of the NATO airstrikes, leaving and then re-entering Kosovo to create the illusion that hundreds of thousands of refugees were fleeing the province." Associated Press news story, 10 May 1999. Astonishingly, the *Truth in Media Bulletin* for 13 May 1999 gave some credence to Matic's story, arguing that NATO's explanations were scarcely better.

14. Anthony Borden, 'Veton Surroi: Sharing the Risks of Democracy' in *Balkan Crisis Reports* 50, 23 June 1999.

15. Shkelzen Maliqi, 'Chaos and Complexities in Kouchner's Kosovo' in Balkan Crisis Reports 107, 14 January 2000.

16. Fron Nazi, 'The Struggle for a Kosovo Authority' in *Balkan Crisis Reports* 85, 18 October 1999.

17. Anthony Borden, 'Surroi: Still Building the New Kosovo' in *Balkan Crisis Reports* 127, 24 March 2000.

17. THE KOSOVO/A CRISIS: CONFLICTING PRINCIPLES, CONFLICTING AGENDAS: THE NATO WAR

C. G. Jacobsen

The alienation of the Albanian majority Kosovars of Kosovo (or Kosova, as they prefer), which erupted into civil violence in Spring 1998, reflects a kaleidoscope of differing historiographies and differing moral prisms.

THE CLAIMS OF KOSOVARS AND SERBS

To Serbs, Kosovo Polje, just outside the capital Pristina, site of their historic defeat by the Ottoman empire, remains a nation-defining beacon, the core of their self-identity. The idea that the province should be ceded because another ethnicity (the Kosovars) is – perhaps temporarily – the majority there seems as alien and preposterous to them as would be, to Americans, the suggestion that the Alamo be returned to Mexico because it is surrounded by a Mexican majority. The Serbs note that only recently have the Albanians constituted a majority, contrary to the case of the Serb majority in Croatia (now 'cleansed', with Western support) or of Srpska regions in Bosnia – both of which have multi-century roots. The Kosovar majority, moreover, is historically contrived, in that it resulted from World War II Nazi and subsequent Titoist policies, which sought to weaken the opposition Serb nationalists by dividing Serb lands among different jurisdictions. In this case the Nazis and Tito encouraged and

Research on Russia and Eastern Europe, Volume 3, pages 321–330
2000 by Elsevier Science Inc.
ISBN: 0-7623-0280-1

even ordered Serb out-migration, while arranging and welcoming an Albanian influx – the impact of which was dramatized through the new population's significantly higher birth rates.

In the 1980s Serb nationalism became a rising tide in response to the failures of a rudder-less ideology (it was preceded by other separatist theologies, such as those of Tudjman and Izetbegovic). Slobodan Milosevic coopted this Serb nationalist movement by crudely revoking Kosovo's autonomous status; the ratifying vote was secured with Yugoslav National Army guns ringing the Assembly. He took other measures also dictated by the politics of divisiveness, in particular banning Albanian language education and driving it underground. So out-of-proportion were his responses that they fueled rather than quelled the separatist threat against which they were aimed.

On the other hand, Milosevic's 1994 split from the harder-line nationalism of his far-right and also most of his 'democratic' opposition became a defining condition for the 1995 Dayton agreements. This switch by Milosevic also facilitated subsequent conciliatory approaches to Ibrahim Rugova, Kosova's clandestinely elected president, who continued to advocate pacifism, compromise, and autonomy rather than separation. But the centerpiece of Milosevic's return to moderation – the agreed resumption of Albanian education – was not immediately implemented, due to Milosevic's continuing need to guard his nationalist flank, and to the increased activism of the Kosovo Liberation Army (KLA), invigorated through arms infusion from civil war-torn Albania, and Rugova's consequent need to guard his nationalist flank.

To Kosovars, be they independistas, limited-sovereignty advocates, or status quo acceptors, the issues of Albanian language facilities and local autonomy/responsibility were quintessential human rights issues. And indeed, they echoed precisely the same issues as defined by the Serbs of Krajina and Srpska. They also echoed the rights guarantees identified by the Professors' Appeal in 1994 as necessary preconditions to the resolution of [Yugoslav] conflicts.[1]

The Contact Group of outside powers (United States, Britain, France, Germany, and Russia), who were the guarantors of the Dayton Accord, were in a quandary over these issues, and so were such anti-Serb crusaders as U.S. Secretary of State Madeleine Albright. America's early support for Kosovo's autonomy, first expressed by U.S. President George Bush, provided a rationale for intervening on behalf of Kosovars. Yet the West's earlier refusal to grant to Serb sub-units of the new states (Krajina and Srpska) the same separatist rights granted former Yugoslav sub-units (such as Tudjman's Croatia and Izetbegovic's Bosnia) were powerful arguments against condoning outright Kosovo independence. The consequence, vis à vis Krajina and Srpska, was moral inconsistency,[2] which again tended to inflame conflict instead of dampen it. While officially opposing the independence option, the Washington-led Contact Group (despite dissent from Moscow and Paris) embraced maximalist demands for Kosovar rights, without imposing strictures against terrorism. At the same time, the Contact Group imposed sanctions on Serbia for the violence of its response, without rewarding its March 1998 agreement to re-institute language education and other rights – thus simultaneously extending the grounds both for Kosovar extremism and for paranoid Serb xenophobia.

THE ONSET OF VIOLENT OFFENSIVES

The KLA then launched an all-out armed struggle, supported by Albania and Iran and indirectly by at least some U.S. and other agencies. It was met by a brutal and largely successful Serbian/Yugoslav counter-offensive that followed through the Summer of 1998, bringing tragedies that were perpetrated by fanatics of both sides. The particular killings that occasioned NATO's early October decision to gather an air armada for punitive attacks against Serbia were notable only for their timing. The Western media representative (an AP correspondent) who was first to see the bodies wrote a measured report, noting that body identifications were uncertain. (One must remember that in the Bosnian wars, all the differing sides would often claim the victims as theirs, slaughtered by one of the others.) In this case too, Serb authorities denied responsibility, promising an investigation and access thereto.[3] Yet the next morning the *New York Times'* front page splashed on its centre top the victims' photo, now labeled as Muslim women and children slaughtered by Serbs.[4] U.S. President Clinton and British Prime Minister Tony Blair issued the call for war. Why the late, sudden rush to judgment? Why then?

   Cynics harked back two weeks, when the U.S. government had chosen the day of Monica Lewinsky's final testimony to the Starr inquiry to launch missile strikes, with no prior Grand Jury charge or indictment, against Osama Bin Laaden's 'terrorist' camp in Afghanistan and Sudan's 'chemical-biological weapons' facility. (The case against Bin Laaden, once ally, now nemesis, rang true, that against Sudan less so: western Ambassadors and specialists visiting the bombed-out plant found no evidence]. The timing of the *New York Times'* splash and the U.S.-British response was similarly serendipitous. In Washington the big news concerned formal impeachment charges and evidence; in Britain, Blair had just been humiliated by left-wing adversaries.[5] The timeliness and focus of the diversion were further fueled by news of hard-line Bosnian election victories – a stinging reversal to Dayton architects and reigning formulas for Bosnia's reconstruction.

Having been threatened, the air strike was averted, seemingly at the 11th hour, through an agreement negotiated between U.S. envoy Richard Holbrook and Slobodan Milosevic specifying that Yugoslav army and special forces units introduced for the offensive would withdraw, with verification to be provided by 2000 civilian OSCE observers and aerial over-flights. Western aid to victims and refugees was to be unhindered. Viewed against the size of the air armada that NATO finally assembled, which was sea- and land-based and which included six B-52s, Milosevic's concessions were minimal. Serbia had already declared victory and the withdrawal of forces before the threat was assembled.[6] Its willingness to reinstate local autonomy provisions had already been announced. Lesser-scope foreign aid agencies, NGO and Western Embassy and media observers were already in place and the more extensive aid and observation presence now decreed would relieve Belgrade of reconstruction costs and burdens. Moreover, the agreement's acceptance of ultimate Serb and Yugoslav sovereignty effectively extended the verifiers' duty to monitor KLA efforts at re-assembling itself, thwart its re-supply ambitions, and legitimize Serb counter-actions. The presence of Russian and other OSCE contingents more sympathetic to Serbia would dictate a degree of non-partisanship.

Indeed, by January 1999 the failure of the first (US-led) OCSE contingent to limit KLA activities became the pretext for renewed Serb mini-offensives and yet more tragedies. Another atrocity attributed to Serbia, and disregard for the fact that the KLA had wrought "more deaths during the cease-fire than the Serbian security forces," brought a renewed NATO threat against Belgrade (just as the Senate impeachment trial formally commenced).[7]

NATO THREATENS, CLIMBS DOWN, THEN STRIKES

Washington's and London's apparent climb-down in October has been attributed to a variety of factors. Russia and China vigorously opposed any assault that was not sanctioned by the United Nations. Russia sent surface-to-air missile (SAM) targeting upgrades to Serbia, while threatening more substantial deliveries and a break to its 'partnership' with NATO.[8] All-NATO approval of the assault was secured, pro forma, but there was manifest internal opposition, from Greece and Italy in particular. Finally, Serbia/Yugoslavia's declared preparation for war and combat brought re-evaluations of the costs and dangers of escalation: missile/air only assault would likely be counter-productive, thereby necessitating a campaign on the ground that would require resources for which there was little political will.[9]

Yet after backing away from an attack in October, by March 1999 NATO unleashed its war as culmination to an ever-starker crescendo of ultimata. It dismissed Belgrade's nearly total acceptance of the [G8's] principal demands, including the call for an armed international guarantor presence. (The only outstanding issue was Yugoslav insistence that this force be mandated by the United Nations rather than by a unilateralist NATO decision.)[10] The decisive influence was apparently Madeleine Albright's conviction that Milosevic would 'cave' within two days of NATO bombing. She dismissed Russia as too dependent on Western/IMF loans, and therefore as effectively impotent. Other factors influencing the decision to intervene ranged from the anti-Serb antipathies (which still resounded eight years after the Croat-funded public relations firms first established the moral parameters of the emergent Yugoslav secession and succession wars) to Washington's assertion of its position as the global security arbiter; to the United States' wish to protect more than $200 billion worth of post-Gulf War reconstruction projects, Middle Eastern oil and gas wealth and arms sales prospects; to Saddam's abiding charge that America's interventionist policies were anti-Islam, and would never be unleashed against a Christian nation.[11]

NATO's de facto embrace of Kosova Liberation Army extremism (notwithstanding Europol's indictment of the KLA for drug and other smuggling activities and the CIA's earlier judgment that it was a 'terrorist' group[12]) legitimized extremist Serb nationalists who previously had found no legitimacy in Serbia/Yugoslavia. Besieged by 'errant' bombs, dropped from 15,000 feet to minimize SAM exposure, and vengeful Serb paramilitaries, most Kosovars fled. (Even before supplies ran out and less precise munitions wrought still more carnage, 'precision bombs' had – presumably accidentally – struck factories, schools, trains, embassies and maternity wards.) Many Kosovars were expelled for military reason, to thwart cell-phone and other targeting communication to NATO bombers. Post-war information would show that killings were in fact rare, belying both NATO and KLA war-time propaganda.[13]

Still, within weeks, a low-level war, which had been a blip on the world horror scale (other arenas, from Sudan and Rwanda to Kurdistan and Tibet offered 2,000 times the death and carnage!),[14] had been transformed into calamity, with most Kosovars in exile. However, their shelter, food, health and education were provided – in stark contrast to the continuing squalor of the still more numerous Serb refugees from the earlier Yugoslav secession and succession wars as well as from Kosovo. Those Serbs received no media attention, only 5% as much aid as that provided to refugees elsewhere, and were (and are still) denied any realistic 'right to return'. Serbs and the Kosovars who did not flee were subject to an ever-escalating barrage. Deadly chemicals from bombed-out petrochemical and pharmaceutical plants seeped into ground water and rivers, affecting neighboring countries also. Depleted uranium munitions left an equally deadly legacy, making this, in effect, a low-level nuclear war.

As the bombing began, Russia broke relations with NATO, noted missiles' re-targetability, sent a Naval intelligence vessel to the Aegean, and even explored possible Yugoslav membership in the Belarus-Russia union. She did accept a mediating role (perhaps, indeed, in part as quid pro quo for IMF loans). But as the war wore on, it was clear that any conflict resolution formula acceptable to the Milosevic regime or any likely successor regime, would resemble the formula Serbia would have accepted at Rambouillet for the sake of avoiding war. This would be a UN-legitimized and helmeted stabilization force, albeit dominated by NATO troops, on the ground in Kosovo/a and with no mandate in remaining Yugoslavia. However, though some other members of NATO showed interest in the formula, the war's primary architects still refused to accept the implication that they had been mistaken or calculated wrong. Instead, they reacted viscerally by escalating the war and demanding total, abject surrender. They took the same stance that had been adopted at Versailles, and which had later spawned World War II. Fifty years of Yugoslav, Serb, Montenegrin, and Kosovo development was laid waste.

Six weeks into the conflict Pope John Paul ll and Rumanian Patriarch Teoctist issued a joint condemnation of 'murderous bombing' and 'forced expulsions'. The dissonance between the finality of the bombings and the distastefulness but non-finality of the forced expulsions stood mute judgment. NATO members had themselves often embraced forced expulsion as a policy in the past, as in the cases of indigenous North American nations, of millions of East European German populations, 900,000 Palestinians, and Serbs from Krajina and Slavonia. Greece's Supreme Council of State indicted and convicted NATO of war crimes, whereas the International War Crimes Tribunal, beholden to Washington and London, indicted Milosevic and other Serb leaders. Bombing runs escalated to nearly 1000 sorties a day, and preparations for ground war proceeded, inexorably.

DE-LEGITIMIZING THE OPPOSITION

Demonization of the antagonist, which always diminishes intellect and diplo-macy, inevitably accompanies conflict – and Milosevic was demonized. This author has long advocated the international law and tribunal that might indict him, though pointing out that such law and tribunal would have to indict also Croatian President Tudjman, Bosnia's Izetbegovic and other world leaders who with or without foresight ordered or condoned analogous atrocities, now includ-ing NATO's. Otherwise it would be little more than a show trial. In fact, Milosevic was under domestic political siege before the bombs, with opponents winning elections in Belgrade, in other larger cities, and in Montenegro. The bombs de-legitimized his opposition, most of which was and is more nationalist than he, and more assertive concerning Serbia's Kosovo heritage. Milosevic had stood as symbol of past failures; now, as long as the bombs fell, he symbolized Serb defiance. Western Yugoslav specialists noted that if the bombs were halted by some sort of compromise solution, domestic Serb opposition would again be legitimate, and likely soon successful. Yet bombing continued, perpetuating both his power and legitimacy. This was a vendetta to destroy a politician on the verge of losing power anyway; for that sake a civilization was reduced to rubble. When the wars first broke out Canada's Former Ambassador, a pre-eminent Yugoslav specialist, called the action 'imbecilic and barbaric'![15]

But the joint statement by the Pope and the Patriarch also reflected a tide of shifting opinion, as images of Yugoslav destruction competed with those of Kosovar refugee plight. Russian and Chinese alienation and anti-Americanism reached ominous levels, but in Western countries too public sanction for war became increasingly questioned. The equation began to resemble early-1968 attitudes towards Vietnam. By May 1999 NATO's war was being interpreted as a contest of hubris versus human lives, with hubris increasingly being challenged, but still the dominant policy.

In the end, shifting polls may have been decisive, along with NATO's calculation about the awesome scale of destruction it was inflicting. All classified intelligence testified to the Yugoslav Army's resilience (its equipment, in Kosovo/a's caves and underground tunnel systems remained 98% intact!) – casting sobering light on the prospective costs of a contested ground war.

The 'victory' declared by NATO on the 79th day proclaimed Serb compliance with all NATO demands. In fact, it constituted acceptance of the compromise Milosevic offered before the war (against the opposition of his more nationalist rivals). In this light the war's lasting devastation of law, lives and civilization provided denouement to a century gone awry. Crucially, the 'victory' terms would interpose an international force, but restrict it to Kosovo/a alone; it would be mandated and thus legitimized by the UN. It would also comprise Russian and other non-NATO contingents – and thus keep alive the possibility of future partition (see below).

The extent of any potential future Serb Kosovo domain would soon appear to shrink dramatically. NATO forces, in stark repudiation of their proclaimed mandate and ambition, stood aside and condoned the subsequent ethnic cleansing of 90% of the Serb population and nearly 100% of Roma and other minority populations. Those who found protection and remained, remained harbingers of a possible future. However, a different future was posited by Yugoslav army threats to return or to sponsor others' (paramilitaries') re-involvement if the terms of their agreed withdrawal continued to be ignored. For Kosovo/a, as for Serbia and rest-Yugoslavia, as well as other Yugoslav successor states, history has clearly NOT ended.

WHAT DID THE WAR PORTEND?

The Washington-London confrontational posture left a disturbing legacy and an ominous precedent. As previously in Croatia and Bosnia, the West's formula was that of a carrot without a stick, this same formula now applied to the KLA. It contrasted sharply to the formula of stick-without-carrot (applied again to Serb interests). Combined, the two approaches were inflammatory and war precipitating.

Let us not forget that this took place in a context where violence in Kosovo/a would always dangerously intertwine with and exacerbate the parallel dynamics of neighboring Macedonian regions (where Albanians constituted a majority) and the clan politics of Albania itself. Any potentially wider conflagration in the area threatened to bring Greek and Bulgarian intervention on the Serb side, Turkish and wider Muslim intervention in support of Albania, a Bosnian Muslim army offensive against Srpska, and (depending on the latter's success) a possible Serb-Croat counter-offensive to finally divide Bosnia. These pan-regional, centrifugal dangers were contained. But the June 1999 agreements did not eliminate their historic portent. That would be accomplished only through inclusive reconstruction, development, security and the generation of hope on a scale such as could, as yet, scarcely be conceived.

The formula for war avoidance in Kosovo/a and beyond, as in Yugoslavia before the wars of succession and secession were unleashed should have rested on generic, equal, human, economic, and civil rights. The rights of minorities, however defined, can be no less than the rights of majorities. Neither's rights can be allowed to negate the other's. Power manipulation, perversion and discrimination can perhaps never be fully eviscerated. But their overt manifestations can be legislated and guarded against. Similarly, history's lesson is unequivocal: lasting conflict resolution is not achieved through partisan diktat; excepting 'permanent solutions' (true genocide), such resolutions always contain the seeds of conflict renewal. Conflict transcendence can evolve only from non-judgmental principles of equality and equity, and the nurturing of reciprocal rights, privileges and opportunities – including reciprocal separation rights, sometimes necessary as a precursor to reconciliation.

In this case, notwithstanding Washington's and London's theological aversion, this would grant majority Kosova genuine, extended autonomy, however defined and secured, but also allow the hiving off to Serbia of Kosovo Polje and northern pro-Serb regions. In regards to conflicts fueled in part by issues of economic deprivation and under-development (i.e. most contemporary conflicts) post-conflict re-building and development aid must also be seen to be non-partisan and equitable, if the solution is not to beget the same problem.

NOTES

1. Signed by Professors Richard Falk, Princeton University, Johan Galtung, Hawaii University & Dir., TRANSCEND, Carl Jacobsen, Carleton University & Dir., Eurasian Security Studies ORU, Jan Øberg, Dir., TFF, and Alex Schiberras Trigona, former Foreign Minister of Malta; distributed by (and available from) TRANSCEND and signatories; first published in *Fedrelandsvennen*, Norway, 7 July 1995, as 'Fredsforslaget', appended to a full-page article on 'Norske Professorer Sentrale I Nytt Fredsforslag For Eks-Jugoslavia'.

2. Thus outrage at the deprivations of 50,000 Kosovar refugees appeared rooted in humanist concern and impulse; yet the glaring lack of concern for or aid to the often equally destitute 527,000 Serb refugees from Bosnia and Croatia (the figure is that provided by the UN, and excludes Serb refugees sheltered in Srpska), jarred assertions of moral purpose. See *The Globe and Mail*, Toronto, 10 November 1998.

3. *CBC Newsworld*, 29 September 1998. Subsequent information, provided by The *Times of London*, 30 September 1998, did indicate Serb (though not necessarily police) culpability – and vengeful cause; the KLA had murdered seven local Serb policemen the night before the killings.

4. *The New York Times*, 30 September 1998.

5. See i.e. *The Times of London*, 28 September 1998.

6. *The Times of London*, 23 September 1998..

7. *The Ottawa Citizen*/Southam Press (with Associated Press files), 21 January 1999.

8. See i.e. Ottawa Citizen, 7 October 1998: 'Serb forces have been supplied with new warheads, fuses and sensors for their mobile SAM 6 missiles...converted it into a faster system with an extended range' [reprinted from The *Daily Telegraph*, with files from *The Times of London* and the Associated Press]. See also *The New York Times*, 7 October 1998, for report on Russian Foreign Minister Igor Ivanov and Defence Minister Igor Sergeyev's visit to Belgrade; and *CBC Newsworld*, same date, for interviews with and reporting on policy positions embraced by other Russian government and Duma leaders.

9. Ibid; see also *New York Times*, 10 October 1998.

10. *Le Monde Diplomatique*, May 1999; see also www.transnational.org and www.transcend.org

11. C.G. Jacobsen, *The New World Order's Defining Crises; the clash of promise and essence*, Aldershot, England: Dartmouth, 1996; see chapters 3 and 4.

12. David Lorge Parnas, "Only the truth can make us free," presentation to McMaster University conference on Globalization and the New World Order', Oct. 7 1999; to be published.

13. John Laughland, "The final death toll in Kosovo? It's sinking fast," *The National Post* [Canada], 4 November 1999 (from *The Spectator*, London): the alleged "biggest

mass grave ever" contained "only seven" bodies; "a whole string of sites where atrocities were allegedly committed have revealed no bodies at all."

14. Was the answer as to why NATO'/Washington's 'moral purpose' did not compel interventions in these arenas: racism, racism, 'Turkey is our ally', 'China has nuclear weapons'; and, if so, what was/is the bottom-line message sent to others...?!

15. James Bissett, 'NATO's brute force an imbecilic policy', *The Ottawa Citizen*, 17 April 1999.

CONCLUSION

18. WHAT ARE THE LESSONS?

Metta Spencer

This book was originally conceived as the report of a Science for Peace conference on 'the lessons of Yugoslavia', held in March 1997. The war in Bosnia had ended by that time and the Dayton Accords were being implemented. It seemed reasonable to suppose that the worst of the catastrophe was over and that the world would not be required to learn additional lessons from yet more new mistakes in the former Yugoslavia. Unfortunately, that was not the case. Surprising events continue to unfold even today, and there can never be any real 'last chapter' to this story. I have continued to collect articles for this book, and to pay attention when each additional scholar, diplomat, humanitarian worker, or war crimes prosecutor has commented on ongoing events in the Balkans. The content of this book is now unlike what had been planned at the time of the conference. Some invited speakers never presented papers for publication, and their places were filled by other authors. With the passage of time and the emergence of new struggles in the former Yugoslavia, new articles were required and some of the articles that had arrived on time required revisions. In any case, now is the time to sum up the lessons that should be taken from all these observations.

I shall pry loose and list some lessons from the various chapters of this book, as well as from other papers and conferences on the same topic. Not every reader will accept my whole list, though most of you probably do share the key values of these authors, at least in regretting the breakup of Yugoslavia and deploring the violence that it brought. My objective is to identify crucial turning points or decisions that made the breakup more likely or made it harder to bring the fighting to a halt.

This book, and the lessons that it proposes, represent only a limited array of interpretations, all of which are grounded on particular values that can be

Research on Russia and Eastern Europe, Volume 3, pages 333–357.
2000 by Elsevier Science Inc.
ISBN: 0-7623-0280-1

– and, during a time of nationalistic warfare, inevitably are – contested. Some members of the audience at the 1997 conference saw the program as ideologically biased against their own group's views. They had internalized a rule from Tito's society: that every conference must invite speakers approximately in proportion to the size of the constituent ethnic groups. These papers are by no means a representative sample of the opinions of ex-Yugoslavs. Instead of selecting speakers and authors by the traditional quota system or for any other ascribed attributes, we invited contributors on the basis of professional or academic standards. Nevertheless, it is true that when we invited people to contribute, I could usually guess what they were likely to conclude and, although there are significant differences of opinion among them on several topics, none of their papers reflect ethno-nationalism, nor do any of them celebrate the break-up of Yugoslavia or the wars through which the new sovereign countries emerged from that break-up. If the papers seem here consistently to derogate separatism, this can be attributed to a biased process of selection only to this extent: We did not invite any contributors whose conclusions were based on ethnic loyalty.

For whose benefit are our so-called 'lessons' compiled? Who is supposed to learn them and live by them? Answer: the lessons are meant primarily for official policy-makers and activists functioning in the international arena, not for the ex-Yugoslavs who occupy leadership roles in the successor states. Regrettably, most of the politicians in the former Yugoslavia remain partial to particularistic interests and continue their fights against old enemies; they will hardly be attracted to the recommendations of this book. Our purpose is to avoid precisely the kinds of disasters that arose from the misguided policies of the political figures who, all too often, remain in power in the year 2000. However, the lessons derived from the south Slavs' experience may perhaps be learned profitably by political leaders in other countries facing fissiparous tendencies similar to those preceding Yugoslavia's downfall – at least if those politicians seek to avert the perils of separatism. (A politician reading this book to learn how to support secession will find no help here.)

When we attempt to extract lessons from the historical record and to learn those lessons, we must generalize about the outcomes of crucial events. However, generalization is not always warranted, for under different circumstances, a given course of action might turn out quite differently. Only insofar as generalizations do hold up can researchers warn decision-makers of the problems with which a particular course of action is typically fraught – but when such warnings are well-founded, they may be exceedingly useful.

PARTITIONING STATES. Let me illustrate the usefulness of a valid generalization that had previously been unrecognized. There is a popular principle that the right to 'self-determination' is fundamental to democracy, and that every 'people' is entitled to secede from the state where its members live, to found a sovereign country of its own. This doctrine can be challenged on a variety of logical and theoretical grounds, but the most important challenge arises from two empirical facts: (a) that almost all secessions result in warfare, and (b) that those conflicts tend to persist for several generations, offsetting the advantages that the separatists had expected to gain. This empirical generalization, which is overwhelmingly supported by historical evidence, is not widely recognized.[1] It is a perfect example of a lesson that the world desperately needs to learn, for only when it is fully recognized can many disastrous decisions be averted. Thus this illustration suggests

> *Lesson 1:* Unless all legitimate stakeholders and minorities within a state agree to do so, avoid dividing a state into two or more sovereign entities as a way of resolving a clash of ethno-nationalistic claims.

This recommendation is, in practice, gradually becoming the prevailing principle in international relations, although officially there are numerous documents that seemingly contradict it by assuring peoples that they are entitled to self-determination. Even now, populists typically assure separatists that they can expect recognition for their claims of national sovereignty, but in reality such claims have rarely been accepted in recent years by the United Nations or by powerful states.

Nevertheless, as with many other rules, Lesson Number One cannot always be applied straightforwardly, for one runs into a dilemma. For example, as we have just noted, the international community justifiably tends to discourage separatists. However, separatists nevertheless are inclined to declare independence unilaterally, over the objections of the larger state, and of minorities and other stakeholders. They may then be victimized by those who reject their secessionist aspirations (Bosnian Muslims and Kosovar Albanians are examples of such separatist victims). If the international community (UN peacekeepers, say, or NATO) intervenes to protect them as victims of aggression, the intervention will de facto constitute a military action in support of the separatists, whether or not it is intended to advance their cause. This happened most conspicuously in Kosova; although there was no international recognition of the Kosovars' claim to sovereignty, by defending their human rights against the attacks of their Serbian enemies, NATO unintentionally established Kosova as a virtually independent state under the control of the Albanian inhabitants.

International decision-makers in such a situation find themselves in a dilemma, in effect unable to avoid taking sides in the dispute. They may defend the weak, thereby fighting on behalf of separatism. Alternatively, they may coldly warn separatists against declaring independence unilaterally (i.e. without the consent of all minorities and legitimate stakeholders) on penalty of forfeiting any intervention on their behalf, even if they become victims of aggression or war crimes. UNPROFOR and NATO faced this dilemma during the wars in Croatia, Bosnia, and especially Kosovo, where they eventually became de facto allies of the KLA, even while opposing that group's separatist goals. Thus certain 'lessons of Yugoslavia' cannot always be applied in a completely consistent way, for each principle has to be considered in the context of the other, possibly contradictory, principles and values to which decision-makers are committed.

I shall discuss the lessons of Yugoslavia by classifying them within the following four headings, which roughly represent the temporal phases of that country's painful drama: (1) structural conduciveness; (2) interventions during the crisis; (3) Dayton and its aftermath; and (4) the Kosovo war.

STRUCTURAL CONDUCIVENESS TO SECESSION AND WAR

When we discuss the 'causes' of any phenomenon, we may refer to any number of antecedent or concurrent conditions that made it more likely or even inevitable. In this section we shall consider some circumstances predating, but increasing the likelihood of the wars over the break-up of Yugoslavia.

Possibly the most contentious issue of the 1997 Lessons of Yugoslavia conference was whether the country's crisis had been caused more by domestic or foreign policy-makers. Ironically, a few Western speakers blamed the West, whereas all the ex-Yugoslavs attributed most of the blame to the political leaders of their own country. Regardless of which position is right, there is plenty of blame to go around. Our purpose is to learn from both the domestic and foreign errors, so as to avoid their recurrence in other countries. I shall consider the factors that were internal to Yugoslav society first, and only later those circumstances that were created predominantly by foreign decision-makers.

Reducing Internal Conditions Conducive To Secession And War

TRUTH-TELLING AND RECONCILIATION. It would be a mistake to conclude that the Balkans are a uniquely violent region in which ancient ethnic hatreds inevitably flare up again and again. During the Tito years, there was little evidence of ethno-nationalistic hatred; on the other hand, the conditions remained latent for the renewal of old enmities dating back to World War II or earlier. Tito had attempted to suppress these conflicts by declaring a new era of 'brotherhood and unity' without allowing people to discuss or even find out what had actually happened during that period, when the conflict had constituted a civil war more than an international one. If there had been a public airing of facts and of culpability instead of a fiat imposing reconciliation by edict, the bitter past might indeed have been surmounted.[2]

> *Lesson 2:* After a war, reconciliation between the former antagonists should be undertaken only in conjunction with a fair judicial inquiry in which culpability is publicly determined for war crimes and crimes against humanity.

DEMOCRATIZATION. According to Sonja Licht, Yugoslavia was in 1990 the only society in which the old issues of World War II remained alive, and this was because the country had not yet seriously begun its democratization. This observation, combined with the many other reasons for promoting democracy, yields the following.

> *Lesson 3:* In order to complete the unfinished business of a war so that a society genuinely reconciles and moves forward, institutionalize the structures of a democratic society, including systems safeguarding the rights of minorities.

CULTURE OF PEACE. One legacy of World War II was that the Partisan victors continued to celebrate their military accomplishments for decades afterward. Moreover, in successfully standing up to the Soviets, the Yugoslavs confirmed a political culture that emphasized the value of military strength. Yugoslavia maintained one of the ten largest armies in Europe, along with a major military-industrial complex. Though understandable, this emphasis on military values was counterproductive when conflicts arose that threatened the viability of the federal government. The leaders of the republics turned immediately to military measures instead of alternative systems of dispute resolution.

Moreover, all multi-ethnic societies need to maintain value systems that foster respect for diversity – especially ethnic diversity. This does not necessarily happen automatically, but may require that conscious attention be paid, especially to the education and media systems. Hence:

> *Lesson 4:* Cultivate a political culture that fosters reconciliation, civic cooperation, and other benign ways of managing conflicts rather than glorifying war or preparing for it. Institutionally, this system should support formal mechanisms for addressing ethnic disputes, as well as educational and media programs fostering respect for diversity.

SPATIAL DISTRIBUTION OF ETHNIC POPULATIONS. Some of the former Yugoslavia's structural problems do date back centuries. Especially problematic was the spatial distribution of ethnic-national groups within the entire country. In certain places (especially Sarajevo) multiple ethnic populations lived side by side in an integrated fashion. In other places, there were ethnic concentrations, and indeed the republics had been designed to give a special territory to each of these ethnic communities. As Dusko Sekulic, Randy Hodson, and Garth Massey have established, intolerance was generally highest among ethnic populations that constituted a local majority within an enclave, surrounded by people predominantly of a different ethnic group. This was the case, for example, in the Krajina, where Serbs were locally a majority within a generally Croatian republic, and in Kosovo, an enclave where the local Albanian majority were surrounded by Serbs.

'Ethnic cleansing' often resulted in the creation of new enclaves instead of mixed, integrated populations. As a policy, the expulsion or internal migration of groups can create new enclaves, exacerbating the inter-group tensions that had caused the problems in the first place. Yet ethnic segregation was not an incidental outcome of the war, but the very raison d'etre for the fighting. Insofar as political leaders can influence the population distributions within their countries in the future, they should become familiar with the results of the research, and be guided by

Lesson 5: Avoid creating ethnic enclaves.

CONSTITUTIONAL AMENDMENTS. The most serious problems for Yugoslavia's survival resulted from constitutional provisions dating back several years. In 1974 the republics were guaranteed the right to secede, though no procedures were spelled out by which this might take place. With the decline of communism world-wide, the demands increased for more democracy. However, there was no built-in amending formula by which constitutional reforms might be accomplished. By the end of the 1980s, political pressures in some republics increased the demand for decentralization,[3] whereas other republics and international financial institutions called for a more centralized and accountable federal government. In the minds of many Croatians and Slovenians, democracy itself seemed logically to entail de-centralization. Accordingly, no new federal constitution could be adopted allowing federal leaders to be elected directly by the citizenry. Only a confederal system would be acceptable to the proponents of decentralization. There existed no constitutional mechanism for resolving this impasse. When a Croatian separatist, Stipe Mesic, was elected to the rotating presidency of the federal government, the proponents of a strong central state kept him from being seated. It was at this point that the break-up of the country became almost inevitable, and all sides began to prepare to fight. This short historical summation points us toward two important conclusions.

Lesson 6: Whenever constitutional changes are introduced, practicable measures should always be specified for amending it in the future and resolving any impasses.

POLITICAL INCENTIVES AND CENTRALIZATION. However, as Timothy Donais has pointed out, every move toward decentralization tends to put a country on a 'slippery slope' toward further decentralization or separatism which can be exceedingly hard to reverse. Hence,

Lesson 7: Constitutional reform of a federal system should therefore attempt to offset any increased incentives to politicians who favor local control by creating ways in which they can benefit from a strong central government.

ECONOMIC DISPARITIES. There were several other internal factors that predisposed the populace toward breaking up Yugoslavia. For example, wide disparities existed between the standards of living of different republics. Despite its commitment to communist ideology, the government had done little to minimize regional economic gaps, the magnitude of which would have been shocking even in a capitalist society. Yet as the federal government became weakened and less able to transfer funds from one republic to another it became even more difficult to overcome those inequalities.

Lesson 8: Economic disparities must be minimized to reduce friction between regions of the country.

INTEGRATION OF INFRASTRUCTURE. As decentralization gathered support, the republics began to prepare for Yugoslavia's break-up by dividing up the infrastructure. Each republic came to expect to own a complete and self-sufficient industrial system, whether or not it was economically supportable. This trend should have been opposed.

Lesson 9: Do not permit the infrastructure of the federal state to be fragmented and reallocated among the various territorial entities that constitute it.

ELECTORAL LEGITIMATION. Finally, as democracy became an international norm accepted by Yugoslavs, pressures increased for elections to be held at both the federal and republican levels. For the sake of establishing the government's legitimacy, it was urgently necessary for the first free federal elections to be held before the elections in the republics, yet politicians in the separatist-oriented republics kept this from happening.

Lesson 10: Do not hold separate elections at the level of provinces or republics before the federal elections have been held.

To some extent, this mistake in Yugoslavia resulted from foreign pressures, and it is to those international influences that we now turn.

Reducing External Conditions Conducive to Secession and War

During the period preceding the violent break-up of Yugoslavia, a number of unfavorable circumstances were created inadvertently by decisions of international financial institutions, foreign states, and foreign activists. At least, one can reasonably suppose that the harm was done inadvertently, though some participants in the Lessons of Yugoslavia conference seemed to believe that it was a deliberate policy of the U.S. State Department to promote Yugoslavia's dissolution. We need not adopt any such Machiavellian theory here in order to recognize that negative effects sometimes did take place as a result of foreign – or even specifically Western – policies.

DIVERSE SOURCES OF INFORMATION. One such unfortunate effect has been pointed out by George Urban, who long headed Radio Free Europe (RFE).[4] Throughout the cold war, RFE broadcast news and analyses to the socialist countries of Central and Eastern Europe. The Soviet Union sometimes tried to jam those short-wave broadcasts, but in general the messages got through and were heard by audiences of millions, despite the dangers that they incurred by listening. Political dissidents usually tried to convey their messages to RFE or to one of the other Western radio systems, which spread the word back to their own countries.[5] In this way, a significant portion of each socialist society became familiar with the ideological debates of the day and, when the time came, were ready and eager to adopt democratic practices.

However, because the Western countries long courted Tito's favor, they did not permit any RFE broadcasts to be directed into Yugoslavia. According to Urban, this meant that the populace had no access to a free press or to opinions contrary to the official governmental analyses. When communism began to crumble, the Yugoslavs were seriously unfamiliar with democratic politics or with the very notion of pluralistic governance. Even since the wars in Croatia, Bosnia, and Kosova, the mass media throughout the former Yugoslavia tend to be controlled by particular political organs, and in Serbia by the regime itself. Journalism as a profession has been underdeveloped in Yugoslavia, with a few unusual exceptions, and the content is grossly biased, while the mass audiences have not acquired sufficiently critical attitudes to claim freedom of expression as a right.

Lesson 11: When dealing with a government that manipulates the press, democratic states and non-governmental organizations should foster the diversity of news sources and political debates and provide extensive support and technical assistance to the mass media. If the government prohibits such support of a free press within the country, alternative sources of information should be broadcast into the country from outside, as for example by Radio Free Europe/Radio Liberty. In the current period, access to the Internet can also be supported.

FOREIGN ECONOMIC INFLUENCES. Some Western participants in the Lessons of Yugoslavia conference blamed the United States in particular for Yugoslavia's demise. Thus Margarita Papandreou maintained that it was the U.S. Congress that required free elections to be held within six months in six Yugoslav republics; this was before the federal elections could be held. Along the same lines, Michel Chossudovsky attributed the worst of Yugoslavia's problems to the macroeconomic reforms imposed by the U.S.-controlled International Monetary Fund. He also claimed that Yugoslavia's federal government, led by Premier Ante Markovic, had been prevented by President George Bush from transferring funds to the republics and autonomous provinces – a change that exacerbated the political fracturing of the federal state. Most Yugoslav analysts, on the other hand, expressed great admiration for Markovic's reforms, which they say were generally accepted throughout the country as bitter, but necessary, medicine. They noted that the main problem with the transfer of funds was that some republics stopped transferring money to the federal government rather than the other way around. Moreover, the republics had been spending money on a lavish scale without even informing the federal government. The international monetary institutions began to insist on greater accountability.

Mihailo Crnobrnja, who had been Yugoslavia's ambassador to the European Community (which I will abbreviate as EU, since it is now the European Union), argued that it was Europeans who made the most serious mistakes affecting his country.[6] By the end of the 1980s more than two-thirds of Yugoslavia's economic relations were with the EU, making it Yugoslavia's most important economic partner – one capable of exerting powerful leverage. Since 1988, Yugoslavia had been moving away from non-alignment and had applied for an association agreement with the EU. The EU responded cautiously, requiring Yugoslavia to go through the same formal process as the other applicants. In 1990, Prime Minister Markovic requested support for his economic program. Although the EU praised his economic reforms, they turned his request down because political conditions had not been met. These political shortcomings were the worrisome human rights conditions in Kosovo and Slovenia's refusal to allow multi-party federal elections to be held. According to Crnobrnja,

> The EU let Markovic go back to Yugoslavia empty-handed, much to the delight of particularistic ethno-nationalist elites who saw him as a clear danger to their divisive plans and

strategies. The federal government's economic program soon collapsed and with it the last realistic rallying point to hold Yugoslavia together. The EU had lost a golden opportunity to influence more decisively the evolution of internal politics, before war broke out.

Markovic was expected to deal successfully with rising ethno-nationalism first, and the EU would 'reward' Yugoslavia after he had dealt with the problem. It was not seen clearly that Markovic needed the economic and political support of the EU to battle the rising nationalistic tide and proceed with reforms which, without the aid of the EU, were doomed.

While it is by no means certain that the support for Prime Minister Markovic's program would have avoided the breakup of Yugoslavia and the bloody confrontations that ensued, it is, in retrospect, absolutely certain that the type of conditionality imposed by the EU on Yugoslavia at the time that this program was rejected was ineffective, to say the least.[7]

The EU implicitly admitted as much only 11 months later. When the EU presidents, the two Jacques – Santer and Delors – paid a visit to Yugoslavia on the eve of its implosion, they had on offer a substantial economic package, if only the belligerents would calm down their divisive rhetoric and sit down to discuss outstanding issues in a rational and democratic way. The political situation in Yugoslavia, which prevented EU's assistance to Markovic only a year earlier, was infinitely worse at this time and yet the offer was made. . .

Crnobrnja maintained that the EU had simply used the wrong instruments for influencing Yugoslavia. "The EU," he said, "is an economic giant and a political pygmy. It tried to build up its political strength. Its economic strength probably would have been more effective as an instrument to pacify." From the EU's mistake we can derive

Lesson 12: If and when conditions are attached to the disbursement of foreign financial aid, the donor country should do so only from a well-informed position involving considerable knowledge of the context and implications of the conditions being stipulated. More specifically, when a country is facing internal separatist movements, its federal government should usually continue to be supported while it deals with that internal crisis, unless there are compelling reasons for doing otherwise.

ECONOMIC SANCTIONS. On the other hand, Lesson Twelve is a troubling and controversial admonition that requires at least a second thought. Whereas most people would agree in theory to the use of conditionalities and other economic pressures to induce reforms and prevent the break-up of a country, there is little agreement about the circumstances – if any – when such pressures may appropriately be applied. In 1992, after Milosevic's Serbia had incurred widespread opprobrium, economic sanctions were applied to the country, and were tightened during the wars over Bosnia and Kosovo. These sanctions were not observed by all countries (Russia, for example, continued to supply oil to Serbia) but some limited success can be attributed to them in terms of changes in Milosevic's policies. Nevertheless, in most cases, economic sanctions injure innocent civilians at random, while the political leaders are barely affected. Melica Bookman wrote that the sanctions

> caused the sudden closing of an otherwise internationally oriented economy. Sanctions resulted in the loss of foreign markets and rapidly outmoded the existing technology. That translated into decreased production, unemployment and poverty. Sanctions also decreased personal income and thus decreased the demand for goods and services. At the same time, sanctions decreased the supply of goods and services, especially basics such as food, heating, housing, transportation.... [T]he jobs that were cut have been from the sectors in which women are highly represented. Under conditions of economic and financial crises, the social policies that women have come to depend on were negatively affected (policies such as maternity leave, child supplements, free education, social help, etc.).

Thus one must question the ethics of economic sanctions as a weapon. There have been occasions – as in South Africa during apartheid – when the very population that will suffer most from sanctions nevertheless requests that they be imposed. On other occasions, however – notably in Iraq after the Gulf War – most ordinary people complain and plead for the sanctions to be lifted. Sometimes they are counterproductive, stimulating the populace to support the dictator who rules them instead of rising up against him. Thus if we can formulate any conclusion on this subject, it will have to be a guarded one such as this:

Lesson 13: If economic sanctions are to be imposed on a state to penalize the policies of an irresponsible or aggressive government, those sanctions must be chosen carefully to minimize the harm done to ordinary citizens.

In any case, economic sanctions are not normally imposed before the crisis and hence are not a matter of structural conduciveness but rather are one of the interventions attempted during the crisis. Let us consider that crisis phase now.

INTERVENTIONS DURING THE FIGHTING

As soon as organized fighting begins, the options available for resolving the crisis change markedly. At the same time, more outsiders may attempt to intervene, hoping to reduce the bloodshed and bring the antagonists to some kind of resolution. In this section we shall consider lessons that can be gained from the Yugoslav experiences of four different groups: (1) diplomats and officials conducting international relations and making decisions about military aid; (2) the United Nations peacekeepers; (3) non-governmental peace organizations; and (4) victims and displaced persons. Each of these categories of actors attempted to manage the catastrophe with the specific resources available to them. None was conspicuously successful.

Arms Transfers

Wars cannot be fought without weapons. A simplistic corollary of this fact is that all wars might be ended simply by abolishing all weapons everywhere. However desirable this proposal may appear, it is of course not achievable within any foreseeable time, for a variety of reasons that are worth considering.

First, in the case of Yugoslavia, considerable weaponry had been manufactured domestically, so that the JNA had access to supplies for much of the war without appealing to foreign powers to sell or donate weapons.

Second, even in principle most people would agree that a sovereign state is legitimately entitled to maintain an army or militia, whereas other political groups within the country are not entitled to do so. Indeed, a familiar definition of the state is: that organization within a territory that has a monopoly over the legitimate use of force. By and large such a monopoly of violence is beneficial, reducing the occasion for people to react with dangerous vigilante justice. However, such a definition calls attention to the problematic aspect of concentrating power: that the monopoly of violence may too easily be used against the citizenry, since they cannot defend themselves effectively. Hence even the Romans posed the question: "Who will guard the guards?"

In the Yugoslav case, in view of the dangers involved in having a unified JNA, the army was divided into seven units – one for each republic. Still, the JNA was disproportionately controlled by Serbs, and the seceding republics sometimes had to procure weapons from other sources so as to defend their citizens from the JNA. In some cases they were able to do so; in other cases not.

There is a third reason why the arms trade cannot easily be abolished: It is a generally accepted principle that sovereign states may legitimately determine their own foreign relations, including their military support or opposition to other specific governments abroad. Although the United Nations Charter restricts arms transfers, this is not closely observed, since each country actually remains jealous of its sovereign right to supply arms to friendly foreign powers. The arming of dissident factions within a foreign country is quite a different matter; the legitimacy of actively undermining a another government is more questionable in terms of the rules of international relations, but nevertheless such interference happens frequently and cannot be prevented.

Fourth, the transfer of arms is by no means under the exclusive control of any single state or any group of states. Private weapons manufacturers necessarily seek profits and therefore usually sell their products without much regard for the legitimacy of the buyers' political aspirations. Moreover, under certain circumstances a supply of arms may be seized or otherwise appropriated without payment. For example, when the Albanian government was in virtual collapse, its arms depots were raided and a huge supply of machine guns were stolen, then sold to the KLA in Kosovo.

For all four of these reasons, the availability of weapons cannot easily be reduced or regulated by universal standards. Each nation sells or donates arms for its own reasons without necessarily consulting its allies. In this respect, there is no 'international community', but only international anarchy. The United Nations has attempted to establish a degree of transparency regarding arms transfers by creating a registry, but compliance is limited.

The result is that particular states and international alliances face dilemmas that sometimes induce them to violate some of their own policies. For example, many states are, in principle, unwilling to support separatist movements. However, when a separatist group finds itself under seige and unable to defend its territory from an oppressive central government, there is considerable basis for sympathizing with the victimized group, even if their goals are not supportable. For instance, when people in Yugoslav enclaves (including Sarajevo, Srebrenica, Krajina, and Kosovo) could not defend themselves because of a lack of weaponry, many people who opposed the victims' goals nevertheless regarded themselves as obligated to provide weapons to them. They reasoned that the victims should at least be able to defend themselves on an equal footing with their adversaries. Along the same lines, there was a widely shared sense of guilt because the world had failed to defend the vulnerable, quite apart from whether their political policies were wrong or right.

Given these realities it is hard to identify particular lessons about arms transfers to learn from the Yugoslav tragedy. The following three principles are logical but may seem almost utopian within the current world order. Nevertheless, they should be entertained seriously as long-term policies.

Lesson 14: Minimize the level of militarism, wherever possible, for the sake of: (a) reducing the risk of armed confrontations, (b) permitting each economy to develop in a socially beneficial direction, and (c) avoiding the grave environmental effects of producing and using weaponry.

Lesson 15: Because Lesson 14 cannot always be put into effect without jeopardizing the security and human rights of a victimized group that lacks military defences, it must be pursued in conjunction with two other basic principles, which will be elaborated elsewhere, namely:

(a) International law and systems of enforcing it must be developed so that the rights and security of vulnerable populations can be protected when they are not armed.

(b) Voluntary civilian monitors and peacekeeprs, and sometimes also lightly armed military peacekeepers under UN direction should be encouraged to enter sites of likely confrontation to prevent the outbreak of fighting.

Lesson 16: States should, whenever possible, maximize the consistency between their willingness to arm a vulnerable population and their support for, or opposition to, its political goals. Each party to a conflict (whether it be the established government or a group claiming independence for itself) should know in advance what the international responses will be to the various options that it is considering.

International Diplomatic Interventions

The outbreak of any war is evidence that political officials have failed in their most important duty: to manage a looming dispute by implementing policies acceptable to all the parties involved. As the failure of Yugoslav politicians became apparent, the officials of foreign countries and international organizations stepped in, but they were able neither to arrive at any common definition of the problem nor to apply to it any accepted rules of governance. Indeed, even after the wars, all of the issues remain unresolved and are likely to lead to new conflicts in other countries unless the international community answers them definitively.

DIPLOMATIC RECOGNITION OF SECESSION. The most serious issues are these: Is self-determination a legitimate right that can properly be invoked to justify secession? If so, who is entitled to self-determination – a unit of a federal government (e.g. a republic or province) or a 'people'? Since Yugoslavia was ethnically mixed (as are almost all countries) if a *republic* secedes, this will automatically remove numerous members of a minority 'people' from their homeland against their will. On the other hand, if every *people* is entitled to self-determination, then all enclaves where an ethnic group locally constitutes a majority may secede, and this will require that international borders be changed. There is disagreement over the legitimacy of moving borders.

To manage this conflict, the international community has generally declared that secession should not be recognized diplomatically except when carried out with full protection for the rights of minorities. It was hoped that this rule would persuade minorities to remain within the new breakaway states. However, those in Yugoslavia were not persuaded – nor indeed were the conditions generally met before the breakaway states gained recognition. Logically one might well argue that if the Croats and Bosniaks had a right to secede from Yugoslavia, then the Serbs in Eastern Slavonia, the Krajina, and Bosnia-Herzegovina should equally have the right to secede from Croatia and Bosnia. And further, when the Bosnian Serbs boycotted the referendum, they had some constitutional basis for protest, since all decisions of constitutional import were supposed to be made consensually in Bosnia-Herzegovina.

While Secretary-General Perez de Cuellar was still in office he had anticipated this impasse before his term had expired, and had proposed the following three rules: (1) Do not recognize any breakaway republic or region in Yugoslavia before making abolutely sure that the minority problem there has been solved. (2) Offer 'symmetric recognition' instead of recognizing your fevorite groups while denying recognition to others. (3) Have a plan for Yugoslavia as a whole. His recommendations were ignored.

After having been recognized as independent states, Croatia and Bosnia-Herzegovina invoked the principle of 'territorial integrity' and insisted then that their borders were unchangeable. However, local pockets of minorities within each of those states demanded the right to rejoin their nationality group's own republic. The Western countries accepted the principle of territorial integrity for Bosnia and Croatia, over the objections of the minority populations, who then fought for their own 'self-determination' – the right to secede from the newly independent state and, ideally, to attach their territory to the titular republic of their ethnic community.

The conference speakers in Toronto agreed that international recognition of Slovenia and Croatia had been given prematurely, if indeed it should ever have been extended at all, given the lack of commitment in those new states (especially Croatia) to the protection of internal minorities. The participants also noted that the sale of arms by all parties violated the UN Charter and contributed directly to the onset of warfare. In view of this consensus, we can propose the following principle:

> *Lesson 17:* No separatist state should be recognized by the international community until all its constituent minorities are satisfied with the terms of the partition, or until an impartial body accountable to the United Nations has ascertained that the minorities' human rights will be secure within the new regime. One (but not the only) way of determining this is to assure that a majority of voters from each significant ethnic population must consent in a referendum to the partition before it can be carried out, and that the partition then be supervised by the United Nations.

THE CHANGING OF BORDERS. On one related issue, it is impossible to say what lesson should have been learned, since the speakers in Toronto held diverging opinions: There was no consensus as to whether the international community should, as a principle, resist demands that borders be changed.

Possibly this question would be one of several resolved by adoption of the following recommendation.

> *Lesson 18:* Clarification of international law is urgently needed to standardize the conditions of legitimate secession. An appropriate method of clarification might be for the General Assembly of the United Nations to request an opinion on these matters from the International Court of Justice.

This recommendation, which should be assigned highest priority, is consistent with the proposal advanced by the International Commission on the Balkans:

> The commission recommends the development of an international judicial institution to elaborate on the meaning of the right to 'self-determination' of 'peoples' as expressed in the UN Charter. There is an inherent tension between that principle and the no less important international commitment to the inviolability of borders. All the Balkan protagonists have different interpretations on these matters. There is a clear need for a tribunal on the limits to self-determination. This need not be a new institution. One obvious candidate would be the present World Court; another could be the European Commission and Court on Human Rights.[8]

NON-MILITARY PEACEKEEPING INTERVENTIONS. As the crisis deepened and casualties began to mount, many observers came to regard as shameful the reluctance of the international community to intervene early enough to stop the warfare before it cost so many lives. On this point there was, however, no obvious consensus among the Toronto conference participants – at least as regard military intervention. On the other hand, many people agreed upon the desirability of using diplomatic and non-violent civilian interventions even before the organized violence began. Some of the suggestions ran along the following lines: Act early. Combat hate speech whenever it appears. Train local citizens in civil disobedience to reduce the power of demagogues. Send in volunteer corps of social workers and peace brigades wearing white helmets and carrying camcorders to document violations of human rights. Send in astute conflict journalists who can analyze the conflicts underlying the fighting, instead of simply reporting on the progress of the war. Civilian peacebuilders should learn the local language, preferably before arriving in the conflict zone, and should commit to stay at least one or two years. Listen to both sides. Give people a voice so they can express themselves and regulate their own affairs. Early mediation costs only thousands of dollars, as compared to billions for the war that will otherwise follow.

> *Lesson 19:* Once blood has flowed it is harder to resolve conflicts. Therefore, when there are strong indications that a conflict may result in warfare, the international community should organize diplomatic and civilian peacebuilding services as early as possible to forestall violence. Such teams should

go into an area only with the invitation of local NGOs and activists, and should work with grassroots members of all ethnic groups, instead of with the parties in power. Instead of apportioning blame, they should identify the nature of problems and how these can be solved, paying systematic attention to the human dimensions of conflict.

MILITARY PEACEKEEPING INTERVENTIONS. By mid-1992 the United Nations forces were assigned tasks for which resources were not available and rules of engagement were not clearly specified. The result was confusion, in which military protection was offered to civilians but then could not be effectively provided. UNPROFOR's mission was hampered by lack of consent, lack of cooperation, and frequent mutually hostile actions by all three factions.

Lesson 20: International military peacekeepers should never be sent to an area of conflict without a clear mandate and adequate resources. If the international community is going to become involved in unsettled civil wars by sending military units to areas where there is limited consent to their presence and participation, they should either "go big or stay home."

AFTER DAYTON

When the heads of all the contending factions of the former Yugoslavia finally met at Dayton, Ohio and agreed to terms ending the war, their breakthrough was hardly voluntary. Considerable pressure – including direct bombing campaigns – had been placed upon them, especially by the United States, to conclude the fighting. Even so, the conference succeeded only barely and at the last moment. In fact, many knowledgeable people do not call the agreement that emerged from the effort successful at all. Perhaps something is better than nothing, but only in that minimal sense can the Dayton Accords be regarded as a satisfactory conclusion to the conflict. Nevertheless, Dayton was a turning point; the problems that had to be handled after it were fundamentally political and economic in nature, in contrast to the military problems that had preceded that conference. I will not deal here with the controversial processes by which the Accord was wrought, but only with the issues that emerged later, as it was being implemented.

AMBIGUITY. The war was fought to settle one issue: whether Bosnia would remain a united, multi-ethnic country or be partitioned along ethnic lines. Unfortunately, the Dayton Accord in effect left that question unresolved. The continuing problems that followed result largely from the very ambiguity of the agreement – though there would have been no agreement at all if its terms had actually established a clear decision concerning the crucial issue. Thus we must make one important recommendation, even while we recognize that it may be unattainable.

> *Lesson 21:* The terms established in settling a war should clearly resolve the key conflicts over which the war was fought – though this objective may in practice be waived for the sake of reaching even an ostensible agreement. Dayton contained enormous ambiguities which have not been – and apparently cannot be – settled definitively even five years later.

HUMAN SECURITY. The international community, if not all the various former Yugoslav communities, agree that the ending of a war should enable the inhabitants of the region to regain their personal security and the basic freedoms to which all human beings are entitled.

> *Lesson 22:* Human security must be re-established in the war region. This includes the following objectives: Provide police services that are not managed by people who have themselves committed war crimes. Provide stable, but soft, borders that people can cross readily. Facilitate the return of refugees and displaced persons to their homes. Disarm the fighters instead of rearming them.

Most of these obvious principles have not been implemented at all in post-Dayton Bosnia-Herzegovina and barely more so in Croatia or Serbia.

NATIONALIST LEADERS. Some of the politicians and military officers who precipitated the war have been indicted for war crimes by the Tribunal in The Hague. Many others would be indicted if it were not so difficult to gather evidence strong enough for a criminal prosecution (i.e. 'beyond a reasonable doubt'). All states and United Nations personnel are obliged to cooperate with the court by executing its arrest warrants, yet some of the indicted persons have not been arrested, but continue to exercise considerable political influence. Other nationalist authoritarian politicians indeed were re-elected to the same high offices they held before the war, positions in which they brought disaster upon their own citizenry. This is partly because the individuals were not held accountable for their own actions and partly because a flawed electoral structure was not improved by the Dayton Accords to create a democratic system in which ethnic divisions are not permanently divisive. Two lessons need to be recognized in this connection.

> *Lesson 23:* Those who are indicted for war crimes or crimes against humanity should be arrested and brought to trial.

> *Lesson 24:* Nationalist demagogues should not be allowed to run for political office during the first years of transition, and their political leadership should not be accepted as legitimate.

DEMOCRATIZATION AND POWER-SHARING. Motivated by a praise-worthy intention to establish democracy in Bosnia-Herzegovina, the international diplomats who negotiated the terms of the Dayton Accord insisted that elections be held very soon after the hostilities ceased. This hasty arrangement did not address some of the key preconditions for genuine democracy – such as the establishment of (a) a free press over a long enough period to enable an informed public opinion to take shape and express itself politically, and (b) electoral systems that do not reinforce, but instead counter, the existing nationalist political cleavages. Until such changes are made, any elections that are held may bring back to power the same demagogues who ruled throughout the war – and, worse yet, cloak them with a kind of legitimacy which would have eluded them in the absence of elections. This predictable outcome is what actually happened. We offer, then, another recommendation.

Lesson 25: Elections should not be held until the preconditions for democracy have been well established. In particular, a free press should be established for a period of time in which political controversies are exposed to open and extensive debate and in which journalists force politicians to account publicly for their actions.

In ethnically diverse states, the transition to democratic governance magnifies the political relevance of ethnic difference. Minority ethnic groups living in a state dominated by another ethnic group may feel anxious and attempt to defend themselves by concentrating their communities geographically and mobilizing them to vote for parties that will seek their own nationalistic advantages. These tendencies should be anticipated and the electoral system should be restructured so as to foster power-sharing, along the following lines.[9]

Lesson 26: At the executive level, no significant group should be completely excluded from power.

Lesson 27: On all issues of common concern, decisions should be made jointly by the different groups or their representatives; on all other issues, decisions should be made by and for each separate group. Federalism is one embodiment of this principle but other non-territorially based forms of constituencies can also be created.

Lesson 28: Where a minority group is included within a power-sharing government, it could be outvoted on all major issues. Hence a minority veto is an institutional arrangement that minorities may need to protect their vital interests.

Lesson 29: Political appointments, public funds, and political representation should be divided among major groups according to their share of the overall population. This suggests that proportional representation electoral systems are superior to plurality or 'first-past-the-post' systems.

Lesson 30: Rather than the current system (in which candidates can be elected solely on the support of their own ethnic group) the electoral system should require successful candidates to appeal to voters of all ethnicities, thereby rewarding moderation and discouraging ethnic extremism. For example, since the joint presidency is made up of one member from each of Bosnia's main ethnic groups, it would be an improvement to allow all Bosnians a vote for each of the three presidency positions. Thus Bosniaks would be able to choose not only among the Bosniak presidential candidates, but among the Serb and Croat candidates as well.

Theoretically, this last proposal, which was advanced by the International Crisis Group, could be implemented quite easily. In reality, however, the Dayton constitution has entrenched some of the very principles that most need to be eliminated by reforming the electoral system. Moreover, the nationalists, having been elected to office, gained sufficient power to prevent any changes that would threaten their own hold over the Bosnian political system. For these reasons, disappointed reformers have concluded that the only realistic way in which nationalism may be soon curtailed is not through electoral reforms, but rather through building up the institutions of civil society: unions, an independent media, and other non-state organizations.

POST-WAR FAMILIES AND ECONOMIC RECONSTRUCTION. In the aftermath of the fighting, large numbers of families are now headed by women. Refugee and displaced families are particularly likely to constitute women and children, yet many of these women have few skills or tools. The economic reconstruction is aimed predominantly at the demobilized soldiers, at the expense of households headed by women. The stringent circumstances of the economy generally tend to load extra burdens on mothers by cutting back support for children's schooling, day care, health services, and related social benefits. Women's educational opportunities have been reduced, while infant mortality has increased, along with death due to complications in childbirth.

Lesson 31: During the economic reconstruction following a war, the needs of women and children in female-headed families should be particularly noted, for they increase in numbers and face grave disadvantages.

THE KOSOVO WAR

It was Kosovo's problems that launched the events resulting in the breakup of Yugoslavia and the accompanying wars of secession. Kosovo's problems were not resolved at all during those conflicts, yet the international community persisted in ignoring those problems until war finally became inevitable. An earlier and well-coordinated international policy might have prevented the whole catastrophe by vigorously recognizing and supporting the Kosovars' elected Gandhian leader, Ibrahim Rugova. Instead, Rugova and his LDK party were never treated as full participants in any of the conferences designed to resolve the ongoing crisis. The London conference of 1992 relegated all these elected LDK personnel to an outer room instead of seating them as delegates. Rugova was not invited to Dayton or to Rambouillet. Had he been treated as a legitimate statesman from the beginning – truly a party to the negotiations – the KLA would not have emerged and there would have been no war in Kosovo. Instead, the behavior of the great powers gave resentful Kosovars ample grounds for concluding that nonviolence does not work, and that violence pays. Clearly this was a mistake.

> *Lesson 32:* If a non-violent, indigenous, legitimately representative leadership (such as Rugova's) exists within a group that is at odds with an oppressive regime (such as Milosevic's pro-Serbian nationalistic regime in Belgrade), foreign governments should include the nonviolent movement in all negotiations involving its interests. Moral support and encouragement should be accorded, and, when appropriate, material assistance should be provided as well.

On the other hand, one must acknowledge certain circumstances that made it difficult for Western countries to apply Lesson 32 to Rugova's movement, as one can see in retrospect that they should have done. For one thing, if Clinton had invited Rugova to Dayton, Milosevic would not have attended and the war in Bosnia would have proceeded.

Other inhibiting circumstances resulted from flaws in the LDK itself. The chief problem is that the Kosovar group indicated early in the conflict that its goal was independence. If Rugova had been willing to consider some greater measure of autonomy, short of sovereignty, the friendly foreign states might have been less apprehensive about acknowledging his authority. Instead, for someone who was so remarkably committed to nonviolence, Rugova's political approach was incongruously inflexible. Even within civil society, he discouraged contact between Kosovars and moderate Serbs. Kosovars at the grass-roots level sometimes regarded his methods of reaching decisions as undemocratic. He avoided taking serious risks with confrontational tactics such as street demonstrations, but chose instead to appeal to foreign governments for support. When no such support was manifested – but only humiliating exclusion from the locus of real negotiations – his timidity outraged bolder Kosovars, who deserted the LDK in favor of armed struggle, further diminishing Rugova's authority. During the fighting, the KLA may have wielded more authority within the Kosovar population than the LDK, though most people recognized the criminal links and excessively violent methods of the guerrilla army. Thus there is a lesson here for the leadership of a resistance movement.

> *Lesson 33:* Leaders of non-violent movements seeking independence or autonomy will benefit from keeping in good contact with the more radical members of their constituencies, employing confrontational methods upon occasion so as never to seem craven, and to encourage dialogue and nego-tiation at all levels of society with minority groups and moderate members of the adversary's community. Even if the leader's ultimate goal remains that of complete independence, he or she can probably win greater support from foreign statesmen by displaying willingness to discuss other options short of the ultimate goal.

If we turn now to examine the opportunities missed by the world's leaders, the failures on their part seem almost limitless, beginning in the early 1990s. The instances of short-sightedness are too numerous to name here, but it is easy to see what they all had in common: a lack of thoroughness in the analysis that informed the decisions of the international community.

The war in Kosovo illustrates the regularity of this shortcoming. The great powers developed no long-range strategy for dealing with Milosevic and even-tually announced that they had exhausted all options except bombing. In fact, there had been many non-violent (or less violent) alternatives earlier that had not even been considered. (I will mention only three examples: The opposition movement in Serbia in the winter of 1997 could have been supported. Radio and television coverage could have been broadcast from abroad to inform rural Serbians about the atrocities being committed in their name. The foreign bank accounts of Serbia's corrupt leaders – especially in Cyprus – might have been impounded.)

Moreover, the international community failed to anticipate the consequences of their own policies, and even failed to carry out a rational analysis of the relative costs and benefits of the options available to them. A wise strategy with long vision would have recognized, for example, that the bombing of Serbia and Kosovo would, at least temporarily: (a) solidify Milosevic's support among the Serbian people and undermine the credibility of the move-ments that had been opposing his regime, and (b) give Milosevic a method of disguising his plan for ethnic cleansing, whereby most Kosovars would be expelled from their country after the bombing began. That is, he could plausibly claim that they were not being forced out but rather were fleeing from NATO's bombs.

If NATO leaders had used the information readily available to them that pointed to Milosevic's probable expulsion of a whole nation of people, they might have reached a different conclusion about the relative costs and benefits of carrying out their bombing campaign. When a war is fought to protect a group of people, but it actually inflicts far more harm on those people than they would have experienced otherwise, it is irrational to launch such a war. The intention of the bombing was to protect Kosovars (and to be fair, most Kosovars are glad that it was waged because the they have a better chance now to win independence and the Serbs have mostly fled from Kosovo, leaving it in their control) but in no sense were they 'protected' by the bombing. A large peacekeeping force on the ground might have been able to offer some protection, but even this is questionable.

Finally, as an example of short-sightedness, one must recognize the counterproductive effects for ethnic diversity and tolerance of the bombing. After the cease-fire, the peacekeepers could not keep the Albanian population from exacting retribution on the remaining Serbs upon their return to Kosovo. Most of the Serbs fled, so that the country is now overwhelmingly Albanian. The outcome ran completely counter to NATO's original intention to prevent 'ethnic cleansing' and to create ethnic harmony between the Kosovars and Serbs.

The leaders of the NATO countries continue to argue that the bombing took place because "something had to be done," and there was "no alternative." This is largely true. By the time the Western countries had made all the errors and miscalculations that they had reached throughout a whole decade or more, there were indeed very few feasible alternatives that would have protected the human rights of the Kosovar population. Nevertheless, that does not mean that the bombing was a sound policy. By then, all of NATO's options were bad; bombing was probably the worst one of all. It was counterproductive, yielding results that were more harmful than beneficial to the Kosovar people themselves, not to mention the Serbs.

What is the lesson here? Simply this: think before acting! Think hard. Think long. Think far, far ahead.

Lesson 34: World leaders confronting nationalistic demagogues should cultivate long-range vision, look for non-violent alternatives that may not be obvious, and constantly think reasonably about the relative costs and benefits of various courses of action. The fact that all other options besides force have seemingly been exhausted does not make war a rational choice. Sometimes doing something is counterproductive – even worse than doing nothing. But leaders who look beyond the momentary situation to consider future possibilities can usually identify numerous promising options far ahead of time. This kind of strategic planning would have prevented the bungling that resulted in the catastrophe that befell the former Yugoslavia.

NOTES

1. Having edited a sizeable book that compares empirically the outcomes of numerous separatist movements, I am not inclined to repeat much of that evidence here, but will only cite it: Metta Spencer, ed. *Separatism: Democracy and Disintegration* (Lanham, Md: Rowman and Littlefield, 1998). See also Robert K Schaeffer, *Warpaths: The Politics of Partition* (New York: Hill and Wang, 1990).

2. In 2000 the same kind of situation exists as forty years before: the recently guilty have not been brought to justice and the historical record remains cloudy concerning the wars of the 1990s. Indeed, most citizens of the former Yugoslavia generally remain skeptical and uncooperative toward the International War Crimes Tribunal that is still going on in The Hague. Probably such cynical attitudes will continue until it is demonstrated to these people that war criminals will indeed be brought to justice within the framework of a fair rule of law. There is a debate as to whether this is best handled through procedures similar to South Africa's Truth and Reconciliation Commission (in which amnesty is granted to perpetrators who confess fully) or whether the only adequate approach is the tribunal system, which will be institutionalized in the new International Criminal Court. See 'Prosecuting War Criminals', an interview with Madam Justice Louise Arbour of the Canadian Supreme Court, shortly after she left her position as chief prosecutor of the War Crimes Tribunal of Yugoslavia and Rwanda. Justice Norman Dyson and Metta Spencer, *Peace Magazine*, Spring 2000. Madam Justice Arbour favors the stronger system, which is based on international standards of criminal justice.

3. For their part, many citizens of Western democracies also believed that democracy was strengthened by decentralization. If this was true of the Soviet Union, it was not true of Yugoslavia, which was already too decentralized to be politically and economically coordinated well.

4. George R. Urban, *Radio Free Europe and the Pursuit of Democracy: My War Within the Cold War* (New Haven: Yale University Press, 1997).

5. In an interview, the dissident physicist Yuri Orlov, who had been confined in Siberia, told me that even there villagers managed to get the news from RFE/RL by going fishing on a lake, taking their short-wave radio along. He mentioned meeting a woman in the village who recognized his name and who told him that when she was in secondary school in Irkutsk, she and other girls had listened to RFE/RL in the washroom, where she became familiar with Orlov's dissident activities.

6. My account here is a much-abbreviated version of Crnobrnja's paper, delivered at the Lessons of Yugoslavia conference but not published in this collection.

7. However, this negative result may be contrasted to a comparable situation in which the EU has attached conditions on Turkey's admission the the EU, evidently with results that are constructive for human rights.

8. Leo Tindemans et al. *Unfinished Peace: Report of the International Commission on the Balkans* (Aspen Institute Berlin, Carnegie Endowment for International Peace), p. 162.

9. The power-sharing arrangements proposed in lessons 23–26 are directly adopted from Arend Lipjhart, 'The Power-Sharing Approach', in Joseph V. Montville (ed.) *Conflict and Peacemaking in Multiethnic Societies* (Lexington: Lexington Books, 1990), pp. 493–94. See the extensive discussion by Timothy Donais in this volume.

APPENDIX:
YUGOSLAVIA TIMELINE

1908. Bulgaria declares its independence from the Ottoman Empire; the Austro-Hungarian Empire takes control of Bosnia and Herzegovina.

1912. The First Balkan War. Bulgaria, Greece, Serbia and Montenegro jointly drive the Turks out of Macedonia and northern Greece.

1913. Greece, Serbia and Romania fight Bulgaria in a Second Balkan War over territory.

1914. In Sarajevo a Serbian nationalist assassinates Archduke Franz Ferdinand, heir to the Austro-Hungarian throne. Austria retaliates against Serbia, drawing in other European powers and igniting World War I.

1918. Germany and the Ottoman and Austro-Hungarian empires are defeated in World War I. The two empires are dissolved, and new Balkan political borders are drawn up by victorious allied powers. The Kingdom of the Serbs, Croats, and Slovenes is formed. Croatia, Slovenia, and Bosnia and Herzegovina had been part of the fallen Austro-Hungarian empire; Serbia and Montenegro had been an independent state. Macedonia was then part of Serbia. In the war's aftermath, both Albanians and Serbs lay claim to Kosovo. The newly created Kingdom of Serbs, Croats and Slovenes regains control of Kosovo. As a minority, Albanians are promised extensive rights by minority rights treaties. The Albanians, however, claim the guarantees are never implemented and that the Serbs engage in widespread massacres and repression in the 1920s. The Serbs also accuse Albania of fomenting discontent in Kosovo.

Research on Russia and Eastern Europe, Volume 3, pages 359–374.
2000 by Elsevier Science Inc.
ISBN: 0-7623-0280-1

1929. King proclaims his personal dictatorship and abolishes the constitution. The monarchy's name is changed to Yugoslavia.

1931. The King grants a constitution to Yugoslavia, reducing the power of the parliament.

1934. King Aleksandr is assassinated.

1941. Yugoslav Croats join with the Nazi side after Germany invades. Josef Broz Tito begins a Serb-led partisan war against the Germans and Croats.

1945. The monarchy becomes a communist republic after World War II, under Prime Minister Tito, and is called the Federal People's Republic of Yugoslavia. It is composed of six republics: Serbia, Croatia, Bosnia and Herzegovina, Macedonia, Slovenia, and Montenegro, as well as two provinces of Serbia – Kosovo and Vojvodina.

1948. Stalin ends the special relationship between Yugoslavia and the Soviet Union.

1963. A new constitution introduces 'socialist democracy', self-management.

1971 'Croatian Spring', is a political reform movement, emerges in Zagreb, then other democratic reformists emerge among students in Belgrade.

1974. A new constitution is adopted for the federal government. It grants autonomy to Serbia's Kosovo and Vojvodina provinces, emphasizes ethnic pluralism.

1980. Tito keeps ethnic tensions in check until his death in 1980, when without his pan-Slavic influence, ethnic and nationalist differences become tense. • May: Death of Tito.

1981. Students rebel in Kosovo, with local Albanian nationalist support. • The federal constitution is amended, but without providing mechanisms for the democratic management of ethnic relations.

1987. Serb nationalist Slobodan Milosevic becomes leader of Yugoslavia, carries on a chauvinistic nationalist movement through 1989.

1989. The Yugoslav government rescinds Kosovo autonomy. • May: Milosevic becomes president of the Republic of Serbia. • June: A million people meet in Kosovo to hear Milosevic proclaim the land sacred to Serbs.

1990. September: Serbia adopts a new constitution. • April–December: First multiparty elections in six republics of former Yugoslavia. Serbian Communist Party leader Slobodan Milosevic is elected Serbian President.

1991. June: US Secretary of State James Baker visits Belgrade, meets with political leaders. • With 90% of its population ethnic Slovenians, Slovenia is able to break away with only brief fighting. Because 12% of Croatia's population is Serbian, however, rump Yugoslavia fights against its secession for the next four years. As Croatia moves towards independence, it evicts most of its Serbian population. July: Yugoslav army announces withdrawal from Slovenia. • June: Slovenia and Croatia each declare independence. • 'Mothers Movement' forms spontaneously in Serbia with the outbreak of war in Slovenia; its members demand that their sons in the army be brought back from Slovenia. • July: Genscher begins pushing the EC to recognize Croatia and Slovenia. • August 25: Fight begins over Vukovar, will last 86 days. • Throughout summer: Serb–Croat skirmishes going on since early 1991 escalate into war in Croatia between Croats and rebel Serbs, backed by the Yugoslav army. • September: United Nations imposes arms embargo on all of former Yugoslavia, including Bosnia • Kosovo's clandestine parliament declares Kosovo a sovereign and independent state. A month later, a national referendum sees overwhelming approval from the Albanians for the decision.• October: Serbs shell Dubrovnik. • November: Cyrus Vance is negotiating a truce between Croatia and Yugoslavia's army. • December 23: Germany recognizes Croatia and Slovenia without waiting for the decision of the Badinter Commission. The European Community, under pressure from Germany, also agrees to do so.

1992. During this year, 100,000 to 150,000 professionals leave Serbia. • Albanians organize multiparty elections which are declared illegal by the Serbs. The Democratic League wins 96 out of 140 seats and Rugova is elected president. He opts for passive resistance to Serb rule warning his fellow citizens not to provide the Serbs with a pretext for a violent crackdown in Kosovo. • Macedonia declares independence. • January: The truce negotiated by UN mediator Cyrus Vance is signed; it will prove lasting. UN peacekeepers will patrol it, with headquarters in Sarajevo, in attempt to prevent war in Bosnia. • Feb. to March 1992: Croatians, originally fighting with the Muslims against the Serbs, start their own 'ethnic cleansing' campaign. • The UN Security Council sends

14,000 peacekeeping troops to Croatia. • March 3: Bosnia declares itself an independent nation. Bosnian Serbs demand that Bosnia withdraw its declaration of independence. When this does not happen, fighting begins. • April: Rock concert is held in Serbia to show popular solidarity with Sarajevo. • April: Bosnia and Herzegovina declares independence. It is 43.7% Muslim, 31.4% Serbian, and 17.3% Croatian. It erupts into war. By 1995, the country has been partitioned into three areas, each governed by one of the three ethnic groups and made up of roughly 90% of its own ethnic group. • Serbia and Montenegro form the Federal Republic of Yugoslavia, with Slobodan Milosevic as its leader. This new government is not recognized by the United States as the successor state to the former Yugoslavia. • The U.S. and European Union recognize Bosnia as independent state. • Nationalist Serb snipers fire on peaceful demonstrators in Sarajevo, marking the beginning of the war. Bosnian Serb soldiers are formally discharged from the Yugoslav army, but allowed to keep all of their weapons. • Intense fighting in Bosnia. • May 27: A mortar shell fired from a Serb position in the hills of Sarajevo kills 16 people waiting in line for bread. • UN imposes sanctions on Serb-led Yugoslavia. • May 3: Bosnia's Muslim president, Alija Izetbegovic, is taken hostage by Yugoslav troops on return from peace talks in Lisbon, freed the following day. • Yugoslav army relinquishes command of its estimated 100,000 troops in Bosnia, effectively creating a Bosnian Serb army. • May 30: United Nations imposes sanctions on a new, smaller Yugoslavia made up of Serbia and Montenegro, for fomenting war in Bosnia and Croatia. • Summer: There are reports of 'ethnic cleansing', a policy of slaughtering Muslim inhabitants of towns or driving them away, in order to create an ethnically pure region. Reports of concentration camps, mass rapes. • June 29: Peacekeepers hoist UN flag at Sarajevo airport after Serbs leave. • July 3: International airlift begins to Sarajevo. • August: Major international conference on Yugoslavia in London. Agreements on aid, cease-fire, never implemented. • Sept. 19: UN Security Council drops Yugoslavia from General Assembly. • Nov. 16: UN Security Council authorizes naval blockade of Serbia and Montenegro.

Winter 1992–93. Gas, water and electricity service are at best sporadic in Sarajevo. UN humanitarian convoys to Muslim enclaves in central Bosnia crowded with refugees are blocked by Serb forces, leading to acute shortages of food, fuel, and medicine. UN declares several Bosnian cities 'safe areas,' to no one's relief. • Pres. Clinton orders humanitarian aid and food to be air-lifted to those places.

1993. Real income per capital in Yugoslavia drops to 44% of its 1989 level, with inflation among the highest in world economic history. Unemployment rate is

15%. • Jan. 2: International mediators Cyrus Vance and Lord Owen unveil plan to divide Bosnia into 10 provinces, mostly along ethnic lines. • Feb. 22: Security Council sets up a war crimes tribunal for former Yugoslavia. • March 25: Izetbegovic signs Vance-Owen peace plan in New York. • March: Bosnian Croats and Muslims begin fighting over the 30% of Bosnia not seized by Bosnian Serbs. • April 12: NATO jets begin to enforce UN no-fly zone over Bosnia. • April 26: Tighter UN trade sanctions against Yugoslavia. • April and May: Croatian side, openly supported by Croatian regular military units, attacks the Bosniacs, who had been their allies until then. The Croatian leaders intend to divide Bosnia with the Serbs, creating Greater Croatia. • Following Serb assault on Srebrenica and dramatic crisis of refugees arriving in Tuzla, Security Council declares six 'safe areas' for Bosnian Muslims: Sarajevo, Tuzla, Bihac, Srebrenica, Zepa and Gorazde. • May 2: Bosnian Serb leader Radovan Karadzic signs Vance-Owen plan in Greece, but his assembly rejects it. • May 15–16: In a referendum, Bosnian Serbs overwhelmingly reject Vance-Owen plan in favor of an independent Bosnian Serb state. • May 31: Yugoslav federal Parliament ousts Dobrica Cosic, seen as too peaceable by Milosevic, as Yugoslav federal president. Thousands demonstrate, clash with police in Belgrade. • June 16: Mediators meet with Milosevic, Izetbegovic, Croatian President Franjo Tudjman and Bosnian leaders in Geneva. Plan emerges to split Bosnia three ways. Izetbegovic walks out. • July 30: Warring sides reach preliminary agreement in Geneva on Union of Republics of Bosnia and Herzegovina with three states and three peoples. August: Izetbegovic walks out after Serbs violate cease-fire. • Fall: Bosnian Government army makes some territorial gains against Croatian separatists, reputedly with the arms supplied by the Serbs. Both Yugoslav and Croatian army regulars are observed fighting in Bosnia. • The breakaway Serb republic of Bosnia orders a general mobilization among all the Bosnian Serb refugees, planning for an all out assault that will lead to the end of war. • Bosnian government rejects the Owen-Stoltenberg Plan, which would have maintained the state of Bosnia-Herzegovina, but divided it internally along ethnic lines. • Mortar barrages on Sarajevo lighten up, and Serbs withdraw from some strategic positions, when US and NATO threaten air strikes. Firing resumes when it becomes obvious that no action will be taken.

1994. January: France, which has the most UN troops in Bosnia, calls for NATO to use air strikes to relieve the humanitarian crisis in Bosnia. French intellectuals start a party 'Europe Begins at Sarajevo', for the elections for the European Parliament. Its platform is that Europe's humanity and civility is challenged by its inactivity in the Bosnia crisis. • Feb. 4: The market place massacre, which leaves 68 people dead and over 200 wounded in Sarajevo leads NATO to issue ultimatum for Serbs to withdraw their artillery to 20 km from Sarajevo, and for all warring parties to hand over their heavy weapons to UN observers. • Feb. 9: NATO gives Bosnian Serbs 10 days to withdraw heavy guns from Sarajevo region or face air strikes. • Feb. 17: Karadzic agrees to remove guns from around Sarajevo if soldiers from Russia join peacekeeping mission. • Feb. 20: Russian peacekeepers arrive. NATO deadline expires; UN says it is satisfied heavy guns are being removed. • Feb. 28: U.S.F-16 fighters, flying for NATO, down four Bosnian Serb warplanes violating ''no-fly'' zone. The shots are the first fired by NATO. • March 18: In Washington, Bosnia's Muslim-led government and Bosnian Croats sign a U.S.-brokered accord, ending a year-long war. • April 22: After two airstrikes against Serbs advancing on Gorazde, NATO delivers fresh ultimatum to Serbs to stop firing and pull back or face air strikes. • April 27: UN says the Serbs have mostly complied with NATO ultimatum. • May: The major powers form the Contact Group, an informal Security Council of the UN in which Germay replaces China. It announces a new peace plan, including a four-month cease-fire and eventual partition of Bosnia. • Summer: Bosnian Government army makes successful advances against separatist Serbs, recapturing some of the territory around Bihac, in Bosnia's North-East corner. • July: Croats accept the Contact Group plan outright, Muslims reluctantly, Bosnian Serb reject it. • Aug. 4: Milosevic cuts ties with Bosnian Serbs for rejecting plan. • Fall: Cease fire around Sarajevo is spotty, but holding. Bosnian Serb forces are reinforced by Croatian Serb forces from the neighboring Krajina region, press against Bosnian governemnt, re-recapturing the region around Bihac, which is shelled and bombed relentlessly. NATO 'strikes back' and bombs the runways in the Serb held airport in Krajina from which bombing raids are flown. Serbs hold over 300 UN troops hostage against further air raids. • Oct. 29: Bosnian government forces score their biggest victory of the war around Bihac, northwest Bosnia. Fierce Serb counterattack a week later. • Nov. 21: NATO launches its largest action ever, about 50 jets and support planes attacking Serb airfield, but fail to take out Serb jets attacking Bihac. • Nov. 25: Serbs detain 55 Canadian peacekeepers against further air strikes. Eventually more than 400 peacekeepers held. NATO attempts air strike on Serbs near Bihac. Mission called off after UN fails to pinpoint targets. • Dec. 20: Former U.S. President Jimmy Carter ends mediating mission with announcement of Bosnian cease-fire. Cease fire does not affect Croat Serbs who continue the siege of Bihac. Despite early problems with violence, the cease-fire lasts four months.

1995. Jan. 1: Four-month, nationwide truce takes effect. Bihac is never quiet; elsewhere, fighting dies down or stops. • Jan. 28: 1000th day of the siege of Sarajevo. • Feb: Cease-fire violations by Bosnian Serbs are increasingly common. UN monitors observe helicopters crossing from Serbia to Bosnia, presumably to resupply the Bosnian Serbs, a breach of promise by Milosevic to put them under an internal embargo. • Feb. 13: United Nations tribunal on human rights violation in the Balkans charges 21 Bosnian Serb commanders with genocide and crimes against humanity. • Feb. 15 – 22: Under the pressures from European allies, U.S. agrees to loosen economic sanctions against Yugoslavia, in return for Milosevic's recognition of territorial integrity of Croatia and Bosnia-Herzegovina. Milosevic refuses. • Mar. 9: According to New York Times, a CIA report has concluded that 90% of the acts of 'ethnic cleansing' were carried out by Serbs and that leading Serbian politicians almost certainly played a role in the crimes. • April 8: U.S. aid plane hit by gunfire, all UN aid flights to Sarajevo canceled. • May 1, 1995: Fighting renews as Carter's four month cease-fire ends in Bosnia. • Croatian government begins a new offensive against Croatian Serbs. • UN efforts to extend the truce fail. • Croatia launches blitz offensive to recapture chunk of land from rebel Serbs. Serbs retaliate by rocketing Zagreb; six killed, nearly 200 wounded. • May 24: UN orders Serbs to return heavy weapons to UN control and remove all heavy weapons around Sarajevo. The UN commander in Bosnia threatens to use strikes if heavy weapons in Sarajevo are not silenced within 24 hours. • May 25: Serbs ignore UN order. NATO attacks Serb ammunition depot. Serbs respond by shelling 'safe areas', including Tuzla, where 71 people are killed and over 150 injured. • May 26: Bosnian Serbs seize UN peacekeeping troops, using them as human shields against NATO airstrikes. All hostages are subsequently released. Several days later, British Prime Minister John Major says it may be necessary to remove British troops from Bosnia if the risk becomes too great. • NATO warplanes attack more ammunition depots. Eventually more than 370 UN peacekeepers are seized. • May 28: France, Britain and United States send thousands more troops toward Bosnia. • June 2: Serbs shoot down U.S.F-16 over northern Bosnia, release 121 UN hostages. • June 3: NATO defense chiefs, meeting in Paris, agree on rapid reaction force to bolster UN peacekeepers in Bosnia. • June 6: U.S. envoy Robert Frasure fails to agree after weeks of talks with Milosevic on Serbia recognizing Bosnia. • June: Serbs release 111 more UN hostages. • June 8: U.S. Marines rescue downed pilot of U.S.F-16. • NATO approves new rapid reaction force, but also says peacekeepers will leave Bosnia by fall if rebel Serbs don't accept new force. Complex evacuation plan approved. • June 14: All but last 26 UN hostages released. • June 15: Bosnian government launches offensive to break siege of Sarajevo. Offensive gradually stalls; Serbs step up shelling of Sarajevo and other 'safe areas'. • June 18: Last 26 UN hostages released. • June 30: Bosnian government, increasingly bitter, demands review of UN mission. • German parliament approves deployment of fighter jets for rapid reaction force. • July: United Nations peacekeepers undertake 'Operation Active Presence', deploying troops so as to deter a Croatian attack in the Krajina; it does not succeed. • July 2: French peacekeepers use 120 mm mortar on lone road into Sarajevo. • July 6: Gen. Mladic's forces begin shelling Srebrenica. • July 10: Serbs capture Srebrenica and the Dutch peacekeepers, launch biggest mass murder in Europe since World War II. • July 11: Last-minute NATO airstrikes fail to stop Serb advance. Serb forces sweep into the UN safe area, causing a massive exodus of civilians. Dutch peacekeepers call in air strikes by U.S. and Dutch warplanes, but the defensive effort fails and the peacekeepers withdraw. • July 12–13: Some 20,000 Muslim women, children and elderly expelled to Tuzla, bringing tales of atrocities. • July 16–17: Some 4,000 Muslim men who marched through Serb-held land reach government-held Tuzla; another 11,000 thought missing. • July 18: Bosnian government troops threaten to take UN peacekeepers hostage unless the UN orders air strikes to prevent the fall of Zepa. The Bosnian Serbs, close to capturing the town, say they'll respond to air strikes by shelling eight Ukrainian peacekeepers, who are in a UN base near Zepa. • July 21: NATO threatens to use air strikes. After international military leaders meet in London, NATO threatens air strikes to protect the safe area of Gorazde, early use of Rapid Reaction Force. • July 23: UN commanders in Bosnia order the Rapid Reaction Force to send artillery units to Sarajevo. Part of a 12,000-member contingent of mostly French and British soldiers, the special combat group settles in at Mount Igman overlooking the city. A day later, UNPROFOR spokesman Lt. Col. Chris Vernon warns of escalation in the Bosnian conflict • Serbs kill two French peacekeepers; UN threatens punishment from Rapid Reaction Force. • July 25: Safe area of Zepa crumbles before advancing Bosnian Serb forces. Many Muslim refugees are packed onto evacuation buses by Bosnian Serbs. After executing the Muslim commander of the government forces, the Serbs burn the town. • War crimes tribunal indicts Karadzic, Mladic for genocide, crimes against humanity. Martic charged with war crimes for bombing Zagreb. • July 26: The US Senate votes to lift the arms embargo against Bosnia. Imposed on all of former Yugoslavia in September 1991, the embargo weighs heaviest on Bosnian government forces because Serbs inherited weapons from the Serb-led Yugoslav army. • July 28: The war widens as Croatia sends thousands of troops into Bosnia. They cut Serbian supply lines and overtake the towns of Glamoc and Grahavo in southwestern Bosnia. The days to come will bring more Croatian gains. July 31: Croats shell outskirts of Knin. • August 1: The U.S. Congress votes to lift the arms embargo against Bosnia. President Clinton warns that this will involve US troops in an evacuation of UN peacekeepers.• NATO extends its threat of anti-Serb air strikes to protect UN safe areas beyond Gorazde. • Aug. 3: Offer by rebel Serbs to bow to some Croatian authority rejected by government. Serbs shell Dubrovnik area. • August 4: Less than 36 hours after starting their advance, Croatian forces recapture the rebel Serb 'capital' of Knin, shelling UN peacekeepers and civilians. Recapture most of Serb-held lands in four days. • Thousands of Serb civilians beginning stream toward Bosnia. Eventually more than 180,000 flee their homes. • NATO warplanes fire missiles at Croatian Serb radar site after being threatened by surface-to-air missiles. • Aug. 7: Column of Serb refugees attacked by military jet; at least five killed. • Aug. 9: Mobs of Croats batter Serb refugees with bricks, chunks of concrete in Sisak. • Aug. 10: U.S. ambassador to UN calls for war crimes tribunal investigation after spy photographs show evidence of mass graves of executed Bosnian Muslims. • Aug. 18: U.S. diplomats shuttle between Serb and Croat leaders with peace plan. Peacekeepers begin pull out from Gorazde. • Aug. 19: Three key diplomats for US peace initiative, Robert Frasure, Joseph Kruzel and Nelson Drew, killed when armored personnel carrier slips off Mount Igman road. Three other Americans and three French injured. • Aug. 20: Human rights investigators suspect at least four mass graves exist around Knin. • Aug. 22: Serbs shell Sarajevo region, killing six and wounding 38, including six Egyptian peacekeepers, after government shells Serb arms factory. • Aug. 28: Bosnian Serbs fire shell into a busy Sarajevo market area, killing 37 and wounding scores. • UN secretly pulls out last peacekeepers of Gorazde enclave. • Aug. 30: NATO planes launch massive airstrikes to silence Serb guns around Sarajevo. Serbs shell Sarajevo in response. • Sept. 1: NATO suspends attacks; U.S. announces that hostile parties agree to a discuss permanent peace. • Sept. 5: NATO resumes attacks to force withdrawal of Serb guns around Sarajevo. • Sept. 8: Warring factions agree to formally maintain Bosnia but sub-divide it into Serb and Muslim-Croat sections. • Sept. 13: Croats and Muslims advance on Serbs in central and western Bosnia. • Sept. 14: NATO suspends attacks. Milosevic pledges that Bosnian Serbs will withdraw guns from around Sarajevo. Red Cross says about 8,000 Muslims from Srebrenica missing and unaccounted for. • Sept. 15: Serbs let Sarajevo airport reopen for the first time in five months. • Sept. 26: Bosnian factions agree on basic outlines of peace plan. • Sept. 29: European Union accuses Croatian army of murder, mass looting, arson. • Oct. 3: Rebel Serbs in Croatia agree to give up last swath of territory they hold there. • Oct. 5: Warring Bosnian parties agree to a 60-day cease-fire. • Nov. 1: Bosnian peace talks open in Dayton, Ohio. • Nov. 16: Bosnian Serb leader Radovan Karadzic and Gen. Ratko Mladic, his military commander, indicted for war crimes for their alleged roles in Srebrenica massacres. • Nov. 21: Balkan leaders initial peace accord, granting 51% of Bosnian territory to Muslim-Croat federation; 49% to Serbs. • Nov. 22: Security Council suspends sanctions against Serbia, eases arms embargo against former Yugoslavian states. • Nov. 23: Karadzic accepts peace plan after meeting with Milosevic. • Nov. 30: UN votes to end peacekeeping mission by Jan. 31. •

Dec. 1: NATO authorizes deploying 60,000 troops to Bosnia; appoints Javier Solana NATO secretary general. • Dec. 4: British, U.S. troops land in former Yugoslavia to begin groundwork for peacekeeping mission. • Dec. 5: Polls show majority of Americans oppose sending troops to Bosnia. • Dec. 12: Bosnian Serbs release captured French pilots. • Dec. 13: Senate defeats measure to cut off funds for U.S. troops in Bosnia. • Dec. 14: Presidents of warring parties sign peace plan, setting stage for deployment of 60,000 NATO troops. • Bosnian, Serb governments agree to formal diplomatic recognition. • Dec. 15: UN Security Council transfers peacekeeping duties to NATO. • Dec. 16: Joulwan issues order for 60,000 NATO troops to enter Bosnia. • Dec. 18: Break in fog allows 14 U.S. flights to arrive in Tuzla. • Dec. 19: Assistant Secretary of State Richard C. Holbrooke, chief American negotiator of the Dayton agreement, announces he'll step down, be succeeded by career diplomat John Kornblum. • Dec. 20: NATO takes over command of Bosnia peace mission. • Dec. 22: Thousands of Serbs flee Sarajevo suburbs, many carrying coffins of relatives. • Dec. 24: First American helicopters arrive in Tuzla, while French extend control in Sarajevo. • Dec. 27: Government, rebel Serb troops pull back from area around Sarajevo. • Dec. 31: First U.S. tanks roll across pontoon bridge over Sava.

1996. Jan. 4: Serbs bow to international pressure, free 16 civilians. Italian military engineer injured by sniper in Sarajevo; Italian soldiers return fire to defend him. • Jan. 6: NATO deploys troops, armored vehicles to help keep peace. • Jan. 11: As Sarajevo Serbs torch houses, prepare to flee, their leaders urge American envoy to delay reunification of city. • Jan. 13: President Clinton visits front-line troops in Bosnia, along with Bosnian, Croat and Serb leaders. • Jan. 14: Several thousand Muslim, Croat and Serb troops pull back from confrontation line in British sector, beating deadline by five days. • Jan. 19: Planned prisoner release falls far short of goal, with Croats, Muslims freeing 225 of 900. Serbs renege on promise to release dozens. • Jan. 31: Serb-held Grbavica reattached to Sarajevo with opening of Bridge of Brotherhood and Unity. • Feb. 3: U.S. military sustains first casualty. • Rebel Serbs withdraw forces from Sarajevo suburbs. • Irate Croats surround European Union mission, attack car in response to plan to reunify Mostar. • Feb. 8: Bosnian Serb army breaks off contacts with NATO over detention of suspected war criminals, bans civilians in Serb territory from crossing into federation lands, threatens to arrest Muslims and Croats crossing into Serb territory. • Feb. 18: Leaders at Rome summit agree to reunify Sarajevo and Mostar and to conform to procedures for arresting suspected war criminals. Bosnian Serbs agree to resume contact with NATO. • Feb. 20: Some Bosnian Serb leaders organize mass exodus from suburbs, while moderates urge residents to remain. • Feb. 27: Security Council lifts sanctions against Bosnian Serbs. • Feb. 29: Sarajevo siege officially ends. • March 2: Slobodan Milosevic is overwhelmingly re-elected head of Socialist party. • March 9: 20,000 in Belgrade rally against Milosevic. • March 11: U.S. pledges $100 million to rearm Bosnia, draws criticism from European leaders. • March 13: Gangs from Sarajevo terrorize Serb suburbs after handover. • March 14: Top Croatian politicians fly to Sarajevo to discuss federation as NATO, others fear federation is falling apart. • March 19: Sarajevo reunited. • March 23: Bosnian government releases 109 Serb prisoners. • War crimes investigators find human remains and other evidence of mass grave 18 miles from Srebrenica. • April 5: Mass grave in northern Bosnia contains 181 bodies; thought to be Serbs killed by Croats. • April 8: U.S. officials admit Clinton knew of illegal arms shipments from Iran to Bosnia. • April 13: Nations pledge $1.23 billion to rebuild federation-held section of Bosnia. Offer little aid to Bosnian Serbs until suspected war criminals are turned over to tribunal. federation work. • Persian Gulf countries to donate $100 million to help with upgrade of Bosnian government forces. • April 18: NATO says all sides miss deadline to pull back weapons and soldiers, despite efforts to comply. • April 26: Pentagon says that even if the NATO-led peace mission in Bosnia ends as scheduled in December, a substantial number of U.S. troops will remain at least until January. • May 15: Bosnian Serb leader Radovan Karadzic fires moderate premier. • May 23: Bosnian Croats and Muslims agree to postpone Mostar elections. • May 29: Yugoslav war crimes tribunal issues its first indictment for Srebrenica massacres. • June 6: Adm. Leighton W. Smith will be replaced as commander of NATO-led troops in Bosnia. NATO-led troops will remain in Bosnia past end of mission on Dec. 20. • September: Elections are held in Bosnia, returning nationalist groups to power in each ethnic region.

1997. Growing frustrated with the pace of change under Rugova's rule, some Albanians choose violence to force concessions from Belgrade. A shadowy group calling itself the Kosovo Liberation Army (KLA) emerges. • January: Opposition coalition Zajedno ('together') holds marches daily in Belgrade to protest against Milosevic's failure to recognize election returns. • Patriarch Pavle repeated the Holy Synod's recent condemnation of the Milosevic regime. The Yugoslav army will not oppose the student demonstrators. • Jan. 3: Bosnia's new government convenes for the first time in Serb-run Lukavica, near Sarajevo. Deputies in the lower house of the Bosnian parliament approves the government and the nomination of the two joint prime ministers, Boro Bosic, and Serb, and Haris Silajdzic, a Muslim. • The Party of Democratic Action (SDA) headed by President Izetbegovic, confirms that it received $500,000 from Iran in mid–1996. The LA Times had reported that Iran gave the money for use in the run-up to the September elections. • Spring: OSCE supervises voter registrations throughout Bosnia, registering 2.5 million. • Jan. 4: Kosovo human rights activist Adem Demaci is elected chairman of the Parliamentary Party of Kosovo. Demaci is expected to compete with shadow-state president Ibrahim Rugova in upcoming presidential elections. • Jan. 8: German Foreign Minister Klaus Kinkel endorses the demands of Serbian opposition group, Zajedno. • Fifty-two of the 160 members of the Serbian Academy of Sciences warns the government to recognize all opposition victories. • Bosnian Serb President Plavsic says in a letter to UN Secretary-General Kofi Annan that the Bosnian Serbs will not hand over Radovan Karadzic or Ratko Mladic, both of whom are indicted war criminals. • Jan. 27: The UNHCR-sponsored plan to return Muslim families to their home village has suspended because of well-organized mob violence by Serb civilian crowds with the apparent complicity of the Republika Srpska police. • Feb. 5: The Bosnian federal defense minister, Ante Jelavic, and other top defense official met with diplomats from Turkey and Egypt, which are supporting the U.S.-sponsored 'Train and Equip' program for the Bosnian military. • Feb. 13: Alija Izetbegovic and Haris Silajdzic agree to give the UN police increased powers to control Mostar. • March: Breakdown of law and order in Albania provides a source of weapons for Kosovar guerrilla fighters. • Federal Yugoslavia gets a new government on 20 March, stemming from the 3 November legislative elections. The cabinet appointments reinforce the belief that Milosevic is building up the federal government before assuming the federal presidency later in 1997. • July 22: President Plavsic wins a major victory when Bosnia's constitutional court overrules the cabinet's objections to her decision to dissolve parliament and call for new elections. • Montenegro's governing party candidate, Milo Djukanovic, insists that the country will continue to be part of federal Yugoslavia, but only as Serbia's fully equal partner. • War Crimes Tribunal Prosecutor Louise Arbour accepts the Montenegrin prosecutor's invitation to visit Podgorica. • July 31: OSCE representatives slams Bosnian Serb TV for engaging in propaganda in 'gross violation' of the Dayton rules. German Foreign Minister Klaus Kinkel has called on NATO to jam the broadcasts. • August 25: Kiro Gligorov says that all the ethnic Albanian parties in Macedonia want to secede from that state. • August 27: In Doboj, Karadzic's police retake control of a TV relay tower that Plavsic's backers had seized the previous day. • Aug. 28: Bosnian Serbs assaul SFOR troops and UN police in Brcko; NATO troops fire back with tear gas, then evacuate their personnel. • State Department officials say those Bosnian Serb authorities who harbor war criminals will not receive any financial assistance. • Carlos Westendorp, High Representative in Bosnia, urges UN Security Council to pass a resolution freezing all bank accounts belonging to indicted war criminals and the confiscation of their property. Westendorp also establishes a commission to investigate corruption in the Bosnian government. • September 13–14: Municipal elections are held in Bosnia after being postponed four times. • Winter: Bonn Peace Implementation Council grants the High Representative increased powers to impose decisions in Bosnia and remove obstructionist local officials. • UN removes some of its observers on the border between Albania and Macedonia. All three of its posts are scheduled for closure, as are three of the six on the Serbian border.

1998. January: Moderate democrat Milorad Dodik becomes prime minister of Republika Srpska. • Feb.: Serbians crack down on Kosovar citizens, many of whom begin leaving the province. • March: Rugova and his party appeal to international community – especially the U.S. and EU – to put pressure on Belgrade to end violence in Kosovo. U.S. State Department, Russia, and the EU condemn the violent repressions there. The Albanian parliament asks for NATO presence in the western Balkan region. The KLA announces it will seek revenge on Serbian security forces in villages around Drenica. • March 19: Some 40,000 Kosovars take to the streets of Pristina to protest against Serbian government repression. Hours later, 50,000 Serbs began a counterprotest. • Summer: Almost all Bosnian municipal assemblies outside of Srebrenica have been certified as having complied with the results of the municipal elections. • Milosevic sends troops to Kosovo to quash unrest in the province. A guerrilla war breaks out, replacing the passive protests of Albanians until that time. • The initial commitment of SFOR troops to Bosnia was set to expire in mid–1998, but is extended indefinitely. • Sept.: Bosnia's second round of elections after Dayton take place, mostly resulting in a continuation of the same political trends that marked the 1996 elections. • Oct.: Serbian Deputy Prime Minister Vopjislav Seselj launches repression against Serbian radio and TV stations rebroadcasting 'anti-patriotic' messages and the programs of Western countries such as Radio Free Europe. • Oct. 12: After repeated threats of NATO air strikes, Milosevic agrees at the last minute to a truce calling for the removal of Serbian troops from Kosovo. Despite the agreement between him and Ambassador Holbrooke, fighting continues in the region. • Fall: Some 2,000 unarmed civilians begin arriving in Kosova under a mandate from the OSCE to monitor the uneasy truce. In neighboring Macedonia, a 1,700-strong, French-led NATO rapid reaction force begin assembling in order to evacuate the monitors if they ran into danger. • Dec. 23: Ibrahim Rugova said that the Serbian forces "will be able to exterminate [the Kosovars] in the spring in a couple of days if they want to." Rugova defended his long-standing policy of non-violence.

1999. Jan.: The killing of 45 ethnic Albanians by Serb forces in the town of Racak leads to international pleas for peace once again. • Serb and ethnic Albanian representatives meet at Rambouillet, France to discuss options for peace. Both sides are reluctant to compromise, and the talks break up without a plan in place. • Tribunal Prosecutor Louise Arbour attempts to enter Kosovo to investigate possible war crimes. She is refused entry. • Feb. 23: The terms of a proposed peace agreement for Kosovo are made public. • Mar: Republika Srpska President Nikola Poplasen is dismissed for obstructing implementation of the Dayton Accord. • March 15: Fighting continues in Kosovo, even as a second round of talks gets underway in Paris. • March 18: The ethnic Albanians finally take a step toward peace, signing a deal that calls for interim autonomy and a NATO force of 28,000 to monitor the region. Milosevic responds by reiterating Serbian disapproval and the talks are again suspended without an agreement. The Serbs return home under the threat of NATO airstrikes. • March 22: Richard Holbrooke visits Belgrade in a final, unsuccessful attempt to convince Milosevic to agree to NATO's terms. • March 24: NATO begins launching air strikes in an attempt to force Serbia to cease hostilities and allow ethnic Albanian refugees to return to their homes in Kosovo. • Serb troops force thousands of ethnic Albanians out of the town of Djakovica. At least 47 men are believed to have been rounded up and shot. Serb forces are also accused of raping women and destroying many ethnic Albanians' citizenship papers. • April 14: NATO bombs accidentally hit two convoys of ethnic Albanian refugees being escorted by Serb police. Yugoslav officials put the death toll at more than 60. • April 27: The UN and Human Rights Watch report that Serbian troops killed 200–300 men in the village of Meja. Witnesses tell of Serb troops clearing and burning villages, then separating men ages 18 to 65 from their families and shooting them. • May 2: In Belgrade, Rev. Jesse Jackson succeeds in winning the release of the three American hostages after negotiating with Milosevic. • May 5: An Apache helicopter crashes in Albania while on a training mission. • Arbitration panel decides that Brcko will become a neutral community under international supervision, rather than part of Republika Srpska. • International authorities dismiss Bosnian Serb President Nikolal Poplasen after he tried to get rid of the pro-Dayton Serb prime minister, Milorad Dodik, who then resigns in protest against the Brcko ruling. • May 6: The first ethnic Albanian refugees from Kosovo arrives in the US. Hundreds of thousands of Kosovars abandoned their homes, forming a refugee crisis that primarily affected Albania and Macedonia. Western countries offer refuge to some of the 860,000 ethnic Albanians who left Kosovo. • May 7: Three Chinese journalists are killed in Belgrade when NATO accidentally bombs the Chinese Embassy. NATO attributes the mistake to outdated maps. Massive protests erupt in Beijing. •

May 13: More than 80 ethnic Albanians are killed and at least 100 are injured when NATO bombs a village believed to have been a Serb military post. NATO claims the victims were being used by Serb troops as human shields. • May 21: NATO again hits an unintended target – a KLA stronghold. • May 27: The UN's International War Crimes Tribunal formally indicts Milosevic and four other Yugoslav officials for crimes against humanity. They are accused of being responsible for the deportation of 740,000 ethnic Albanians from Kosovo this year as well as the murder of more than 340 identified victims. • June 3: Milosevic and the Serbian parliament accept a proposal drawn up by representatives from Russia, the EU, and the U.S. • June 9: Yugoslavia and Western nations sign a formal agreement calling for the withdrawal of Serb troops from Kosovo and a subsequent halting of NATO's air campaign. An international peacekeeping force headed by NATO is to monitor Kosovo and the return of the refugees. Russia's role in the operation remains ambiguous. • June 10: The UN Security Council approves a resolution that authorizes the plan for peace in Kosovo by a vote of 14–0. China abstains. • June 11: As peacekeeping forces prepare to enter Kosovo, an uninvited Russian convoy also heads to the region. Western officials try to figure out how to maintain control over the peace-keeping efforts without stirring up further conflict with Russia. • June 28: Patriarch Pavle and other Serbian leaders appeal to the UN secretary-general for better protection by UN peacekeepers, KFOR, of the local Serbs in Kosovo. • July: Some 40 world leaders and 17 international organizations held a summit in Sarajevo to launch the Stability Pact for Southeast Europe. Bodo Hombach of Germany is appointed by the European Union as coordinator, and will begin his work in Brussels. Serbia will be excluded from this 'new Balkan order' as long as Milosevic remains in power. • Oct.: In the aftermath of the Kosova war, polls show that 90% of Albanian Kosovars would vote for the pacifist leader Ibrahim Rugova instead of the leaders of the KLA, if an election were held now. • Nov.: Mrs. Plavsic says she agrees with the decision of Carlos Westendorp to oust Nikola Poplasen, her successor, for not respecting the Dayton agreement. • Nov.: Crime statistics show that levels of violence in Kosovo are the same as before the war. But many of the victims are Serbs and the perpe-trators are Albanians. • Serbian paramilitaries forced out of Kosovo have moved into Montenegro, some to find a hiding place and some to join the police. • November. Montenegro offers amnesty to those who refused to fight for Milosevic in the recent war. • December: Franjo Tudjman dies of cancer.

2000. Stipe Mesic is elected successor to Croatia's Tudjman. The elections also brought a more liberal government to Croatia. The HDZ has split. • French peacekeepers arrest Bosnian Serb leader Momcilo Krajisnik, who has been indicted for war crimes. • March: Pledges to the Security Pact projects at a meeting in Brussels were much higher than expected – $1.8 billion. • April: Macedonia's parliament votes to return property expropriated by the Communists over 50 years ago. • Serbian opposition leaders, though feuding among themselves, managed to stage a demonstration in Belgrade with up to 200,000 participants. • President Djukanovic of Montenegro is in serious conflict with Milosevic and his government has expressed an intention to secede from Yugoslavia. • In Bosnia, the Social Democratic Party (formerly communist) wins victories everywhere except in Republika Srpska, where the recent arrest of Momcile Krajisnik had infuriated many voters, so that they display their continuing commitment to nationalism. • May: Disaffection is growing in the Serbian army. Some of the military's ablest troops have left because of deteriorating conditions.

NOTES ON CONTRIBUTORS

Milica Z. Bookman
Milica Z. Bookman is professor of economics at St. Joseph's University in Philadelphia. She has authored six books and numerous articles, many of them dealing with the economic conditions in her native Yugoslavia.

Michel J. Chossudovsky
Michel Chossudovsky is Professor of Economics at the University of Ottawa. He has acted as economic adviser to governments of developing countries and has worked as a consultant for several national and international organizations. His research focuses on the debt crisis and the impact of IMF-World Bank sponsored macro-economic reforms. He has also undertaken a number of country studies, including Yugoslavia and its successor states, Rwanda, and Somalia.

Timothy Donais
Timothy Donais is a PhD Candidate in Political Science at York University in Toronto, and his current research is focussed on the democratization process in post-war Bosnia and Herzegovina. Between 1996 and 2000, he has worked in various capacities with the OSCE Mission to Bosnia and Herzegovina, including press and public information officer and elections supervisor.

Carl G. Jacobsen
Carl Jacobsen is a professor of political science at Carleton University where he administers research programs in post-communist studies. He has recently been engaged in teaching peace research and co-authored a book with Johan Galtung: *Searching for Peace: The Road to Transcend* (London: Pluto Press, 2000)

David Last
Major David Last is a professor in the Royal Military College of Canada at Kingston, Ontario. He developed training programs for Canadian peacekeepers in Cyprus and other locations and later served with the peacekeeping forces in Croatia and Bosnia during the wars there, working closely with local leaders in outlying regions, especially during the transition to NATO's peacekeeping efforts.

Sonja Licht

Sonja Licht is President of the Open Society Fund in Belgrade. She has long been an activist in Yugoslavia's democratic opposition, and was co-founder of the Helsinki Citizens Assembly. After receiving a M.A. in sociology at the University of Belgrade in 1971, she worked as a researcher at the Institute of International Politics and Economics and at the Centre for Cultural Development. From 1990-92, she worked at the Institute for European Studies She has written numerous articles on nationalism and pluralism, and on the impact of economic sanctions and the role of the peace movement.

Corey Levine

Corey Levine has her M.A. in Human Rights from the University of Essex. Her particular areas of interest include post-conflict and peacebuilding, gender, democratisation and civil society issues. She was recently on secondment with the OSCE Mission in Kosovo.

Srećko Mihailović

Srecko Mihailovic is a researcher in the Center For Political Studies and Public Opinion Research, Institute of Social Sciences, University of Belgrade. He is particularly interested in studying the effects of the mass media in his country.

Jan Øberg

Jan Øberg is a co-founder and director of TFF, the Transnational Foundation in Lund, Sweden. In his own research and practice he specializes in nonviolence, conflict-mitigation, ethics, and world order. Between September 1991 and 1996, TFF had 22 missions visiting the former Yugoslavia and conducting more than 1200 interviews, with the objective of providing background reports and analyses for use by humanitarian organization, diplomats, the press, and the general public.

Margarita Papandreou

Margarita Papandreou was born in the United States but has lived in Athens for many years. Initially trained as a social worker at the University of Minnesota, she became more interested in political affairs during the lengthy period when, as a member of a political family, she was exiled from Greece and lived in Canada. She analyzed that period in her book, *Nightmare in Athens* (Englewood Cliffs: Prentice-Hall, 1970). It was later, while she was first lady of Greece, that she founded where she founded an important organization that was initially called Women for a Meaningful Summit, now Women for Mutual Security. Ms. Papandreou continues to function as the global coordinator of that group.

Robert Schaeffer

Robert Schaeffer is a Professor of Global Sociology at Kansas State University. He is the author of *Severed States: Dilemmas of Democracy in a Divided World* (1999) and *Power to the People: Democratization Around the World* (1997).

Dusko Sekulic

Professor Sekulic teaches in the sciology department in The Flinders University of South Australia. For the past several years he has worked with several colleagues on a study involving survey analysis regarding nationalism in the various regions of the former Yugoslavia.

Darko Silovic

Darko Silovic was the last Ambassador and Permanent Representative of former Yugoslavia to the United Nations in New Yok (1990-1992). In that capacity he was an immediate witness to the outbreak of the open conflict and an active participant in the efforts of the U.N. and the international community to stop it.

Mr. Silovic has worked for thirty years as a career diplomat in the foreign service of former Yugoslavia, mainly in multilateral, conference diplomacy. After relinquishing his post because of serious disagreements with the policies of the parties that led to the conflict, he held various positions in the United Nations, at Headquarters in New York and in the field in peacekeeping, political, and humanitarian affairs. He was until recently the head of the U.N. peacekeeping mission in Tajikistan.

Kenneth Simons

Kenneth Simons was editor of *Peace News* in London England for six years and, as a journalist and an activist in sponsoring organization, War Resisters International, traveled to Yugoslavia and Kosovo, as well as to various other areas where nonviolent resistance required support. He is now a Toronto-based researcher who works most closely with *Peace Magazine*.

Metta Spencer

Metta Spencer is Professor Emeritus of Sociology, University of Toronto and Editor in Chief of *Peace Magazine*. She authored nine editions of an introductory sociology textbook, *Foundations of Modern Sociology*, and for 14 years directed a program in peace and conflict studies at the Mississauga campus of the university. Her most recent book is *Separatism: Democracy and Disintegration* (Lanham, Maryland: Rowman and Littlefield, 1998). She is vice-president of Science for Peace.

Dorie Wilsnack

Dorie Wilsnack worked for 25 years in the U.S. peace movement, concentrating from 1991–1998 on building American support for peace and human rights groups in former Yugoslavia. She has visited the region of former Yugoslavia often and prepared many resource materials for American audiences, including Working for Peace in the Balkans, a directory carrying descriptions of over 200 U.S. NGOs working on human rights and humanitarian aid. She worked in Belgrade and Pristina for three months with the Balkan Peace Team and currently works as a Coordinator of the BPT in Minden, Germany.

Mitja Žagar

Mitja Zagar holds a doctorate in law and is a senior research fellow at the Institute for Ethnic Studies, Ljubljana and an associate professor of social sciences. His research primarily concerns ethnic relations and comparative constitutional law. He is also an expert adviser and research coordinator at the Ministry of Science and Technology of the Republic of Slovenia, and a member of an advisory council of Slovenia's minister of foreign affairs. As Visiting Fulbright Associate Professor at Wayne State University he taught courses on ethnic conflict in the Balkans.

Professor Zagar participated in drafting the agreement between the Republic of Hungary and the Republic of Slovenia on ethnic minorities. He has published extensively on topics related to nation states; sovereignty and supranationality; processes of international integration; Human Rights and Rights of Minorities.